Nutrition and Feeding in Animal Production

Nutrition and Feeding in Animal Production

Edited by Arden Barnes

SYRAWOOD
PUBLISHING HOUSE
New York

Published by Syrawood Publishing House,
750 Third Avenue, 9th Floor,
New York, NY 10017, USA
www.syrawoodpublishinghouse.com

Nutrition and Feeding in Animal Production
Edited by Arden Barnes

International Standard Book Number: 978-1-64740-411-6 (Hardback)

Cataloging-in-publication Data

Nutrition and feeding in animal production / edited by Arden Barnes.
p. cm.
Includes bibliographical references and index.
ISBN 978-1-64740-411-6
1. Animal feeding. 2. Animal nutrition. 3. Animal culture. I. Barnes, Arden.
SF95 .N88 2023
636.085 2--dc23

TABLE OF CONTENTS

PREFACE

Animals get their nutrition either directly from plants they eat or indirectly by consuming other animals that have consumed plants. Some of the mechanisms through which nutrition can be obtained are suction feeding, fluid feeding, filter feeding, bulk feeding and deposit feeding. Good animal feeding is essential to ensure the welfare and health of animals, as well as the production of high quality and safe animal-derived products. Appropriate animal feeding and nutrition is also essential for the proper development and reproduction of animals. Furthermore, appropriate feeding helps the animals in developing strong immunity to fight infections. It also encourages better health in animals, which leads to increased productivity of products such as eggs, milk and meat. This book contains some path-breaking studies on the role of nutrition and feeding in improving animal production. It includes contributions of experts and scientists, which will provide innovative insights into this field. The book is a complete source of knowledge on the present status of this important area of study.

The researches compiled throughout the book are authentic and of high quality, combining several disciplines and from very diverse regions from around the world. Drawing on the contributions of many researchers from diverse countries, the book's objective is to provide the readers with the latest achievements in the area of research. This book will surely be a source of knowledge to all interested and researching the field.

In the end, I would like to express my deep sense of gratitude to all the authors for meeting the set deadlines in completing and submitting their research chapters. I would also like to thank the publisher for the support offered to us throughout the course of the book. Finally, I extend my sincere thanks to my family for being a constant source of inspiration and encouragement.

Editor

1

Effects of Long-Term Supplementation with Brown Seaweeds and Polyphenols in Rabbit on Meat Quality Parameters

Raffaella Rossi [1,*], Francesco Vizzarri [2], Sabrina Ratti [1], Marisa Palazzo [3], Donato Casamassima [3] and Carlo Corino [1]

1 Dipartimento di Medicina Veterinaria, Università Degli Studi di Milano, Via dell'Università 6, 26900 Lodi, Italy; sabrinaratti@hotmail.com (S.R.); carlo.corino@unimi.it (C.C.)
2 Department of Agricultural and Environmental Science, Università di Bari Aldo Moro, Via G. Amendola 165/A, 70126 Bari, Italy; francesco.vizzarri@uniba.it
3 Department of Agricultural, Environmental and Food Sciences, Università Degli Studi del Molise, Via F. De Sanctis 1, 86100 Campobasso, Italy; m.palazzo@unimol.it (M.P.); casamassima.d@unimol.it (D.C.)
* Correspondence: raffaella.rossi@unimi.it

Simple Summary: Safe and natural feed supplement are required to enhance animal health and welfare, answering to the agri-food system and consumers' needs. The research for sustainable feed additives for enhancing rabbit health and meat quality parameters are needed. Plants received considerable consideration as safe feed supplements in animal nutrition. Recent studies indicated that seaweed can be considered a sustainable dietary supplement that improve animals' health and meat quality parameters. Moreover, plant polyphenols have been studied as antioxidants and cholesterol-lowering agents in meat. No data on the effects of long term dietary supplementation of a mixture composed of brown seaweed and plant polyphenols has been available. For this reason, we decided to test the effects of dietary natural supplements in rabbits, based on performance and meat quality parameters.

Abstract: The objective of the present study was to evaluate the effects of dam and offspring dietary supplementation with a natural feed additive on the growth performance and meat quality parameters of growing rabbits. The growing rabbits are selected from lactating does receiving a control diet (C) or diets supplemented with 0.3% (SP1) and 0.6% (SP2) of feed additive containing brown seaweeds (*Laminaria* spp.) and plant extracts. In the postweaning phase, the growing rabbits remained in the treatment group defined by their does and the trial lasted 42 days. The average daily feed intake and feed conversion ratio were improved in the rabbit fed 0.6% of the natural feed additive. The cholesterol content tended to be lower in *Longissimus lumborum* (LL) muscle and decrease in *Semimembranosus* (SM) muscle (in SP2 −41.36% than controls). The α tocopherol and retinol content were enhanced in both muscles of rabbit fed the natural mixture (SP1 and SP2 groups). An improvement of sensory attributes of texture was observed in both muscles from rabbit fed natural mixture. In conclusion, long term supplementation of both lactating does and offspring with the high dosage of brown seaweed and plant polyphenols improves growth performance and enhances meat nutritional and sensory parameters.

Keywords: brown seaweed; meat quality parameters; plant polyphenols; rabbit

1. Introduction

The EU is the second highest world producer of meat rabbits. The majority (83%) of production is in Spain (48.5 million rabbits slaughtered), followed by France (29 million rabbits slaughtered) and Italy (24.5 million rabbits slaughtered). Rabbit meat represents a traditional dietary practice in the Mediterranean area and a new source of consumption in China and Mexico [1–3].

Even if the global rabbit meat consumption per capita is low if compared with other meat (0.19 kg/per capita/year), in EU countries the consumption is about 0.51 kg/per capita/year (Spain 1.09 kg; Italy 0.91 kg; France 0.75 kg) [3].

In the last decades, the genetic selection has turned towards improvement of the offspring number, growth rate and muscle mass. This causes some problems related to global farm efficiency [4,5]. In rabbit farming several diseases occur; digestive disorders are the main problem, affecting antibiotics consumption [6].

Considering that the EU has been committed for years towards a rational use of antibiotics in livestock, nutrition can be one potential efficient strategy for improving rabbit health and meat quality. Additional research that explores the effects of dietary sustainable additives and alternative feed ingredients in rabbit are required.

Moreover, in recent years, the demand for healthy foods that respect the environment and animal welfare has increased [7]. For products of animal origin, the idea of healthy and safe is linked not only to the product nutritional composition but also to the diet, additives, and antibiotics that the animals have taken. In fact, a strong consumers' preference for natural food, free from synthetic additives and antibiotics was observed [8].

Several studies reported that dietary integration with plant extracts containing polyphenols in rabbit improves health, increases nutrient digestibility and growth performances and enhances meat quality parameters [9,10]. Plant extracts offer a considerable range of activities such as antioxidant, anti-inflammatory, antiviral and antimicrobial effects, positively influencing feed digestibility and microbial ecology [6].

Other studies showed that dietary brown seaweeds in livestock had a positive effect on health, growth performance, and meat quality due to their nutraceuticals properties and content of sulfated polysaccharides, phlorotannins, diterpenes, omega-3 polyunsaturated fatty acids, minerals and vitamins [11,12]. Moreover, brown seaweeds represent a renewable and sustainable feed ingredient for their high productivity if compared to other conventional ingredients.

Our previous study reported the effects of a mixture of brown seaweeds and plant polyphenols on the performance of lactating does [13]. Moreover, we also investigated the effect of the same natural supplementation in growing rabbits [14]. In literature, no previous study reported the effects of long term supplementation in rabbit with brown seaweed and plant polyphenols from lactating does to rabbit slaughter. Therefore, we wanted to evaluate if dietary treatments during the entire rabbit rearing period may positively affect the growing rabbit's productive performance and meat quality parameters.

2. Materials and Methods

2.1. Animals and Dietary Treatments

Rabbits were handled following the guidelines of EU Directive 2010/63/EU and national guidelines for the care and use of animals [15]. All experimental procedures involving animals were approved by the National Agricultural and Food Centre ethical committee (No. NPPC 18-10-2016).

The dietary supplementation began in the lactation phase in which does (n = 60) were divided into three experimental groups, homogeneous for body weight (4.83 ± 0.19 kg) and parity order (second), receiving a control diet (C) or diets supplemented with 0.3% (SP1) and 0.6% (SP2) of feed additive containing prebiotic polysaccharides from brown seaweeds (*Laminaria* spp.) and plant extracts containing phenolic acid, hydroxycinnamic acids, tannins, and flavonoids. In the postweaning period, all rabbits remained in the treatment group defined by their does. The trial lasted 42 days.

The dosages of the supplement were chosen based on an in vitro study on the minimal inhibitory concentration (MIC) against *Clostridium* spp., *Staphylococcus* spp., and *Escherichia coli* spp. [16].

No anticoccidials, antibiotics or other medications were included in the diet. The feed additive was included in the mashed diets, then the diets were pelleted.

At weaning (35 days), 144 New Zealand White rabbits, balanced for body weight and sex were housed at the National Agricultural and Food Centre, Nitra (Slovak Republic).

Rabbits were randomly allotted into three experimental groups (48 rabbits per treatment) and were housed in cages (2 rabbits/cage). The cages were equipped with a hopper for feed and an automatic nipple drinking system. The lighting cycle throughout the trial was 16 h of light and 8 h of dark. The building temperature was maintained within 18 ± 4 °C, with a relative humidity of 70 ± 5%.

The animals were fed ad libitum and had free access to water. Table 1 reports the ingredients and chemical composition of the experimental diets.

Table 1. Ingredients (g/kg) and chemical composition of experimental diets.

Ingredients	C	Diet [a] SP1	SP2
Maize	282	279	276
Alfalfa hay	305	305	305
Sunflower meal	135	135	135
Carob bean meal	90	90	90
Wheat	80	80	80
Oat	53	53	53
Cane molasses	20	20	20
Palm seed oil	8	8	8
Soybean oil	7	7	7
Calcium carbonate	7	7	7
Sodium Chloride	3	3	3
DL-Methionine (99%)	2.5	2.5	2.5
Vitamin and mineral premix [b]	2.5	2.5	2.5
Dicalcium phosphate	2	2	2
L-Lysine HCl (78.5%)	1.6	1.6	1.6
Choline (75%)	1.4	1.4	1.4
Dietary supplement	0	3	6
Chemical composition [c]			
Crude protein	184	183.6	183.5
Ether extract	37.5	35.5	35.5
Crude fiber	187	186.8	187
Ash	86	85.7	85.8
Nitrogen free extract	507	507.1	506.9
NDF	302.1	301.5	301.7
ADF	195.8	195.4	195.3
ADL	39.9	39.5	39.5

[a] C = control diet; SP1 = diet supplemented with 0.3% of brown seaweed and plant polyphenols; SP2 = diet supplemented with 0.6% of brown seaweed and plant polyphenols. [b] Supplied per kg diet: 13,500 I.U. vitamin A (trans-retinyl acetate); 800 I.U. vitamin D3 (cholecalciferol); 35 mg vitamin E (α-tocopherol min 91%), 35 mg copper (cupric sulphate pentahydrate). [c] Determined in triplicate.

The chemical composition was analyzed according to the methods of the Association of Analytical Chemists [17], following the recommendations of the European group on rabbit nutrition [18].

Table 2 reports the chemical composition, phenolic composition and carotenoid content of the feed supplement. The supplement phenolic compounds were determined using HPLC-UV-DAD [19]. The beta-carotene content of the feed supplement was performed according to Rakusa et al. [20].

Table 2. Chemical composition and polyphenols content of the dietary supplement.

Item	
	% DM
Dry matter	93.6 ± 5.05
Crude protein	7.2 ± 0.99
Ether extract	0.32 ± 0.01
Crude fiber	11.2 ± 1.02
Carbohydrates	49.6 ± 3.18
Ash	32.7 ± 1.38
Chemical compounds: [a]	mg/kg DM
β-Carotene	402 ± 30.89
Phenolic acid:	
Syringic acid	1059.8 ± 62.82
Hydroxycinnamic acids:	
Neochlorogenic acid	7979.2 ± 468.11
Rosmarinic acid	126.5 ± 8.67
Trans sinapic acid	105.5 ± 8.09
Chlorogenic acid	21.4 ± 3.65
Tannins:	
Ellagic acid	2440.9 ± 148.29
Rutin	272.4 ± 20.82
Flavonoids:	
Myricetin	53.9 ± 5.68

[a] Values are expressed as means (n = 4) ± standard deviation.

2.2. Sampling

The animals were weighed at 0 and 42 days of the experimental trial and were monitored daily to assess the health conditions. Daily feed intake was recorded (feed offered and refused weekly). The data were used to calculate the average daily gain (ADG), average daily feed intake (ADFI) and feed conversion ratio (FCR).

At the end of the trial the rabbits (77 days old) were weighted and after a fasting period of 6 h 1 male rabbit per cage (12 animals/group) were randomly selected and slaughtered. Rabbits were subjected to electrical stunning (100 V, 50 Hz, 2–3 s) and sacrificed by bleeding according to the guidelines established by the European Community (1099/2009/EC) for the protection of animals during slaughter [21].

Carcasses were chilled for 24 h at +4 °C and then dissected, according to the recommendations of the WRSA [22], by removing the skin, distal part of the limbs, genital organs, bladder and the gastrointestinal tract. Carcasses were then weighted, and the dressing percentages were determined. Samples of whole *Longissimus lumborum* (LL) from the 7th thoracic vertebra to the 7th lumbar vertebra and whole thighs were collected. The physical parameters were evaluated on freshly cut LL muscle and *Semimembranosus* (SM) muscle. The whole muscles were then sampled for chemical and sensory analysis, vacuum packed and stored at −20 °C. The chemical parameters were performed on LL and SM muscles and the sensory evaluation was assessed on the LL muscle and whole thighs.

2.3. Physical and Chemical Parameters

All the analyses were performed on LL muscle and *Semimembranosus* (SM) muscle. The pH and color parameters were measured 24 h after slaughter. The pH measurement was performed using a pH meter (HI98191 microcomputer; Hanna Instruments, Vila do Conde, Portugal), calibrated with a standard buffer of pH 4.0 and 7.0.

Meat color parameters were evaluated at 24 h after slaughter on the fresh cut surface of LL and SM muscles, after 20 min of blooming, using a chromameter (Chromameter CR 300 Minolta Ltd., Osaka, Japan) equipped with a D65 light source and a 0° viewing angle geometry according to the reflectance coordinates (L*, a*, b*). The instrument was calibrated using a white calibration plate (Calibration Plate CR-A43; Minolta Camera Co.). The color determination was assessed in triplicate.

The chemical composition of both muscles was determined according to the AOAC methods [17].

The cholesterol content was determined following the method of Du and Ahn [23], using a gas chromatographic method. The cholesterol was identified based on the retention time of the standard (Sigma Aldrich, St. Louis, CA, USA) and quantified with the Chrom Card Data System (version 1.17) software. All samples were analyzed in triplicate.

2.4. Vitamin Content

The α-tocopherol and retinol content in LL and SM muscle were determined using the modified procedure of Zaspel and Csallany [24]. The samples were analyzed using an HPLC system (Kontron Instruments, Milan, Italy) with an autosampler (HPLC autosampler 360, Kontron Instruments, Milan, Italy) with a loop of 20 μL, a high-pressure pump and a C18 column 5 μm, 150 × 4.6 mm (Phenomenex, Torrance, CA, USA). The mobile phase was acetonitrile and methanol (75:25 *v*/*v*) and the flow rate was 1 mL per min. The α-tocopherol and retinol were identified using a fluorimeter detector, comparing the retention time with the standards (Sigma Aldrich, St. Louis, CA, USA). The quantification was performed using the Geminyx system (version 1.91).

2.5. Sensory Analysis

The sample preparation was performed after thawing thighs and LL muscle for 24 h at 4 °C. The whole samples of LL and thighs were prepared in an uncovered stainless-steel dish in a conventional oven (REX, Milano, Italy) at 180 °C. A thermocouple (Pentronic AB, Gunnebobruk, Sweden) was used to register the sample core temperature and samples were removed at 79–80 °C. The roasted samples were then cut into 1.5 cm thick slices (Electrolux 50, 220–24, kW0.2) and the slices were warmed to 60 °C before the evaluation. A trained sensory panel, involving eight members familiar with descriptive analysis procedures [25], was employed. All evaluations were performed in a sensory laboratory equipped according to EN ISO 8598 recommendations [26]. The descriptors, definitions, and standards are reported in Palazzo et al. [27].

The sensory profile was assessed according to EN ISO 13,299 [25] and the panel evaluated the two samples (thigh and LL) on different days in triplicate. Within each session the design was balanced for order and carry-over effects [28]. During training and sampling, the panelists had access to unlimited water and unsalted crackers. They were requested to evaluate the intensity of each attribute by assigning a score between 1 (absence of the sensation) and 9 (extremely intense).

2.6. Statistical Analysis

Data were analyzed using SPSS software (SPSS, PC Statistics 25.0 SPSS Inc., IBM, Segrate, MI, Italy). Productive performance and slaughter and meat quality parameters were analyzed using one-way analysis of variance (ANOVA), with diet as the fixed effect. The sensory data were submitted to ANOVA with samples, judges, replicates, and their interactions as effects (EN ISO 13299, 2010). The significance of these effects was tested with F tests. Post-hoc pairwise contrasts were evaluated by Duncan's test. The cage was considered as the experimental unit for growth performance and the rabbit for the meat quality parameters. Data are reported as mean ± SEM. Differences among treatments were considered significant at $p < 0.05$.

3. Results

3.1. Productive Parameters and Carcass Characteristics

The growth performances and carcass parameters of rabbit fed experimental diets are reported in Table 3. The final weight of the rabbits did not differ among experimental groups ($p > 0.05$), even if the ADG tended to be higher ($p = 0.09$) in both the SP1 and SP2 groups. The ADFI result was lower

($p < 0.05$) in the SP1 group than the results in the other groups. The feed conversion ratio was positively affected ($p < 0.001$) by both the dosage of brown seaweed and plant polyphenols (SP1 and SP2).

Table 3. Growth performances of rabbits fed the control diet (C) and diet supplemented with 3 g/kg (SP1) and 6 g/kg (SP2) of brown seaweed and plant polyphenols.

Item	C	Diet SP1	SP2	*p*-Value
Initial weight, g	829.1 ± 0.21	838.7 ± 0.22	818.8 ± 0.17	0.686
Final weight, g	2657.2 ± 58.62	2757.9 ± 41.54	2864.1 ± 35.51	0.220
ADG, g/d	42.5 ± 1.16	45.7 ± 1.04	46.3 ± 0.73	0.090
ADFI, g/d	172.1 ± 2.45 ^{a}	148.8 ± 3.17 ^{b}	161.9 ± 1.66 ^{a}	0.014
FC	3.95 ± 0.04 ^{A}	3.25 ± 0.06 ^{B}	3.49± 0.03 ^{B}	0.001

ADG, average daily gain; ADFI, average daily feed intake; FC, feed conversion ratio. Data are reported as means ± standard error of the means (SEM); n = 48 ^{a,b} values in the same row are different at $p < 0.05$. ^{A,B} values in the same row are different at $p < 0.01$.

The slaughter weight (2791.3 ± 69.2 g C vs. 3001.2 ± 59.4 g SP1 vs. 2986.1 ± 42.4 g SP2; $p = 0.021$) and the carcass weight (1473.4 ± 38.6 g C vs. 1585.2 ± 33.4 g SP1 vs. 1610.4 ± 19.7 g SP2; $p = 0.010$) of the sampled rabbit (n = 12) resulted higher in groups fed brown seaweed and polyphenols (SP1 and SP2) than in the control. The dressing percentage resulted higher in SP2 group than the others (52.8 ± 0.37% C vs. 52.4 ± 0.32% SP1 vs. 53.9 ± 0.41% SP2; $p = 0.025$).

3.2. Meat Physical and Chemical Parameters

The chemical composition of LL and SM muscle are reported in Tables 4 and 5, respectively. The physical parameters in both muscles were not affected ($p > 0.05$) by long term supplementation of both lactating does and the offspring with low and high dosage of seaweed and polyphenols. The nutritional parameters moisture, crude protein, ether extract and ash were also unaffected by dietary treatments ($p > 0.05$). The cholesterol content in LL muscle tended to be lower ($p = 0.059$) in SP2 than SP1 and the control groups. The cholesterol content of the SM muscle resulted lower ($p < 0.01$) in the group that received the high dosage of seaweed and polyphenols (SP2) than SP1 and the control groups. The cholesterol content reduction was about −40.3% than control and −36.4% than SP1 groups.

Table 4. Physical and chemical parameters of *Longissimus lumborum* muscle of rabbits fed the control diet (C) and diet supplemented with 3 g/kg (SP1) and 6 g/kg (SP2) of brown seaweed and plant polyphenols.

Item	C	Diet SP1	SP2	*p*-Value
pH 24 h	5.85 ± 0.25	5.91± 0.21	5.88 ± 0.16	0.222
Color parameters				
L*	55.85 ± 0.39	56.30 ± 1.01	56.44 ± 0.81	0.873
a*	3.89 ± 0.71	3.99 ± 0.32	4.28 ± 0.42	0.911
b*	11.92 ± 0.67	11.00 ± 0.52	11.28 ± 0.26	0.484
Chemical composition:				
Moisture, %	73.23 ± 0.18	73.02 ± 0.21	72.84 ± 0.18	0.383
Crude Protein, %	24.33 ± 0.14	24.52 ± 0.17	24.65 ± 0.21	0.404
Ether extract, %	1.17 ± 0.06	1.05 ± 0.06	1.19 ± 0.10	0.547
Ash, %	1.22 ± 0.01	1.36 ± 0.08	1.39 ± 0.06	0.132
Cholesterol, mg/100 g	33.23 ± 2.98	28.18 ± 1.24	25.07 ± 1.18	0.059

Data are reported as means ± standard error of the means (SEM); n = 12.

Table 5. Physical and chemical parameters of *Semimembranosus* muscle of rabbits fed the control diet (C) and diet supplemented with 3 g/kg (SP1) and 6 g/kg (SP2) of brown seaweed and plant polyphenols.

Item	C	Diet SP1	SP2	*p*-Value
pH 24 h	5.77 ± 0.20	5.79± 0.18	5.82± 0.21	0.853
Color parameters				
L*	63.92 ± 1.59	63.84 ± 2.01	64.41 ± 1.94	0.951
a*	6.01 ± 0.64	5.94 ± 0.56	6.11 ± 0.74	0.894
b*	3.92 ± 0.32	4.02 ± 0.49	3.78 ± 0.56	0.831
Chemical composition:				
Moisture, %	74.29 ± 0.86	74.22 ± 0.78	74.17 ± 0.72	0.594
Crude Protein, %	23.12 ± 0.53	23.37 ± 0.73	23.61 ± 0.62	0.338
Ether extract, %	1.66 ± 0.46	1.92 ± 0.38	1.51 ± 0.43	0.958
Ash, %	1.21 ± 0.06	1.19 ± 0.09	1.15 ± 0.07	0.329
Cholesterol, mg/100 g	39.85 ± 4.48 ^{A}	37.39 ± 2.07 ^{A}	23.78 ± 1.17 ^{B}	0.002

Data are reported as means ± standard error of the means (SEM); n = 12. ^{A,B} values in the same row are different at $p < 0.01$.

3.3. Muscle Vitamin Content

Figure 1 shows the α-tocopherol (A) and retinol (B) content of both muscles in relation to dietary treatments. The α-tocopherol content resulted higher ($p < 0.001$) in the LL and SM muscles of rabbit fed both dosages of brown seaweed and polyphenols (SP1 and SP2) than in the control. The retinol content was higher ($p < 0.001$) in both muscles of rabbit fed brown seaweed and plant polyphenols (SP1 and SP2) than in the control group.

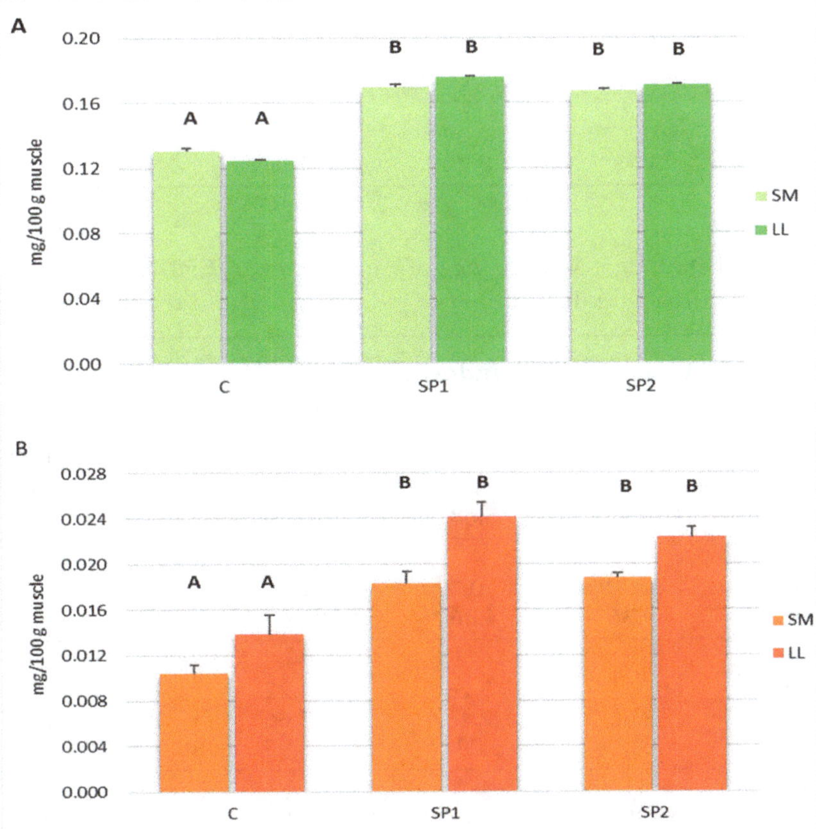

Figure 1. α-tocopherol (**A**) and retinol (**B**) content of *Longissimus lumborum* (LL) and *Semimembranosus* muscle (SM) of rabbits fed control diet (C) and diet supplemented with 3 g/kg (SP1) and 6 g/kg (SP2) of brown seaweed and plant polyphenols. Data are reported as means ± standard error of the means (SEM); n = 12. ^{A,B} values with different superscript letters are different at $p < 0.001$.

3.4. Sensory Profile

The F values for the measured parameters of the LL muscle and thigh sensory profile are reported in Tables 6 and 7, respectively. The data show that dietary supplementation with brown seaweed and plant polyphenols mixture affected ($p < 0.05$) the flavor and texture of the LL muscle. The F values for replicates and interactions were not affected ($p > 0.05$) for all the descriptors. In thigh, taste, flavor and texture were affected ($p < 0.05$) by dietary treatments. Judges presented differences ($p < 0.05$) for aroma, taste, flavor and texture as observed for LL muscle. The judges and the interaction of the treatments per judge revealed differences ($p < 0.05$) for aroma, taste, flavor and texture.

Table 6. The F value and statistical significance of treatments (C, SP1, SP2) and their interaction for sensory descriptors of *Longissumus lumborum* muscle, where the number of judges, n = 8, and the number of replicates, n = 3.

Descriptors	Treatments	Judges	F Value Replicates	T*J	T*R	J*R
Aroma						
Rabbit	2.27	4.52	0.26	1.68	1.55	0.85
Liver	0.05	1.73	1.37	0.70	1.75	1.03
Metallic	2.27	4.83 **	0.08	1.63	1.41	0.83
Taste						
Sweet	0.71	2.67 *	1.33	1.64	0.46	1.23
Salty	2.12	2.83 *	0.63	1.10	1.16	1.22
Flavor						
Rabbit	4.61 *	1.37	0.73	0.95	2.25	1.45
Liver	12.88 ***	5.56 ***	0.30	3.41 **	0.77	1.47
Metallic	5.18 *	5.83 ***	0.45	2.19 *	0.30	1.00
Texture						
Tender	8.53 **	7.80 ***	0.81	2.98 **	0.48	1.12
Juicy	20.31 ***	2.69 *	2.04	3.84 **	1.00	1.12
Stringy	3.56 *	3.26 *	1.04	4.83 ***	1.48	0.77

*** $p < 0.0001$; ** $p < 0.001$; * $p < 0.05$.

Table 7. The F value and statistical significance of treatments (C, SP1, SP2) and their interaction for sensory descriptors of thigh, where the number of judges, n = 8, and the number of replicates, n = 3.

Descriptors	Treatments	Judges	F Value Replicates	T*J	T*R	J*R
Aroma						
Rabbit	1.91	2.58 *	0.09	1.02	1.28	0.80
Liver	0.11	1.35	1.15	0.65	1.07	1.09
Metallic	035	5.68 **	0.89	2.07 *	0.56	1.10
Taste						
Sweet	3.97 *	0.70	2.90	1.29	2.42	0.81
Salty	0.36	4.06 **	1.02	2.18 *	0.37	0.71
Flavor						
Rabbit	6.51 **	3.26 *	1.22	1.26	0.58	2.42
Liver	2.66	2.46 *	1.66	1.47	1.42	1.43
Metallic	2.97 *	5.09 ***	0.95	1.84	1.13	1.17
Texture						
Tender	16.95 ***	4.44 **	0.93	5.23 ***	0.97	0.46
Juicy	12.92 ***	1.47	0.57	4.92 ***	0.86	0.71
Stringy	7.45 **	14.44 ***	0.39	9.57 ***	0.39	0.84

*** $p < 0.0001$; ** $p < 0.001$; * $p < 0.05$.

Figure 2 reports the least squares mean of the different attributes for the LL muscle (A) and thighs (B). In LL muscle the rabbit, liver and metallic flavors were affected by the high dosage of brown seaweed and plant polyphenols, resulting higher ($p < 0.05$) in the SP2 group than the SP1 group and the control groups. The texture parameters were also enhanced in muscle from rabbit fed the natural mixture. In particular, tenderness and juiciness resulted higher ($p < 0.05$) in both groups receiving brown seaweed and plant polyphenols (SP1 and SP2) than the controls. The stringiness resulted lower ($p < 0.05$) in the control and SP2 groups than results in the SP1 group. The attributes related to aroma and taste were unaffected ($p > 0.05$) by dietary treatments. In tights the rabbit and metallic flavors were affected by brown seaweed and plant polyphenols, resulting higher ($p < 0.05$) in both groups receiving the supplement (SP1 and SP2) than in the controls. The sweety taste was positively affected ($p < 0.05$) by the high dosage of natural supplement, in fact, thigh from SP2 group resulted sweetness higher than controls. Tenderness and juiciness resulted higher ($p < 0.05$) in groups receiving the high dosage of brown seaweed and plant polyphenols (SP2) than the SP1 group and the controls. The stringiness resulted lower ($p < 0.05$) in the control group than the SP1 group. The attributes related to aroma were unaffected ($p > 0.05$) by dietary treatments.

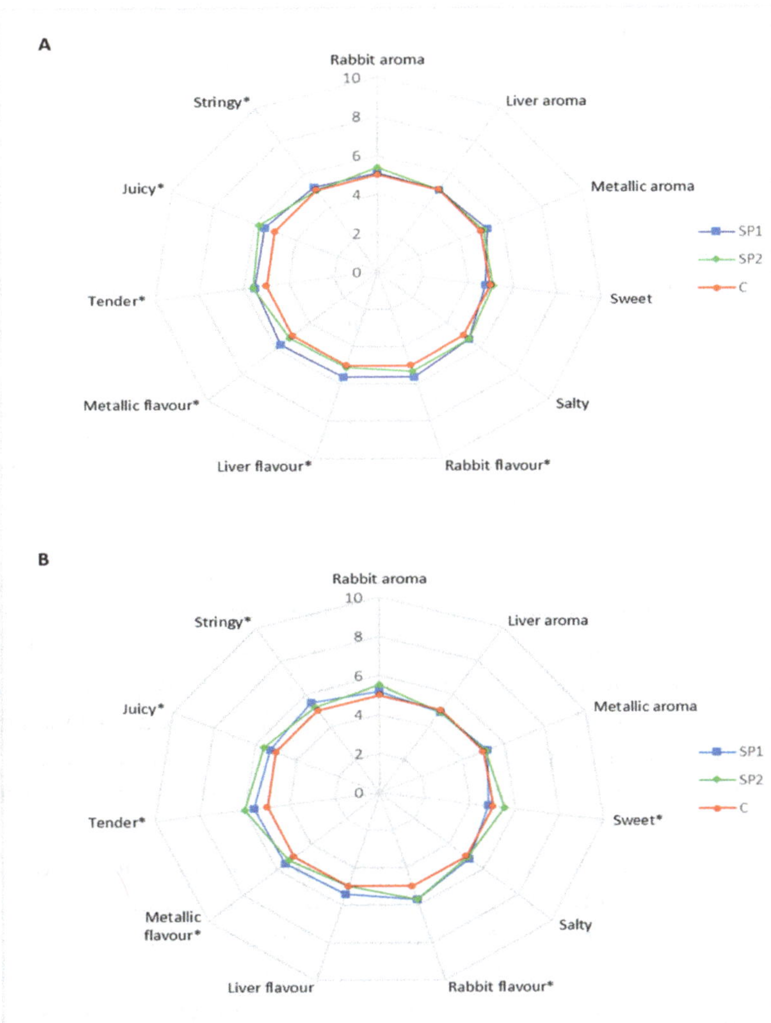

Figure 2. Spider plot of the sensory profile of *Longissimus lumborum* muscle (**A**) and thigh (**B**) of rabbit fed the control diet (C) and the diet supplemented with 3 g/kg (SP1) and 6 g/kg (SP2) of brown seaweed and plant polyphenols. * Significant at $p < 0.05$.

4. Discussion

In the recent years, several studies reported the effect of dietary supplementation with plant polyphenols in rabbit to improve growth performances and meat quality parameters [9,29]. However, there are only few studies on dietary supplementation with seaweed in rabbit. Previous study reported that dietary green seaweed *Ulva Lactuca* in does and male rabbit improve reproductive parameters [30,31]. Moreover, Dalle Zotte et al. [32] reported that a mixture containing dietary microalga *Spirulina* and *Thymus vulgaris*, did not affect growth performance and energy or nutrients digestibility.

To the best of our knowledge, the first study describing the effect of dietary brown seaweed and plant polyphenols in rabbit is our recent study limited to the growing phase [14], but the effects of long term supplementation of this mixture from lactating does to rabbit slaughter has been not yet studied. The present data showed that ADG tended to be higher in rabbit fed the natural mixture and the feed conversion ratio was enhanced in rabbit fed the dietary supplement, suggesting a better feed conversion due to the probiotic, antibacterial and antioxidant effects of the natural mixture used [11]. An enhancement of nutrient digestibility was also observed in several animal species fed seaweed [12]. The present data are in line with studies in rabbit showing that natural extract dietary supplementation positively affected rabbit's growth performances [6,10].

The data on carcass characteristics showed an enhancement in rabbit fed the natural dietary supplement. Previous studies reported no effects of dietary supplementation with polyphenols on carcass characteristics in pigs and rabbit [9,33]. Other studies in pigs reported an enhancement of carcass characteristics due to polyphenols supplementation [34,35]. These different results are probably due to the length and dosage of dietary supplementation and the active principles content of the plant extract used.

The present data show that dietary supplementation with polyphenols and brown seaweed did not affect pH and color parameters in both muscles, in agreement with our previous study in rabbit fed the same mixture [14]. The data fall in the means of the values reported in other studies for LL and SM muscles [36,37].

Likewise, the chemical composition of the LL and SM muscles was not affected by the dietary treatments, except for the cholesterol content. Comparing with our study in rabbit fed the same mixture [14], the dietary supplementation of both the does and the offspring reduces the cholesterol content in both muscles, and this reduction compared to controls was relevant in the SP2 groups (−24.5% and −40.3% in LD and SM muscle, respectively).

Previous studies reported a lowering cholesterol effect of seaweed in livestock and humans. Al-Harthi and El-Deek [38] reported that dietary supplementation with 3–6% of brown seaweed in laying hens improved egg quality, and decreased yolk cholesterol content. In humans, seaweed intake was positively linked to a modulation of blood glucose and cholesterol [39]. This mechanism is probably linked to the activity of the seaweed polysaccharide fucoidan that decreases cholesterol absorption and increases its excretion, modulating reverse cholesterol transport-related protein expression [40]. A cholesterol lowering effect was also observed in rabbit and hares fed plant extract containing polyphenols [41,42].

Our results show that the dietary mixture of brown seaweed and plant polyphenols enhances the content of α tocopherol and retinol in both LL and SM muscles. Seaweeds show antioxidant properties due to the phenols, carotenoid fucoxanthin, tannins, phlorotannins and polysaccharides content. Furthermore, seaweeds are an excellent source of vitamins and, above all, brown seaweeds contain a large amount of α tocopherol and carotenoid [11]. In fact, previous studies in pork reported that seaweed dietary supplementation enhances oxidative stability of LD muscle due to higher muscle α tocopherol content [43,44].

Moreover, plant extract containing polyphenols increased α tocopherol and retinol levels in several tissues, reducing the oxidative markers [45]. In fact, an earlier study in pork, rabbit and hares reported that natural extract containing polyphenols enhanced muscle α tocopherol and retinol [27,46,47].

The data on sensory profile revealed that the dietary mixture of brown seaweed and plant polyphenols affected the flavor in both LL muscle and thigh, even if the differences are hardly perceived. The texture attributes related to tenderness, juiciness and stringiness was improved in in both LL muscle and thigh from rabbit fed the natural mixture. The judges and the interaction of the treatments per judges revealed differences for aroma, taste, flavor and texture, but this is common in sensory evaluations, due to the different use of the scale [48]. Even if there is selection and training of judges, some variability always remains, but the interaction treatments per judges was not significant for most descriptors, indicating low variability among the panel members.

In literature no previous study reported the effects of dietary brown seaweed supplementation on meat sensory quality parameters while several studies reported the effects of dietary plant polyphenols on this parameter. Dietary plant polyphenols from *Lippia* spp. improve the texture parameters related to tenderness and juiciness in donkey and horse *Longissimus dorsi* muscle and it is probably related to the protection against the oxidation process [27]. Moreover, Zhao et al. [49] also reported an improvement of tenderness of lamb meat due to dietary supplementation with wine grape pomace, a rich source of polyphenols. Probably, the protection of the proteolytic enzymes μ-calpain and m-calpain from the oxidative process, increases their functionality, enhancing meat tenderness [50,51].

5. Conclusions

Our results show that dietary supplementation with a high dosage of brown seaweeds and plant polyphenols of both lactating does and the offspring have a positive effect on the average daily feed intake and feed conversion ratio, without affecting other meat physical parameters.

Nutritional characteristics of both LL and SM muscle were positively affected by the natural extract, with a reduction of the cholesterol content and an increase of α tocopherol and retinol content. Moreover, an improvement of the sensory quality perceived by consumers in terms of texture was observed in both LL muscle and thigh from rabbit fed the high dosage of the natural mixture.

From these data, we can conclude that the long term supplementation of both lactating does and the offspring with a high dosage of brown seaweeds and plant polyphenols is a valid approach to boost rabbits' performance and enhance meat nutritional and sensory parameters.

Author Contributions: Conceptualization, C.C. and D.C.; investigation, F.V., R.R., S.R. and M.P.; data curation, F.V. and R.R.; writing—original draft preparation, R.R.; writing—review and editing, C.C., F.V., R.R., S.R. and M.P.; project administration, supervision C.C. and D.C. All authors have read and agreed to the published version of the manuscript.

Acknowledgments: The authors would like to thank the technical staff of the National Agricultural and Food Centre, Nitra (Slovak Republic) for the availability of rabbit, and technical support during the whole experimental period.

References

1. EFSA Panel on Animal Health and Welfare (AHAW). Health and welfare of rabbit farmed in different production system. *EFSA J.* **2020**, *18*, e05944.
2. Petracci, M.; Cavani, C. Rabbit meat processing: Historical perspective to future directions. *World Rabbit Sci.* **2013**, *21*, 217–226. [CrossRef]
3. Szendrő, K.; Szabó-Szentgróti, E.; Szigeti, O. Consumers' Attitude to Consumption of Rabbit Meat in Eight Countries Depending on the Production Method and Its Purchase Form. *Foods* **2020**, *9*, 654. [CrossRef]
4. Gondret, F.; Larzul, C.; Combes, S.; De Rochambeau, H. Carcass composition, bone mechanical properties, and meat quality traits in relation to growth rate in rabbits. *J. Anim. Sci.* **2005**, *83*, 1526–1535. [CrossRef]
5. Hernandez, P.; Arino, B.; Grimal, A.; Blasco, A. Comparison of carcass and meat characteristics of three rabbit lines selected for litter size or growth rate. *Meat Sci.* **2006**, *73*, 645–650. [CrossRef] [PubMed]
6. Dalle Zotte, A.; Celia, C.; Szendrő, Z.S. Herbs and spices inclusion as feed stuff or additive in growing rabbit diets and as additive in rabbit meat: A review. *Livest. Sci.* **2016**, *189*, 82–90. [CrossRef]

7. Petrescu, D.C.; Petrescu-Mag, R.M. Consumer behaviour related to rabbit meat as functional food. *World Rabbit Sci.* **2018**, *26*, 321–333. [CrossRef]
8. Román Luis, S.; Sánchez-Siles, M.; Siegrist, M. The importance of food naturalness for consumers: Results of a systematic review. *Trends Food Sci. Technol.* **2017**, *67*, 44–57. [CrossRef]
9. Palazzo, M.; Vizzarri, F.; Arvay, J.; D'Alessandro, A.G.; Martemucci, G.; Casamassima, D.; Ratti, S.; Corino, C.; Rossi, R. Dietary effect of dried bay leaves (*Laurus nobilis*) meal on selected productive performances and on quality meat traits in growing rabbits. *Livest. Sci.* **2020**, *242*, 104301. [CrossRef]
10. Miteva, D.; Velikov, K.; Ivanova, S.; Dimov, K. Production of rabbit meat with functional properties. *Agro Life Sci. J.* **2020**, *9*, 221–228.
11. Corino, C.; Modina, S.C.; Di Giancamillo, A.; Chiapparini, S.; Rossi, R. Seaweeds in Pig Nutrition. *Animals* **2019**, *9*, 1126. [CrossRef] [PubMed]
12. Morais, T.; Inácio, A.; Coutinho, T.; Ministro, M.; Cotas, J.; Pereira, L.; Bahcevandziev, K. Seaweed Potential in the Animal Feed: A Review. *J. Mar. Sci. Eng.* **2020**, *8*, 559. [CrossRef]
13. Vizzarri, F.; Chiapparini, S.; Corino, C.; Casamassima, D.; Palazzo, M.; Parkanyi, V.; Ondruska, L.; Rossi, R. Dietary supplementation with natural extracts mixture: Effects on reproductive performances, blood biochemical and antioxidant parameters in rabbit does. *Ann. Anim. Sci.* **2019**. [CrossRef]
14. Rossi, R.; Vizzarri, F.; Chiapparini, S.; Ratti, S.; Casamassima, D.; Palazzo, M.; Corino, C. Effects of dietary levels of brown seaweeds and plant polyphenols on growth and meat quality parameters in growing rabbit. *Meat Sci.* **2020**, *161*, 107987. [CrossRef] [PubMed]
15. European Community. Council Directive 2010/63/EU of 22 September 2010 on the Protection of Animals Used for Scientific Purposes. 2010. Available online: Eurlex.europa.eu/LexUriServ/LexUriServ.do?uri=OJ:L:2010:276:0033:0079:EN:PDF (accessed on 30 September 2020).
16. Tosi, G.; (Experimental Zooprophylactic Institute of the Lombardy and Emilia, Forlì, Italy). Personal communication, 2012.
17. AOAC. *Official Methods of Analysis*, 17th ed.; Association of Official Analytical Chemists: Arlington, VA, USA, 2002.
18. European Group on Rabbit Nutrition (EGRN). Technical note: Attempts to harmonize chemical analyses of feeds and faeces for rabbit feeding evaluation. *World Rabbit Sci.* **2001**, *9*, 57–64.
19. Russo, R.; Pucci, L.; Giorgetti, L.; Árvay, J.; Vizzarri, F.; Longo, V.; Pozzo, L. Polyphenolic characterisation of plant mixture (Lisosan® reduction) and its hypocholesterolaemic effect in high fat diet-fed mice. *Nat. Prod. Res.* **2017**, *33*, 651–658. [CrossRef] [PubMed]
20. Rakusa, Z.T.; Srecnik, E.; Roskar, R. Novel HPLC-UV method for simultaneous determination of fat-soluble vitamins and coenzyme Q10 in medicines and supplements. *Acta Chim. Slov.* **2017**, *64*, 523–529. [CrossRef] [PubMed]
21. Council of the European Union. Council Regulation (EC) No. 1099/2009 of 24 September 2009 on the Protection of Animals at the Time of Killing. 2009. Available online: Eur-lex.europa.eu/LexUriServ/LexUriServ.do?uri=OJ:L:2009:303:0001:0030:EN:PDF (accessed on 30 September 2020).
22. Blasco, A.; Ouhayoun, J. Harmonization of criteria and terminology in rabbit meat research. Revised proposal. *World Rabbit Sci.* **1996**, *4*, 93–99. [CrossRef]
23. Du, M.; Ahn, D.U. Simultaneous analysis of tocopherols, cholesterol, and phytosterols using gas chromatography. *J. Food Sci.* **2002**, *67*, 1696–1700. [CrossRef]
24. Zaspel, B.J.; Csallany, A.S. Determination of alpha-tocopherol in tissues and plasma by high-performance liquid chromatography. *Anal. Biochem.* **1983**, *1*, 146–150. [CrossRef]
25. International Organization for Standardization. *Sensory Analysis-Methodology-General Guidance to Establish a Sensory Profile*; ISO 13299; ISO: Geneva, Switzerland, 2010.
26. International Organization for Standardization. *Sensory Analysis: General Guidance for Design of Test Rooms*; ISO 8598; ISO: Geneva, Switzerland, 1989.
27. Palazzo, M.; Vizzarri, F.; Nardoia, M.; Ratti, S.; Pastorelli, G.; Casamassima, D. Dietary *Lippia citriodora* extract in rabbit feeding: Effects on quality of carcass and meat. *Arch. Anim. Breed.* **2015**, *58*, 355–364. [CrossRef]
28. MacFie, H.J.; Bratchell, N.; Greenhoff, K.; Vallis, L.V. Designs to balance the effect of order of presentation and first-order carry-over effect in halls tests. *J. Sens. Stud.* **1989**, *4*, 129–148. [CrossRef]

29. Dabbou, S.; Gai, F.; Renna, M.; Rotolo, L.; Dabbou, S.; Kovitvadhi, C.L.A.; Brugiapaglia, A.; De Marco, M.; Helal, A.N.; Zoccarato, I.; et al. Inclusion of bilberry pomace in rabbit diets: Effects on carcass characteristics and meat quality. *Meat Sci.* **2017**, *124*, 77–83. [CrossRef] [PubMed]
30. El-banna, S.G.; Hassan, A.A.; Okab, A.B.; Koriem, A.A.; Ayoub, M.A. Effect of feeding diets supplemented with seaweed on growth performance and some blood hematological and biochemical characteristics of male Baladi rabbits. In Proceedings of the 4th International Conference on Rabbit Production in Hot Climates, Sharm Elsheikh, Egypt, 24–27 February 2005; pp. 373–382.
31. Okab, A.B.; Samara, E.M.; Abdoun, K.A.; Rafay, J.; Ondruska, L.; Parkanyi, V.; Pivko, J.; Ayoub, M.A.; Al-Haidary, A.A.; Riyadh, S.A.; et al. Effects of dietary seaweed (*Ulva lactuca*) supplementation on the reproductive performance of buck and doe rabbits. *J. Appl. Anim. Res.* **2013**, *41*, 347–355. [CrossRef]
32. Dalle Zotte, A.; Sartori, A.; Bohatir, P.; Remignon, H.; Ricci, R. Effect of dietary supplementation of Spirulina (*Arthrospira platensis*) and thyme (*Thymus vulgaris*) on growth performance, apparent digestibility and health status of companion dwarf rabbits. *Livest. Sci.* **2013**, *152*, 182–191. [CrossRef]
33. Korniewicz, D.; Rózanski, H.; Usydus, Z.; Dobrzanski, Z.; Korniewicz, A.; Kaczmarek, P.; Frankiewicz, A.; Szulc, K. Efficiency of plant extracts (herbiplant cs) in pigs fattening. *Pol. J. Food Nutr. Sci.* **2007**, *57*, 309–315.
34. Rossi, R.; Stella, S.; Ratti, S.; Maghin, F.; Tirloni, E.; Corino, C. Effects of antioxidant mixtures in the diet of finishing pigs on the oxidative status and shelf life of Longissimus dorsi muscle packaged under modified atmosphere. *J. Anim. Sci.* **2017**, *95*, 4986–4997. [CrossRef]
35. Dávila-Ramírez, J.L.; Munguía-Acosta, L.L.; Morales-Coronado, J.G.; García-Salinas, A.D.; González-Ríos, H.; Celaya-Michel, H.; Sosa-Castañeda, J.; Sánchez-Villalba, E.; Anaya-Islas, J.; Barrera-Silva, M.A. Addition of a mixture of plant extracts to diets for growing-finishing pigs on growth performance, blood metabolites, carcass traits, organ weight as a percentage of live weight, quality and sensorial analysis of meat. *Animals* **2020**, *10*, 1229. [CrossRef]
36. Kozioł, K.; Maj, D.; Bieniek, J. Change in the color and pH of rabbit meat in the aging process. *Med. Weter.* **2015**, *71*, 104–108.
37. Tůmová, E.; Bízková, Z.; Skřivanová, V.; Chodová, D.; Martinec, M.; Volek, Z. Comparisons of carcass and meat quality among rabbit breeds of different sizes, and hybrid rabbits. *Livest. Sci.* **2014**, *165*, 8–14. [CrossRef]
38. Al-Harthi, M.A.; El-Deek, A.A. Effect of different dietary concentrations of brown marine algae (*Sargassum dentifebium*) prepared by different ways on plasma and yolk lipid profiles, yolk total carotene and lutein plus zeaxanthin of laying hens. *Ital. J. Anim. Sci.* **2012**, *11*, e64. [CrossRef]
39. Brown, E.M.; Allsopp, P.J.; Magee, P.J.; Gill, C.I.; Nitecki, S.; Strain, C.R.; McSorley, E.M. Seaweed and human health. *Nutr. Rev.* **2014**, *72*, 205–216. [CrossRef] [PubMed]
40. Yang, Z.; Liu, G.; Wang, Y.; Yin, J.; Wang, J.; Xia, B.; Li, T.; Yang, X.; Hou, P.; Hu, S.; et al. Fucoidan A2 from the Brown Seaweed Ascophyllum nodosum Lowers Lipid by Improving Reverse Cholesterol Transport in C57BL/6J Mice Fed a High-Fat Diet. *J. Agric. Food Chem.* **2019**, *67*, 5782–5791. [CrossRef] [PubMed]
41. Vizzarri, F.; Nardoia, M.; Palazzo, M. Effect of dietary Lippia citriodora extract on productive performance and meat quality parameters in hares (*Lepus europaeus* Pall.). *Arch. Tierzucht* **2014**, *57*, 1–7. [CrossRef]
42. Vizzarri, F.; Palazzo, M.; D'Alessandro, A.G.; Casamassima, D. Productive performance and meat quality traits in growing rabbit following the dietary supplementation of Lippia citriodora, Raphanus sativus and *Solanum lycopersicum* extracts. *Livest. Sci.* **2017**, *200*, 53–59. [CrossRef]
43. Rajauria, G.; Draper, J.; McDonnel, M.; O'Doherty, J.V. Effect of dietary seaweed extracts, galacto oligosaccharide and vitamin E supplementation on meat quality parameters in finisher pigs. *Innov. Food Sci. Emerg. Technol.* **2016**, *37*, 269–275. [CrossRef]
44. Moroney, N.C.; O'Grady, M.N.; Robertson, R.C.; Stanton, C.; O'Doherty, J.V.; Kerry, J.P. Influence of level and duration of feeding polysaccharide (laminarin and fucoidan) extracts from brown seaweed (*Laminaria digitata*) on quality indices of fresh pork. *Meat Sci.* **2015**, *99*, 132–141. [CrossRef]
45. Corbi, G.; Conti, V.; Komici, K.; Manzo, V.; Filippelli, A.; Palazzo, M.; Vizzari, F.; Davinelli, S.; Di Costanzo, A.; Scapagnini, G.; et al. Phenolic Plant Extracts Induce Sirt1 Activity and Increase Antioxidant Levels in the Rabbit's Heart and Liver. *Oxid. Med. Cell. Longev.* **2018**, *2018*, 2731289. [CrossRef]
46. Rossi, R.; Pastorelli, G.; Cannata, S.; Tavaniello, S.; Maiorano, G.; Corino, C. Effect of long term dietary supplementation with plant extract on carcass characteristics meat quality and oxidative stability in pork. *Meat Sci.* **2013**, *95*, 542–548. [CrossRef]

47. Palazzo, M.; Schiavitto, M.; Cinone, M.; Vizzarri, F. Rabbit metabolic response and selected meat quality traits. Evaluation of dietary PLX® 23 and LycoBeads® feed supplement. *J. Anim. Physiol. Anim. Nutr.* **2019**, *103*, 383–394. [CrossRef]
48. Lea, P.; Rødbotten, M.; Næs, T. *Analysis of Variance for Sensory Data*; Chirchester John Wiley & Sons Ltd.: New York, NY, USA, 1997.
49. Zhao, J.X.; Li, Q.; Zhang, R.X.; Liu, W.Z.; Ren, Y.S.; Zhang, C.X.; Zhang, J.X. Effect of dietary grape pomace on growth performance, meat quality and antioxidant activity in ram lambs. *Anim. Feed Sci. Technol.* **2018**, *236*, 76–85. [CrossRef]
50. Morán, L.; Andrés, S.; Bodas, R.; Prieto, N.; Giráldez, F.J. Meat texture and antioxidant status are improved when carnosic acid is included in the diet of fattening lambs. *Meat Sci.* **2012**, *91*, 430–434. [CrossRef] [PubMed]
51. Huff–Lonergan, E.; Lonergan, S.M. Mechanisms of water–holding capacity of meat: The role of postmortem biochemical and structural changes. *Meat Sci.* **2005**, *71*, 194–204. [CrossRef] [PubMed]

2

Dietary Application of Tannins as a Potential Mitigation Strategy for Current Challenges in Poultry Production

Janghan Choi and Woo Kyun Kim *

Department of Poultry Science, University of Georgia, Athens, GA 30602, USA; choij@uga.edu
* Correspondence: wkkim@uga.edu

Simple Summary: There are diverse challenges in the poultry production industry that decrease the productivity and efficiency of poultry production, impair animal welfare, and pose issues to public health. Furthermore, the use of antibiotic growth promoters (AGP) in feed, which have been used to improve the growth performance and gut health of chickens, has been restricted in many countries. Tannins, polyphenolic compounds that precipitate proteins, are considered as alternatives for AGP in feed and provide solutions to mitigate challenges in poultry production due to their antimicrobial, antioxidant, anti-inflammatory and gut health promoting effects. However, because high dosages of tannins have antinutritional effects when fed to poultry, determining appropriate dosages of supplemental tannins is critical for their potential implementation as a solution for the challenges faced in poultry production.

Abstract: The poultry industry has an important role in producing sources of protein for the world, and the size of global poultry production continues to increase annually. However, the poultry industry is confronting diverse challenges including bacterial infection (salmonellosis), coccidiosis, oxidative stress, including that caused by heat stress, welfare issues such as food pad dermatitis (FPD) and nitrogen and greenhouse gasses emissions that cumulatively cause food safety issues, reduce the efficacy of poultry production, impair animal welfare, and induce environmental issues. Furthermore, restrictions on the use of AGP have exacerbated several of these negative effects. Tannins, polyphenolic compounds that possess a protein precipitation capacity, have been considered as antinutritional factors in the past because high dosages of tannins can decrease feed intake and negatively affect nutrient digestibility and absorption. However, tannins have been shown to have antimicrobial, antioxidant and anti-inflammatory properties, and as such, have gained interest as promising bioactive compounds to help alleviate the challenges of AGP removal in the poultry industry. In addition, the beneficial effects of tannins can be enhanced by several strategies including heat processing, combining tannins with other bioactive compounds, and encapsulation. As a result, supplementation of tannins alone or in conjunction with the above strategies could be an effective approach to decrease the need of AGP and otherwise improve poultry production efficiency.

Keywords: tannins; chickens; gut health; feed additives; antimicrobials; antioxidants

---

1. Introduction

Poultry products including meat and eggs account for a significant part of global food production and constitute a protein staple throughout the world [1]. The United States Department of Agriculture (USDA) reported in 2016 that global egg production was approximately 70 million tons, and poultry meat production reached more than 100 million tons, which accounted for more than one-third of global meat production [2]. Moreover, global poultry production continues to increase annually [1]. However, there are many challenges in the poultry industry including: bacterial infection (salmonellosis); parasitic infection (coccidiosis); oxidative stress, including that caused by heat stress; welfare issues such as food pad dermatitis (FPD); and nitrogen and greenhouse gas emissions which can cause severe economic losses, threaten food safety and public health, impair animal welfare, and induce environmental pollution [2–5].

Antibiotic growth promoters (AGP) have been supplemented to chicken diets to improve growth performance and gut health, predominantly due to their antimicrobial effects and immunomodulatory functions in chickens [6,7]. However, because of the increased public concern about the transmission of antibiotic-resistant bacteria from poultry products, the use of AGP in poultry production has been banned or restricted in many countries [8–10]. In addition, some producers in the U.S. poultry industry have opted to entirely remove the use of antibiotics and instead raise chickens using "no antibiotics ever (NAE)" or "raised without antibiotics (RWA)" approaches [11,12]. As a consequence, the efficiency of poultry production has decreased due to increases in various bacterial and parasitic infections and reductions in the growth rate of chickens [13]. In addition, because there is no "magic bullet" that can replace AGP, some poultry producers are still using antibiotics in the U.S. and in many other countries, and the use of antibiotics for livestock animals in the world is expected to increase, possibly owing to population growth which is associated with a greater demand for livestock products in middle-income countries [14]. Therefore, it is essential to find alternatives to AGP, which must be cost-effective, eco-friendly and have antimicrobial and growth-promoting effects, without causing side effects (e.g., generation of resistant bacteria) to the animals and humans [15].

Tannins, defined as polyphenolic compounds that can precipitate proteins, are secondary metabolites, which are found in plants, seeds, bark, wood leaves and fruit skins and serve as plant defense mechanisms against predation [16]. High concentrations of tannins have been shown to have antinutritional effects in monogastric animals because tannins can decrease feed intake, nutrient digestibility and growth performance of chickens [17,18]. However, recently in poultry production, tannins have garnered a great deal of attention as an alternative for AGP because of their antimicrobial, antioxidants and anti-inflammation properties [19–21]. In addition, many tannins are considered sustainable feed additives, as they derive from byproducts of plant-based agriculture and industry. For example, chestnut tannins, which are already sometimes supplemented to poultry, are obtained by the distillation of wood that is used in the building industry [22]. However, the effects of tannins on the growth performance and gut ecosystem of the chickens are still inconsistent and their mode of action is unclear. Therefore, it is important to understand the chemical properties and biological effects of tannins to maximize the use of supplemental tannins in chickens. This review is mainly focused on the classification and bioavailability of tannins, the effects of distinct tannins on mitigating the challenges facing poultry production, and strategies to enhance the effects of tannins.

2. Classification and Bioavailability of Tannins

Tannins, defined as polyphenolic compounds that have a protein precipitation capacity, exist in several different types with various molecular weights [23]. Plant tannins are classified into hydrolysable tannins (HT) with tannin derivatives (e.g., gallic acid and ellagic acid) and condensed tannins (CT) [24] (Figure 1). Additionally, phlorotannins (PT) are a third class of tannin unique to brown algae. Tannins have different bioavailabilities (absorbability), and their level of bioavailability varies depending on several factors, including the derivatives of each tannin (e.g., gallic acid and ellagic acid), their affinity to protein, molecular structure and molecular weight. The bioavailability of tannins is an important trait for their functionality and should be considered for tackling different issues in poultry production. Tannins with a low bioavailability potentially have better antimicrobial effects in chickens, whereas highly bioavailable tannins would be more beneficial as antioxidant and anti-inflammatory agents.

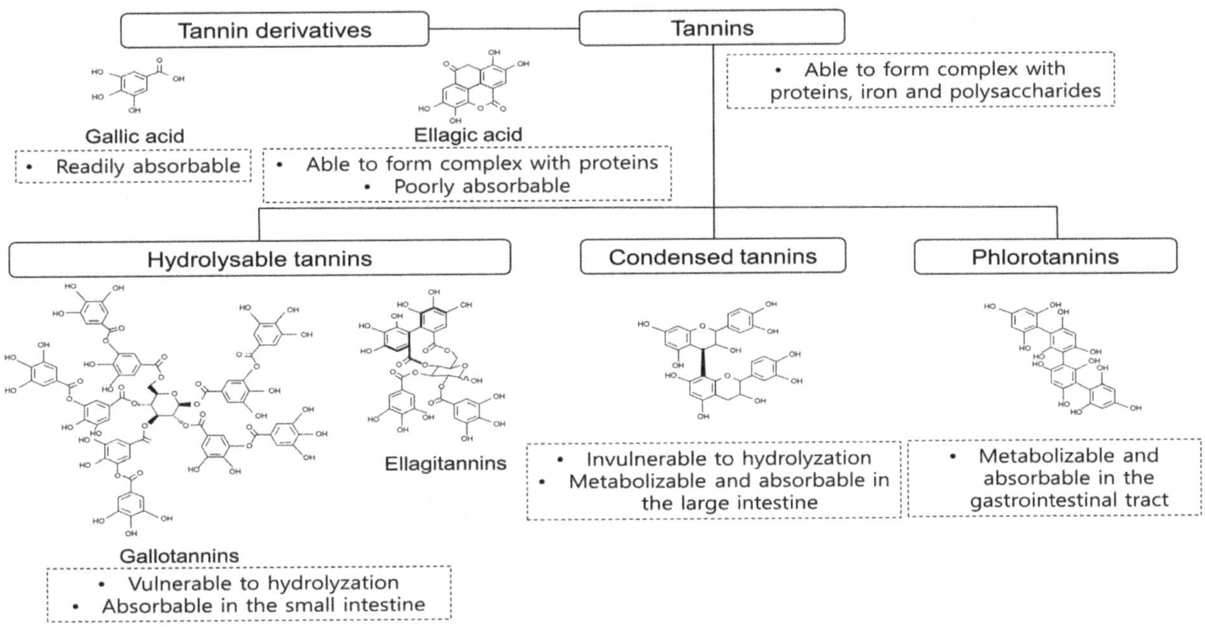

Figure 1. Classification and examples of molecular structure of tannins. Hydrolysable tannins and condensed tannins belong to plant tannins, and phlorotannins are found in brown algae.

The HT including predominately gallotannins and ellagitannins (molecular weight 500 to 3000 Da) contain a polyol (normally glucose) as a central core, which is esterified with phenolic groups (e.g., gallic acid and ellagic acid) [25]. Under certain conditions (e.g., thermal processing, esterification, and acid or base treatment), HT can be hydrolyzed to yield free gallic acid and ellagic acid. Gallotannins produce one molecule of sugar and 9 to 10 molecules of gallic acid when hydrolyzed, whereas ellagitannins yield one molecule of sugar and several molecules of gallic acid and ellagic acid [26]. The unhydrolyzed HT are partially absorbable in the small intestine [27]. Gallic acid is readily absorbable into the blood stream [28], while ellagic acid has low bioavailability due to strong affinity for proteins and poor absorption [29]. Hence, HT include gallotannins and ellagitannins and have different bioavailability depends on the components and structure.

The CT are defined as oligomeric or polymeric flavonoids containing flavan-3-olunits such as catechin, epicatechin, gallocatechin and epigallocatechin [24]. The CT exist in fruits (e.g., berries, pears and apples), forage legumes (e.g., lentils, black-eyed peas, chickpeas and red kidney beans), nuts, red and green grapes (and their juice and wine) [30–32]. The CT have higher molecular weight (1000—20,000 Da) and more complex structure compared to HT [24]. Unlike HT, CT are not vulnerable to hydrolyzation, which may imply low bioavailability in the gastrointestinal tract (GIT) of the chickens. This is because high molecular weight of tannins are hardly absorbable in the intestine [33]. Kahle et al. [34] reported that around 90% procyanidins (CT) were recovered in the distal ileum, which imply that most of CT can reach to the large intestine. The CT in the large intestine can be metabolized and absorbed by epithelial cells [35]. The CT have low bioavailability compared to HT and can be delivered to the large intestine of the chicken.

The PT, algae-derived polyphenols, have molecular sizes ranging 126 Da to 650 kDa, but the majority of them are 10 and 100 kDa [36,37]. By using in-vitro models, Corona et al. [38] demonstrated that PT can be metabolized and absorbed in the upper GIT potentially due to low molecular weight of PT, and limited amounts of biologically active PT were delivered to the colon. In addition, the author also demonstrated that higher molecular weight of PT are less vulnerable to metabolization and absorption in the upper GIT [38]. Nwosu et al. [39] showed that PT can be metabolized and absorbed by the colon cells (e.g., Caco-2). Therefore, PT can be easily metabolized and absorbed in the GIT tracts due to their low molecular weights. However, more future in-vivo experiments are required to study bioavailability of PT in chickens.

Taken together, tannins are classified into HT, CT and PT, and their bioavailability are depending on their components, structure and molecular weight. It is important to understand specific chemical properties each tannin prior to application in poultry diets.

3. Traditional Viewpoints on Tannins as Antinutritional Factors in Poultry Production

While tannins were known as beneficial bioactive compounds in ruminants, they have been considered as antinutritional factors in poultry diets [16]. In the (GIT) of ruminants, tannins bind proteins under rumen pH (pH 5.5–7.0), inhibiting the microbial degradation of dietary proteins. Upon reaching the abomasum (pH 2.5–3.5), the non-covalent linkages between proteins and tannins are broken, and free protein can be absorbed by the host in the distal small intestine (pH 7.5) [40]. As such, tannins are known to increase ruminant's protein utilization and decrease gas emission [41,42]. However, in chickens, tannins were considered as phytotoxins predominately due to their protein binding properties, which impair dietary protein digestion and decrease activity of digestive enzymes [43]. Moreover, tannins bind proline-rich, hydrophobic salivary proteins of chickens, forming complexes that are responsible for an astringent taste of the feed, and in turn lead to low palatability and decreased feed intake of chickens [44]. In addition, a high concentration of tannic acid (25 g/kg) showed toxicity to chickens by showing liver proteolytic activity in broiler chickens [45]. Lee et al. [46] reported that tannic acid impaired growth performance, hematological indices and plasma iron status in weaning pigs potentially because tannic acid can form complex with iron. However, these toxic properties of tannins were shown when tannins were included in the diets more than 7.5 g/kg (based on tannic acid, considered as the standard of HT), and many current studies proved that appropriate amounts of tannins (based on tannic acid; standard of tannins) ranging from to 0.5 g/kg to 5 g/kg in poultry could improve grow rate and gut health due to their potential antimicrobial, antioxidants and anti-inflammatory functions (Table 1). Therefore, while tannins were considered as anti-nutritional factors in the past, tannins at appropriate dosages have potentials to improve growth performance and gut health of chickens.

Table 1. Effects of different dosages of tannic acid on the broiler chickens.

Dosages of Tannic Acid	Outcomes	References
0.5 g/kg	Improved growth performance and immune system in broiler chickens challenged with *Eimeria* spp.	[47]
2 g/kg	Improved foot pad dermatitis of the chickens without affecting growth performance	[48]
5 g/kg	Improved growth performance and decreased lipid oxidation.	[49]
5 g/kg	Increased growth performance, modulated cecal microbial metabolites and decreased cecal pH.	[50]
7.5 and 15 g/kg	Decreased growth performance and did not inhibit the growth of *Salmonella typhimurium* in broiler chickens challenged with *S. typhimurium*.	[51]
10 g/kg	Attenuated fatty acid profile of breast but decreased growth performance in the heat stressed chickens.	[52]
25 g/kg	Decreased growth performance and showed liver toxicity by inducing liver proteolytic activity	[45]

4. Challenges in Poultry Production and Potential Solution by Using Tannins

4.1. *Effects of Tannins on Bacterial Infection (Salmonellosis)*

There are many pathogenic bacteria in chickens and poultry products (meat and eggs), including *Salmonella Typhimurium*, *Escherichia coli* O157:H7, *Campylobacter jejuni*, *Clostridium perfringens*, enterohemorrhagic *Escherichia. coli* (EHEC), *Listeria monocytogenes*, *Arcobacter butzleri*, *Mycobacterium avium* subsp. *Paratuberculosis* and *Aeromonas hydrophila* [53]. These pathogenic bacteria not only impair gut health and growth rate of chickens, but are also a public health threat as foodborne diseases in humans [54]. Salmonellosis, one of the main food-borne diseases from poultry products, is induced by *S. Typhimurium* and *S. Enteritis* and causes, illness, morbidity and mortality in humans [55].

Diverse in-vitro studies showed that tannins and their derivatives showed bacteriostatic (inhibiting bacterial growth) and bactericidal (killing bacteria) effects on *Salmonella* spp. and other pathogens, as shown in Table 2. Potential mechanisms of antibacterial effects of tannins include (1) direct interactions with components in the cell wall to alter morphology of the cell wall and to increase membrane permeability of bacteria [24]; (2) decreasing activities of microbial enzymes [56]; and (3) depriving essential nutrients such as proteins and minerals (e.g., iron) for pathogenic bacteria [57–59]. In addition to bacteriostatic and bactericidal effects of tannins, many in-vitro studies reported that sub-lethal dosages of tannins also restricted pathogenicity of *Salmonella* spp. and other pathogens by inhibiting motility [60], quorum sensing [61] and biofilm formation [62] of pathogenic bacteria.

Table 2. In-vitro antimicrobial effects tannins against diverse pathogenic bacteria.

Tannins Sources	Strains	Results and Conclusions	References
Ellagitannins from Chestnut wood Gallotannins from Tara and Sumach (Gall nuts) Condensed tannins from Quebaracho and *Calliandra calothyrsus* Flavanol gallates from Tea and *Acacia nilotica*	*Salmonella Typhimurium*	All of the tannins inhibited the growth of *S. Typhimurium.*	[63]
Tannic acid Gallic acid	*S. Typhimurium*	Tannic acid and gallic acid had bactericidal effects and gallic acid had higher bactericidal effects than tannic acid	[64]
Condensed tannins extracted from tree leaves viz. babool (*Acacia nilotica*), jamun (*Eugenia jambolana*), peepal (*Ficus religiosa*), subabul (*Leucaenia leucocephala*) and guajava (*Psidium guajava*)	*Escherichia coli* *Staphylococcus aureus* *S. enteritis Enterococcus faecalis*	All of the selected five condensed tannins inhibited the growth of the four pathogenic bacteria	[65]
Tannin extracts from *Cytinus hypocistis* and *C. ruber*	*S. aureus* *S. epidermidis* *E. faecium* *Pseudomonas aeruginosa* *Klebsiella pneumoniae*	Tannin extracts from *C. hypocistis* and *C. ruber* showed antibacterial and antibiofilm activities against gram positive and negative human pathogens.	[66]
Chestnut tannins (80% hydrolysable tannins) Quebracho tannins (75% condensed tannins)	*Clostridium perfringens*	Both hydrolysable and condensed showed antimicrobial effects against *C. perfringens* and neutralized its cytotoxicity.	[67]

In addition to their antibacterial effects, tannins can control systemic infection of *Salmonella* spp. by beneficially modulated components of gut ecosystem. *Salmonella* spp. can enter blood circulation via paracellular and transcellular pathways and use immune cells to enter enterocytes to be distributed in the internal organs and muscles in chickens (Figure 2). Many studies reported that tannins altered expression and functionality of tight junction proteins [68], mucus [69,70], and immune cells [71,72] of chickens. Nevertheless, supplemental tannins have not typically translated to antimicrobial effects against *Salmonella* spp.in in-vivo studies. For example, while Van Parys et al. [73] reported that whereas HT extract of sweet chestnut woods showed strong antimicrobial effects (minimum inhibitory concentration: 25–50 μg/mL and minimum inhibitory concentration: 100 μg/mL) against *S. Typhimurium* isolated from pigs under *in-vitro* conditions, the inclusion of 3000 mg/kg HT in the pig feed did not reduce the *Salmonella* spp. concentration in feces, intestine and internal organs of pigs inoculated 10^7 colony forming units (CFU)/mL of *S. Typhimurium* strain at four days post-inoculum. Similarly, Kubena et al. [51] reported that tannic acid (7.5 or 15 g/kg in the feed) did not modulate the salmonella concentration in cecal content of broiler chickens inoculated with 10^4 CFU of *S. typhimurium.* Potential explanations for such findings include (1) tannins were degraded by host or microbial enzymes or absorbed before they reached to the lower intestine where most of the *Salmonella* spp. and other pathogenic bacteria propagate; (2) tannins formed complexes with components of feedstuffs (polysaccharides and proteins) or endogenous proteins, which inhibits antimicrobial effects of tannins; and (3) experimental factors including the low dosages of tannins, the high dosages of salmonella inoculum or short period of the experiments were possibly obstacles to diminish antimicrobial effects of tannins in in-vivo models [73]. In contrast, Jamroz et al. [74] reported that 1000 mg/kg of sweet chestnut tannins reduced the number of *E. coli* and coliforms bacteria in the small intestine on 28 d; however, growth performance of the tannin-fed birds was decreased in this study.

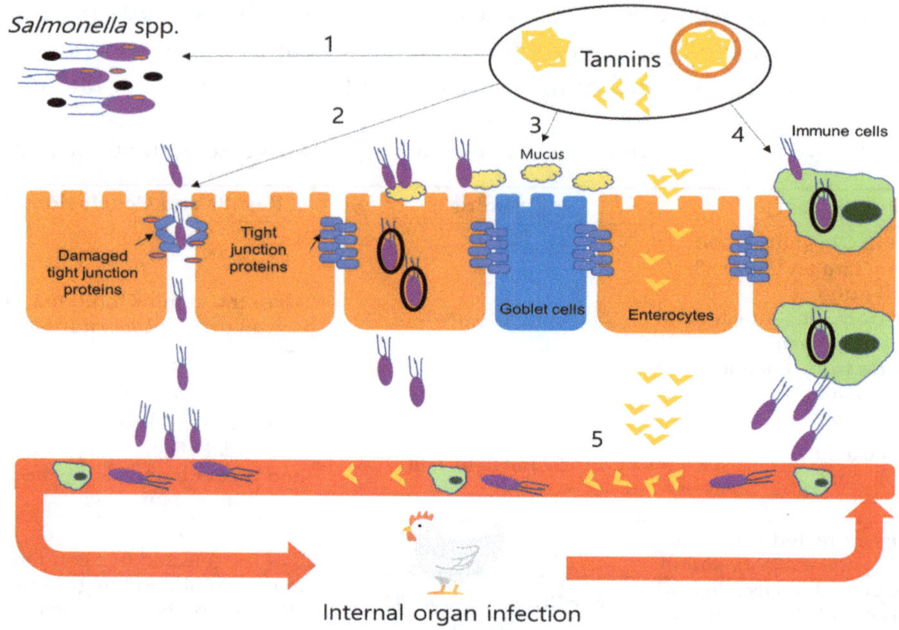

Figure 2. Systemic infection routes of *Salmonella* spp. and potential mechanisms of antibacterial actions of tannins in an in-vivo chicken model. Chickens can orally ingest *Salmonella* spp. from feed or their environment, and subsequently escape the small or large intestine [75]. (1) Tannins and tannin derivatives (tannins, hydrolyzed tannins, and tannin-protein complexes) are known to inhibit the growth of *Salmonella* spp. in the intestine and decrease quorum sensing of the bacteria [61]. Lipopolysaccharides (LPS) of *Salmonella* spp. can impair intestinal barrier function, which allows *Salmonella* spp. to pass the paracellular pathways of the intestine [76,77]. (2) Tannins and tannin derivatives may improve gut barrier integrity by neutralizing LPS or decreasing expression of cytokines, which can impair tight junction proteins [78]. (3) and (4) *Salmonella* spp. also utilized mucus and immune cells to invade epithelial cells [71]. Tannins and their derivates potentially modulate expression and morphology of immune cells and mucus. *Salmonella* spp. can invade epithelial cells in diverse pathways and enter blood circulation and finally colonize in internal organs (e.g., liver, kidney, spleen, etc.) and muscles in chickens [79]. (5) Tannins also can enter the blood circulation and potentially show antimicrobial effects and modulate immune cells to attenuate internal organ infection by *Salmonella* spp. in chickens [80,81].

While many in-vitro studies showed that tannins showed antimicrobial effects against *Salmonella* spp. and other pathogens, the optimal antimicrobial dosages in chickens have not yet been fully determined. Therefore, dose-specific antimicrobial effects of tannins in chickens, as well as their mode of actions, warrant further investigation.

4.2. *Effects of Tannins on Coccidiosis*

Coccidiosis, which is a parasitic disease induced by protozoa of the family *Eimeridae*, is one of the most prevalent and detrimental enteric diseases in poultry production [82]. The nine identified *Eimeria* species in chickens includes *E. acervulina*, *E. brunetti*, *E. maxima*, *E. necatrix*, *E. praecox*, *E. mitis*,

E. tenella, *E. mivati* and *E. hagani* to date [83,84]. *Eimeria* spp. infect and multiply within the mucosal epithelial layers in the different parts of the GIT through the fecal-oral route [85]. Coccidiosis can result in reduced growth rate and gut barrier integrity and induce inflammation, diarrhea, hemorrhaging, and even mortality in broiler chickens, negatively influencing the efficacy of poultry production and welfare [86,87]. The negative effects of coccidiosis on gut health of chickens are closely related to oxidative stress, as *Eimeria* infections cause lipid peroxidation and excessive production of reactive oxygen species (ROS) in chickens [88]. Furthermore, coccidiosis is closely associated with an enteric infectious disease, necrotic enteritis, which is predominately induced by *C. perfringens* with the presence of *Eimeria* spp. [89,90]. To cope with coccidiosis for the poultry industry, prophylactic coccidiostats and anticoccidial drugs have been supplemented in poultry diets [91]. However, resistance for all currently available drugs have been documented, making it imperative to discover novel drug alternatives that induce limited resistance and effectively control coccidiosis in broiler chickens [92]. To find alternatives for prophylactic coccidiostats and anticoccidial drugs, bioactive compounds including prebiotics [93], plant extracts [94], organic acids [95], essential oils [96], lipids (fatty acids) [97], minerals (e.g., zinc) [98] and nitro compounds [99] have been studied in chickens.

Tannins are known to have anticoccidial effects because tannins can form complexes with parasitic enzymes and metal ions, which are essential for *Eimeria* spp. and can stimulate immune system against of the chickens [100–102]. Tonda et al. [47] reported that the dietary supplementation of 500 mg/kg of gallnut tannic acid extract reduced total oocyst number in excreta, and 500 mg/kg of tannic acid or gallnut tannic acid extract decreased intestinal lesion scores in broilers infected with *Eimeria* spp. Furthermore, the authors showed that gallnut tannic acid extract enhanced feed conversion ratio of cocci-vaccinated birds, which possibly implies that gallnut tannic acid extract improved protective immunity following coccidiosis vaccination [47]. Supplemental chestnut HT and quebracho CT tannins reduced *Eimeria* spp. oocyst shedding and parasitic bacterial diarrhea and attenuated negative effects of coccidiosis via immunomodulating and anti-inflammatory effects in rabbits [103]. In broiler chickens, Kaleem et al. [104] showed that administration of *Emblica officinalis* derived tannins improved growth performance and showed immunostimulatory properties and enhanced protective immune system. The beneficial effects of tannins on gut health of chickens infected with *Eimeria* spp. are closely associated with antioxidant properties of tannins, which can restore an *Eimeria*-damaged gastrointestinal [105]. A study by [106] demonstrated that supplementation of grape seed proanthocyanidin extract, rich in CT, enhanced growth performance and attenuated clinical symptoms, potentially by improving antioxidant capacity in chickens infected with *Eimeria tenella*. In contrast, Mansoori and Modirsanei [107] showed that supplemental tannic acid (10 g/kg) numerically increased D-xylose absorption in chickens vaccinated against coccidiosis followed by challenging with *Eimeria* spp., however, supplemental tannic acid increased the total number of oocysts in excreta, indicating that high dosages of tannins can attenuate the efficiency of anticoccidial vaccines and impair appropriate development of immune system against coccidiosis in chickens. The discrepancy may be attributed to different sources and concentrations of tannins and dissimilar experimental conditions (e.g., challenge dosages of *Eimeria* spp.).

Although many studies reported the potential benefits of supplemental tannins in broiler chickens infected *Eimeria* spp., more comprehensive studies are required (1) to study the mechanisms under anticoccidial effects of tannins in chickens; (2) to elucidate mechanisms of the beneficial effects of tannins on the gut health of chickens infected with *Eimeria* spp.; (3) to investigate the effects of tannins in a necrotic enteritis challenge model by using inoculum of *Eimeria* spp. and *C. perfringens*; and (4) to find appropriate concentrations and types of tannins against coccidiosis in chickens.

4.3. Effects of Tannins on Oxidative Stress, Including that Caused by Heat Stress

Heat stress is one of the major obstacles in the poultry industry because heat stress negatively impacts growth performance, gut health, meat quality, and welfare of chickens [108,109]. A potential reason for negative effects of heat stress on chickens is closely associated with excessive production of

ROS via accelerated metabolic reactions due to mitochondrial respiration [110–112]. Under normal conditions, enzymatic and non-enzymatic antioxidants can neutralize ROS and maintain the balance between oxidants and antioxidants [113,114] (Figure 3). However, if there is an imbalance between oxidants and antioxidants in chickens, excessively produced ROS can impair gut health and induce inflammation, which results in decreased growth performance of chickens [115]. Antioxidants including plant extracts [116,117], L-carnitine [118], vitamin C [119,120], vitamin E [121], and selenium [122] have been studied to alleviate heat stress and oxidative stress.

Tannins are believed to relieve or attenuate effects of oxidative stress, including that caused by heat stress by scavenging ROS and modulating enzymatic antioxidants in animals (Figure 3) [116]. The reducing power of tannins including proanthocyanidins, catechins, epicatechin, and procyanidin from grade seeds is approximately 20 times higher than vitamin E and 50 times higher than vitamin C [123]. A study from Sahin et al. [124] reported that 200 or 400 mg/kg of epigallocatechin-3-gallate (the ester of epigallocatechin and gallic acid) from green tea relieved oxidative stress by controlling the hepatic nuclear transcription factors such as nuclear factor κ-light-chain-enhancer of activated B cells (NF-κB) and nuclear factor (erythroid-derived 2)-like 2 (Nrf2) in heat stressed quails. Furthermore, 10 mg/kg supplementation of tannic acid improved fatty acid profile (decreased monosaturated fatty acids) in breast muscle of chickens under heat stress [52]. Ramnath and Rekha [125] showed that supplementation of *Brahma Rasayana* containing various sorts of tannins enhanced activities of enzymatic antioxidants including superoxide dismutase (SOD), glutathione peroxidase (GPx), glutathione reductase (GR), and reduced glutathione (GSH) in blood of chickens raised at a cold temperature. Moreover, the inclusion of grape (*Vitis vinifera*) pomace, rich in CT, enhanced antioxidant enzyme activities (GPx and SOD) and intestinal morphology, and increased relative weight of bursa of Fabricius and thymus in the heat-stressed broiler chickens [126]. In a diquat-induced mouse model (oxidative stress model), tannic acid improved intestinal morphology, activated the antioxidative pathway by reducing protein expression of Kelch like-ECH-associated protein 1 (KEAP1) and enhancing protein expression of nuclear factor erythroid 2-related factor 2 (NRF2), as well as modulated intestinal barrier function in the jejunum. Thus, supplementing appropriate concentrations of tannins potentially would be an effective strategy to attenuate oxidative stress in heat-stressed birds.

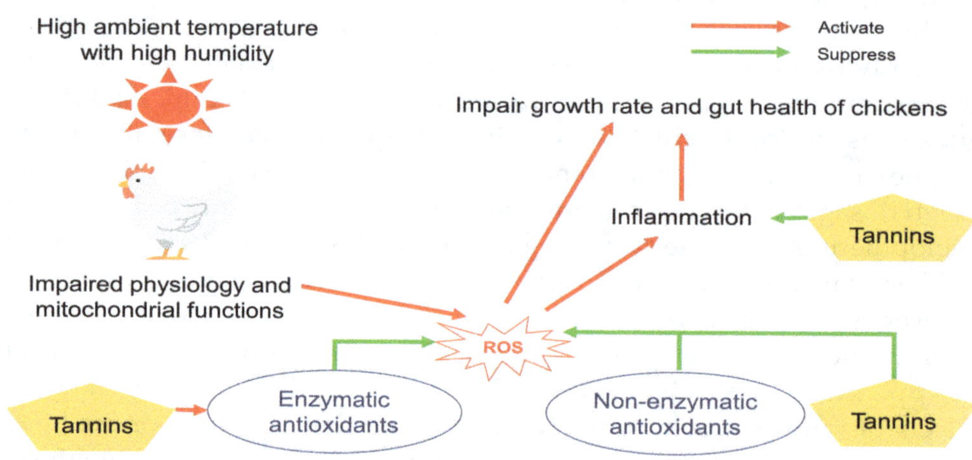

Figure 3. Potential mechanisms for the alleviating effects of tannins in heat-stressed chickens. High ambient temperature with high humidity increase core body temperature of chickens, and this can result in impaired physiology and mitochondrial functions, which lead to excessive reactive oxygen species (ROS) [127]. Tannins, polyphenolic compounds, directly scavenge ROS like non-enzymatic antioxidants [128], and also, tannins can increase activities of antioxidant enzymes such as superoxide dismutase (SOD), glutathione peroxidase (GPX), and glutathione reductase (GR) in broiler chickens. The ROS cause inflammation that can impair growth rate and gut health of chickens and tannins also have anti-inflammatory function [124].

4.4. *Effects of Tannins on Food Pad Dermatitis (FPD)*

Foot pad dermatitis (FPD) is defined as a condition that induces lesions on the plantar surface of the footpads in growing chickens [129]. The FPD causes severe economic losses in poultry production because paws are the third most crucial economic part in broiler chickens, and FPD can impair growth rate, gut health, and welfare of chickens [129,130]. Thøfner et al. [131] showed that FPD and systemic bacterial infections are closely correlated because pathogens can invade to the chickens through damaged epithelium on the foot pads [132]. Litter moisture and litter quality are the most two crucial features that cause FPD among varied factors such as bedding materials and depth, drinkers and nutrient deficiencies [129,133,134]. In addition, increased excreta viscosity has been known to increase the occurrence and severity of FPD in chickens by modulating litter moisture and litter quality [135].

Tannins can relieve the incidence and severity of FPD by enhancing fecal consistency (e.g., fecal dry matter contents) and litter quality [16]. Cengiz et al. [48] reported that 2000 mg/kg supplementation of tannic acid reduced the incidence and severity of FPD in broiler chickens without affecting growth performance, litter quality and intestinal viscosity of chickens. Moreover, 700 mg/kg and 2000 mg/kg of tannin-rich sweet chestnut wood extract increased fecal dry matter contents in chickens [136]. In addition, antimicrobial, antioxidant and anti-inflammatory properties of tannins probably helped to attenuate the incidence and severity of FPD in broiler chickens because oxidative stress and inflammation can exacerbate the severity and consequences of FPD in chickens [137,138]. Thus, supplementation of tannins has potential to reduce severity and incidence of FPD in broiler chickens by enhancing fecal consistency and litter quality.

4.5. *Effects of Tannins on Nitrogen Excretion and Emissions of Noxious and Greenhouse Gases*

Reactive nitrogen species (ammonia, nitrous oxide, and other oxides of nitrogen) and sulfur-containing compounds (hydrogen sulfide and sulfur dioxide) are produced in poultry production facilities and cause environment pollution and greenhouse gases such as carbon monoxide, carbon dioxide and methane [139–142]. Factors affecting the production of nitrogen compounds and detrimental gases include types of feedstuffs, manure conditions, and housing accessories (bedding and heating materials) [143,144]. Nitrogen and greenhouse gases are produced in the livestock animals due to enteric fermentation and manure fermentation, and ruminants contribute most of the livestock greenhouse gas production because of high fermentation rate in the rumen [140]. Although poultry production is not a major contributor of noxious or greenhouse gases, the continued growth of poultry production coupled with higher intensity of excretion and emission per unit compared to other species, more attention is needed to find strategies to mitigate the production of nitrogen excretion and emissions of noxious and greenhouse gases [145].

It is well-established that tannins improve N utilization and decrease methane production in ruminant animals [25,146]. The potential mechanisms underlying effects of tannins on mitigation of emissions of nitrogen and greenhouse gases are inhibiting growth of methanogens, reducing protozoal-associated methane production, and decreasing fiber fermentation [25]. Although chickens (monogastric animals) have distinct GIT from ruminants, tannins may alter gut health and microbiota and improve N utilization in chickens, thereby potentially reducing nitrogen and methane emissions. Ahmed and Yang [147] reported that supplementation of by-products of *Punica granatum* (fruit), which contains HT such as ellagitannins, punicalagin, punicalin and pedunculagin, decreases the emission of ammonia and methanethiol from excreta of broiler chickens. Moreover, a study by Bostami et al. [139] demonstrated that supplementation of fermented pomegranate byproducts, containing ellagitannins, reduced gas emission (ammonia and hydrogen sulfide) from excreta in broiler chickens, potentially via reducing microbial activity and pH of excreta. However, 2000 mg/kg of tannic acid from chestnut wood did not affect ammonia volatilization in broiler chickens. Hence, more studies are required to identify suitable sorts and dosages of tannins to reduce nitrogen excretion and emissions of noxious and greenhouse gases during poultry production.

4.6. Effects of Tannins on Growth Performance, Immune System, Gut Microbiota, Gut Ecosystem and in Chickens Raised under General Conditions

While some authors showed that low concentrations (0.5 g/kg to 5 g/kg) of tannic acid improved growth performance (shown in Table 1), the others reported that supplementation of different sources of tannins at low dosages (0.5 to 5 g/kg) did not affect growth performance [48,136] and even showed negative effects on growth performance of the birds [74]. These differences may be a result of different tannin sources, supplementation period, or specific experiment conditions (e.g., genetics of chickens, temperature and abundance of pathogens in the living conditions).

However, supplemental tannins at appropriate dosages can improve immune system, gut ecosystem and gut microbiota of chickens raised under general conditions. Ramah et al. [148] reported that whereas the high dosage of tannic acid (30 g/kg diet) have negative impacts on immune system by decreasing relative (cluster of differentiation) CD4+, CD8+, CD4+CD8+ and $\gamma\delta$+ cell populations in thymus, spleen, and cecal tonsils and by reducing cytokine mRNA expression in spleen cells, the low dosage of tannic acid (0.5 g/kg) enhanced CD4+CD8+ subpopulations and $\gamma\delta$+ cells in spleen and CD4+CD8+ subpopulations and B cells in cecal tonsils and increasing mRNA expression of IFN-γ in broiler chickens. A study by Karaffová et al. [149], reported that tannins were beneficial to maintain components of mucosal immunity of chickens via upregulating immunoglobulins A and mucin 2. Erlejman et al. [150] also demonstrated that CT can combine with receptors of tumor necrosis factor-α (pro-inflammatory cytokines) to inhibit inflammation, which implies that tannins directly modulate immune system without eliciting antimicrobial and antioxidant properties.

Diaz Carrasco et al. [151] reported that a blend of chestnut (HT) and quebracho tannins (CT) modulated diversity cecal microbiota of chickens and decreased genus *Bacteroides* and increased certain members of order Clostridiales predominately in the families Ruminococcaceae and Lachnospiraceae. A study by Koo and Nyachoti [152] demonstrated that tannic acid positively modulated microbial metabolites in pigs fed oxidized oil. Selective antimicrobial effects of tannins would beneficially modulate microbiome of animals [153]. Viveros et al. [154] also suggested that tannins probably can have prebiotic effects via stimulating the proliferation of the beneficial bacteria.

Microbiome modulating effects of tannins may partially explain the gut health promoting effects of the chickens because microbiota of the chicken is closely associated with the gut ecosystem of chickens [155]. Moreover, an in-vitro study by Brus et al. [19] reported that chestnut tannins stimulated proliferation of enterocytes and enhanced antioxidative properties of the chicken small intestinal epithelial cells. Bilić-Šobot et al. [156] demonstrated that HT decreased production of cell debris in the large intestine of pigs, which leaded to decreased production of skatole, which belongs to the indole family. Together, supplemental tannins have the potential to improve growth performance, gut microbiota and gut ecosystem in broiler chickens, even in the absence of challenge models.

5. Strategies to Maximize the Effects of Supplemental Tannins in the Chickens

5.1. Heat Process on Tannins

Some in-vitro studies showed that heat processed HT had better antimicrobial and antioxidant properties than unprocessed HT [143,144]. This would be because heat processing could partially hydrolyze tannic acid and release gallic acid molecules, and these newly produced gallic acid and galloyl groups had enhanced antimicrobial and antioxidant effects compared to the fresh tannic acid [157]. González et al. [145] also reported that thermal process of *hamamelis virginiana* containing gallotannins and CT improved efficacy of antioxidant properties for inhibiting lipid oxidation. However, because CT are hardly hydrolyzed, enhanced antioxidant of heat-processed *hamamelis virginiana* probably due to hydrolyzation of gallotannins in *hamamelis virginiana* rather than hydrolyzation of CT. Thus, while in-vitro studies found that heat process of tannins could improve their functional properties (e.g., antioxidant and antimicrobial effects) compared to unprocessed HT, it is unknown yet whether heat-processed tannins have more beneficial effects on animal models.

5.2. Co-Supplementation of Tannins with other Bioactive Compounds

Supplementation of tannins with other bioactive compounds could be more beneficial to chickens than supplementing tannins alone for several reasons: (1) complexed form with proteins or polysaccharides of tannins inhibit tannins to form a complex with endogenous and dietary proteins and metal ions; (2) distinct properties of bioactive compounds can show synergistic effects to antimicrobial effects against both gram negative and positive bacteria; (3) different bioactive compounds affect gut health in different ways, which can lead to synergistic effects in animals; and (4) by providing more than two bioactive compounds, pathogenic bacteria are hard to generate resistant system against diverse bioactive compounds. Table 3 shows that tannins have potentials to show synergistic effects with other bioactive compounds.

Probiotics are living microorganisms which beneficially affect the host animals by enhancing animal's intestinal microbial balance [158]. Probiotics may have different mode of actions from tannins to inhibit the growth of pathogenic bacteria and to improve gut health of chickens. Probiotics can improve gut integrity by modulating immune system and maintaining microflora of chickens and tannins, while tannins can show antioxidant and anti-inflammatory properties [159]. However, one of the concerns of using probiotics with tannins could be that tannins may show antimicrobial effects against probiotics. However, Khalil [160] showed that gallic acid and catechin polyphenols did not inhibit the growth of *Streptococcus thermophilus* (probiotics), and Pacheco-Ordaz et al. [153] reported that catechin, gallic, vanillic, ferulic and protocatechuic acids selectively inhibit the growth of pathogenic bacteria without decreasing viability of probiotics. More studies are required to establish synergistic effects and mechanisms of tannins and probiotics in *in-vivo* chicken models.

Table 3. Effects of tannins with other bioactive compounds on the chickens.

Tannins	Other Bioactive Compounds	Outcomes	References
100 mg/kg tannic acid extract	Probiotics (1×10^4 spores/kg *Bacillus coagulans*)	Improved feed conversion ratio of coccidiosis vaccinated broilers.	[47]
240 mg/kg tannic acid	Organic acids (420 mg/kg lactic, 480 mg/kg butyric acid and 480 mg/kg acetic acid)	Decreased *S. enteritis* horizontal transmission in broiler chickens	[161]
Chestnut tannins	Saturated short medium chain fatty acids (C4:0 to C12:0)	Showed strong antimicrobial effects in in-vitro conditions and did not affect growth performance and meat quality of in-vivo chicken models.	[22]

Organic acids, known as strong antimicrobials, are organic compounds with acidic properties. Tannins inhibit the growth of pathogens predominately by inhibiting activities of microbial enzymes and modulating bacterial membrane, but organic acids penetrate bacterial cell wall, and bacteria have to spend a lot of energy to pump out hydrogen molecules, which causes bacterial death [162,163]. Furthermore, organic acids are known to improve intestinal morphology and gut barrier integrity by being energy sources for epithelial cells, which may imply that organic acids with tannins can show synergistic effects [164–166]. Thus, combination of tannins and organic acids can have synergistically increased antimicrobial effects and gut health promoting effects due to different mode of actions.

5.3. Supplementation of Combined or Encapsulated Form of Tannins

If tannins are combined with proteins, polysaccharides and ions before being included in the chickens feed, the tannins in complexes would not bind dietary and endogenous proteins and metal ions in chickens. The tannin complexes would be loose in the high pH (>7.0) in the intestine of chickens, and proteins in the tannin complexes can be degraded by digestive enzymes in the small intestine of chickens [40,167]. However, Lee et al. [46] showed that supplementation of albumin-tannin complexes still decreased growth performance and negatively modulated microbiota, hematological indices and plasma iron status of weaning piglets. The delivery of tannin-protein or polysaccharide complexes in the GIT of chickens, and effects of diverse dosages of supplemental tannin complexes on growth rate and gut health of chickens should be further investigated.

Encapsulation techniques, which offer a physical barrier for bioactive compounds and separate the core material from the environment until their release, have obtained a lot of attention in the livestock industry because encapsulation can maximize the efficacy of feed additives that have stability, cost and environmental issues. Encapsulation has been applied to various vulnerable feed additives such as essential oils [96,168,169], probiotics [170], organic acids [171], bacteriophages [172], zinc [173] and exogenous enzymes [174]. Diverse materials including proteins [96,175], lipids [171], carbohydrates (starch) [169], and polysaccharides [172,176] have been used to encapsulate bioactive compounds.

Encapsulation techniques can be applied to decrease side effects and maximize benefits of tannins in chickens. Encapsulation of tannins can depress the protein binding capacity of tannins, which decreases feed intake by making astringent taste and digestibility of proteins and induces dietary and endogenous protein losses [177]. In addition, more tannins can be delivered to small or large intestine where many pathogens propagate by decreasing bioavailability for absorption in the upper GIT of chickens. Adejoro et al. [178] showed that lipid-encapsulated acacia tannin extracts reduced methane production and enhanced neutral detergent fiber digestibility in sheep. A study by Wang et al. [179] reported that microencapsulated tannic acid improved intestinal morphology in duodenum, increased expression of ileal nutrient transporters (sodium-dependent neutral amino acid transporter; B^0AT1 and peptide transporter 1; PepT1) and modulate microbiota without affecting growth performance of weaned piglets even though ileal maltase activity and gene expression of jejunal sodium-dependent glucose transporter 1 (SGLT1) was reduced. Future studies are required to develop effective encapsulated tannins and to determine appropriate dosages of encapsulated tannins for chickens.

6. Conclusions

There are various kinds of tannins, which have different bioavailability in chickens. Depends on the sorts and dosages of tannins, tannins can elicit detrimental effects on growth performance and gut ecosystem of the broiler chickens or can beneficially modulate growth performance and gut ecosystem of the broiler chickens. In addition, heat processing, co-supplementation with other bioactive compounds and encapsulation potentially enhance the beneficial biological effects of tannins. In conclusion, supplementation of tannins alone or in conjunction with those strategies have large potential to alleviate challenges, replace AGP and improve production efficiency in poultry productions.

Author Contributions: Conceptualization, J.C. and W.K.K.; supervision, W.K.K.; investigation, W.K.K.; Writing—Original draft preparation, J.C.; Writing—Reviewing and editing, W.K.K. and J.C. All authors have read and agreed to the published version of the manuscript.

Acknowledgments: We appreciate Brett Marshall for his help in the manuscript preparation.

References

1. Mottet, A.; Tempio, G. Global poultry production: Current state and future outlook and challenges. *World. Poult. Sci. J.* **2017**, *73*, 245–256. [CrossRef]
2. Pawar, S.S.; Basavaraj, S.; Dhansing, L.V.; Kurade, N.P.; Sahebrao, K.A.; Vitthal, N.A.; Pandit, B.M.; Kumar, B.S. Assessing and Mitigating the Impact of Heat Stress in Poultry. *Adv. Anim. Veter. Sci.* **2016**, *4*, 332–341. [CrossRef]
3. Nkukwana, T. Global poultry production: Current impact and future outlook on the South African poultry industry. *South. Afr. J. Anim. Sci.* **2019**, *48*, 869. [CrossRef]

4. Akil, L.; Ahmad, H.A. Quantitative Risk Assessment Model of Human Salmonellosis Resulting from Consumption of Broiler Chicken. *Diseases* **2019**, *7*. [CrossRef]
5. Hunter, J.M.; Anders, S.A.; Crowe, T.; Korver, D.R.; Bench, C.J. Practical assessment and management of foot pad dermatitis in commercial broiler chickens: A Field Study. *J. Appl. Poult. Res.* **2017**, *26*, 593–604. [CrossRef]
6. Hassan, Y.I.; Lahaye, L.; Gong, M.M.; Peng, J.; Gong, J.; Liu, S.; Gay, C.G.; Yang, C. Innovative drugs, chemicals, and enzymes within the animal production chain. *Veter. Res.* **2018**, *49*, 71. [CrossRef] [PubMed]
7. Kumar, S.; Chen, C.; Indugu, N.; Werlang, G.O.; Singh, M.; Kim, W.K.; Thippareddi, H. Effect of antibiotic withdrawal in feed on chicken gut microbial dynamics, immunity, growth performance and prevalence of foodborne pathogens. *PLoS ONE* **2018**, *13*, e0192450. [CrossRef]
8. Muhammad, J.; Khan, S.; Su, J.Q.; Hesham, A.E.-L.; Ditta, A.; Nawab, J.; Ali, A. Antibiotics in poultry manure and their associated health issues: A systematic review. *J. Soils Sediments* **2020**, *20*, 486–497. [CrossRef]
9. Brown, K.; Uwiera, R.R.; Kalmokoff, M.L.; Brooks, S.P.; Inglis, G.D. Antimicrobial growth promoter use in livestock: A requirement to understand their modes of action to develop effective alternatives. *Int. J. Antimicrob. Agents* **2017**, *49*, 12–24. [CrossRef]
10. Hu, Y.J.; Cowling, B.J. Reducing antibiotic use in livestock, China. *Bull. World Health Organ.* **2020**, *98*, 360. [CrossRef]
11. Karavolias, J.; Salois, M.J.; Baker, K.T.; Watkins, K. Raised without antibiotics: impact on animal welfare and implications for food policy. *Transl. Anim. Sci.* **2018**, *2*, 337–348. [CrossRef]
12. Singer, R.S.; Porter, L.J.; Thomson, D.U.; Gage, M.; Beaudoin, A.; Wishnie, J.K. Raising Animals Without Antibiotics: U.S. Producer and Veter.inarian Experiences and Opinions. *Front. Veter. Sci.* **2019**, *6*. [CrossRef]
13. Echeng, G.; Ehao, H.; Exie, S.; Ewang, X.; Edai, M.; Ehuang, L.; Yuan, Z. Antibiotic alternatives: The substitution of antibiotics in animal husbandry? *Front. Microbiol.* **2014**, *5*. [CrossRef]
14. Van Boeckel, T.P.; Brower, C.; Gilbert, M.; Grenfell, B.T.; Levin, S.A.; Robinson, T.P.; Teillant, A.; Laxminarayan, R. Global trends in antimicrobial use in food animals. *Proc. Natl. Acad. Sci. USA* **2015**, *112*, 5649–5654. [CrossRef]
15. Yang, C.; Chowdhury, M.A.K.; Hou, Y.; Gong, J. Phytogenic Compounds as Alternatives to In-Feed Antibiotics: Potentials and Challenges in Application. *Pathogens* **2015**, *4*, 137–156. [CrossRef]
16. Redondo, L.M.; Chacana, P.A.; Dominguez, J.E.; Miyakawa, M.E.D.F. Perspectives in the use of tannins as alternative to antimicrobial growth promoter factors in poultry. *Front. Microbiol.* **2014**, *5*. [CrossRef]
17. Trevino, J.; Ortiz, L.; Centeno, C. Effect of tannins from faba beans (Vicia faba) on the digestion of starch by growing chicks. *Anim. Feed. Sci. Technol.* **1992**, *37*, 345–349. [CrossRef]
18. Garcia, R.; Mendes, A.; Sartori, J.; Paz, I.C.D.L.A.; Takahashi, S.; Pelícia, K.; Komiyama, C.; Quinteiro, R. Digestibility of feeds containing sorghum, with and without tannin, for broiler chickens submitted to three room temperatures. *Brazil. J. Poultry Sci.* **2004**, *6*, 55–60. [CrossRef]
19. Brus, M.; Gradišnik, L.; Trapečar, M.; Škorjanc, D.; Frangež, R. Beneficial effects of water-soluble chestnut (Castanea sativa Mill.) tannin extract on chicken small intestinal epithelial cell culture. *Poult. Sci.* **2018**, *97*, 1271–1282. [CrossRef]
20. Amirmohammadi, M.; Khajoenia, S.; Bahmani, M.; Rafieian-Kopaei, M.; Eftekhari, Z.; Qorbani, M. In vivo evaluation of antiparasitic effects of Artemisia abrotanum and Salvia officinalis extracts on Syphacia obvelata, Aspiculoris tetrapetra and Hymenolepis nana parasites. *Asian Pac. J. Trop. Dis.* **2014**, *4*, S250–S254. [CrossRef]
21. Daglia, M. Polyphenols as antimicrobial agents. *Curr. Opin. Biotechnol.* **2012**, *23*, 174–181. [CrossRef]
22. Mannelli, F.; Team, H.; Tosi, G.; Secci, G.; Daghio, M.; Massi, P.; Fiorentini, L.; Galigani, I.; Lancini, S.; Rapaccini, S.; et al. Effect of Chestnut Tannins and Short Chain Fatty Acids as Anti-Microbials and as Feeding Supplements in Broilers Rearing and Meat Quality. *Animals* **2019**, *9*. [CrossRef]
23. Haslam, E. *Plant. Polyphenols: Vegetable Tannins Revisited. Australia (UK)*; Cambridge University Press: Cambridge, UK, 1989.
24. Huang, Q.; Liu, X.; Zhao, G.; Hu, T.; Wang, Y. Potential and challenges of tannins as an alternative to in-feed antibiotics for farm animal production. *Anim. Nutr.* **2018**, *4*, 137–150. [CrossRef]
25. Patra, A.K.; Saxena, J. Exploitation of dietary tannins to improve rumen metabolism and ruminant nutrition. *J. Sci. Food Agric.* **2011**, *91*, 24–37. [CrossRef]

26. Lamy, E.; Pinheiro, C.; Rodrigues, L.; Capela-Silva, F.; Lopes, O.; Tavares, S.; Gaspar, R. *Determinants of Tannin-Rich Food and Beverage Consumption: Oral Perception vs. Psychosocial Aspects*; Universidade De Évora: Évora, Portugal, 2016.
27. Kamiloglu, S.; Capanoglu, E.; Bilen, F.D.; Gonzales, G.B.; Grootaert, C.; Van de Wiele, T.; Van Camp, J. Bioaccessibility of polyphenols from plant-processing byproducts of black carrot (Daucus carota L.). *J. Agric. Food Chem.* **2016**, *64*, 2450–2458. [CrossRef]
28. Manach, C.; Williamson, G.; Morand, C.; Scalbert, A.; Rémésy, C. Bioavailability and bioefficacy of polyphenols in humans. I. Review of 97 bioavailability studies. *Am. J. Clin. Nutr.* **2005**, *81*, 230S–242S. [CrossRef] [PubMed]
29. Seeram, N.P.; Lee, R.; Heber, D. Bioavailability of ellagic acid in human plasma after consumption of ellagitannins from pomegranate (Punica granatum L.) juice. *Clin. Chim. Acta* **2004**, *348*, 63–68. [CrossRef]
30. Martinez, K.B.; Mackert, J.D.; McIntosh, M.K. Polyphenols and intestinal health. In *Nutrition and Functional Foods for Healthy Aging*; Elsevier: Amsterdam, The Netherlands, 2017; pp. 191–210. [CrossRef]
31. Selma, M.V.; Espin, J.C.; Tomas-Barberan, F.A. Interaction between phenolics and gut microbiota: Role in human health. *J. Agric. Food Chem.* **2009**, *57*, 6485–6501. [CrossRef]
32. Costain, D.J.; Guha, A.K.; Liwski, R.S.; Lee, T.D. Murine hypodense eosinophils induce tumour cell apoptosis by a granzyme B-dependent mechanism. *Cancer Immunol. Immunother.* **2001**, *50*, 293–299. [CrossRef]
33. Serrano, J.; Puupponen-Pimiä, R.; Dauer, A.; Aura, A.M.; Saura-Calixto, F. Tannins: Current knowledge of food sources, intake, bioavailability and biological effects. *Mol. Nutr. Food Res.* **2009**, *53*, S310–S329. [CrossRef]
34. Kahle, K.; Huemmer, W.; Kempf, M.; Scheppach, W.; Erk, T.; Richling, E. Polyphenols are intensively metabolized in the human gastrointestinal tract after apple juice consumption. *J. Agric. Food Chem.* **2007**, *55*, 10605–10614. [CrossRef]
35. Bawadi, H.; Bansode, R.; Trappey, A., II; Truax, R.; Losso, J. Inhibition of Caco-2 colon, MCF-7 and Hs578T breast, and DU 145 prostatic cancer cell proliferation by water-soluble black bean condensed tannins. *Cancer Lett.* **2005**, *218*, 153–162. [CrossRef]
36. Ragan, M.; Glombitza, K. *Handbook of Physiological Methods*; Cambridge University Press: Cambridge, UK, 1986.
37. Kirke, D.; Smyth, T.; Rai, D.; Kenny, O.; Stengel, D. The chemical and antioxidant stability of isolated low molecular weight phlorotannins. *Food Chem.* **2017**, *221*, 1104–1112. [CrossRef]
38. Corona, G.; Coman, M.M.; Guo, Y.; Hotchkiss, S.; Gill, C.; Yaqoob, P.; Spencer, J.P.; Rowland, I. Effect of simulated gastrointestinal digestion and fermentation on polyphenolic content and bioactivity of brown seaweed phlorotannin-rich extracts. *Mol. Nutr. Food Res.* **2017**, *61*, 1700223. [CrossRef]
39. Nwosu, F.; Morris, J.; Lund, V.A.; Stewart, D.; Ross, H.A.; McDougall, G.J. Anti-proliferative and potential anti-diabetic effects of phenolic-rich extracts from edible marine algae. *Food Chem.* **2011**, *126*, 1006–1012. [CrossRef]
40. Barry, T.; Manley, T.; Duncan, S. The role of condensed tannins in the nutritional value of Lotus pedunculatus for sheep:* 4. Sites of carbohydrate and protein digestion as influenced by dietary reactive tannin concentration. *Br. J. Nutr.* **1986**, *55*, 123–137. [CrossRef]
41. Berard, N.C.; Wang, Y.; Wittenberg, K.; Krause, D.; Coulman, B.; McAllister, T.; Ominski, K. Condensed tannin concentrations found in vegetative and mature forage legumes grown in western Canada. *Can. J. Plant Sci.* **2011**, *91*, 669–675. [CrossRef]
42. Goel, G.; Makkar, H.P. Methane mitigation from ruminants using tannins and saponins. *Trop. Anim. Health Prod.* **2012**, *44*, 729–739. [CrossRef]
43. Medugu, C.; Saleh, B.; Igwebuike, J.; Ndirmbita, R. Strategies to improve the utilization of tannin-rich feed materials by poultry. *Int. J. Poult. Sci.* **2012**, *11*, 417. [CrossRef]
44. Butler, L.G.; Riedl, D.J.; Lebryk, D.; Blytt, H. Interaction of proteins with sorghum tannin: Mechanism, specificity and significance. *J. Am. Oil Chem. Soc.* **1984**, *61*, 916–920. [CrossRef]
45. Marzo, F.; Urdaneta, E.; Santidrian, S. Liver proteolytic activity in tannic acid-fed birds. *Poult. Sci.* **2002**, *81*, 92–94. [CrossRef]
46. Lee, S.; Shinde, P.; Choi, J.; Kwon, I.; Lee, J.; Pak, S.; Cho, W.; Chae, B. Effects of tannic acid supplementation on growth performance, blood hematology, iron status and faecal microflora in weanling pigs. *Livest. Sci.* **2010**, *131*, 281–286. [CrossRef]

47. Tonda, R.; Rubach, J.; Lumpkins, B.; Mathis, G.; Poss, M. Effects of tannic acid extract on performance and intestinal health of broiler chickens following coccidiosis vaccination and/or a mixed-species Eimeria challenge. *Poult. Sci.* **2018**, *97*, 3031–3042. [CrossRef]
48. Cengiz, Ö.; Köksal, B.H.; Tatlı, O.; Sevim, Ö.; Ahsan, U.; Bilgili, S.F.; Önol, A.G. Effect of dietary tannic acid supplementation in corn-or barley-based diets on growth performance, intestinal viscosity, litter quality, and incidence and severity of footpad dermatitis in broiler chickens. *Livest. Sci.* **2017**, *202*, 52–57. [CrossRef]
49. Starčević, K.; Krstulović, L.; Brozić, D.; Maurić, M.; Stojević, Z.; Mikulec, Ž.; Bajić, M.; Mašek, T. Production performance, meat composition and oxidative susceptibility in broiler chicken fed with different phenolic compounds. *J. Sci. Food Agric.* **2015**, *95*, 1172–1178. [CrossRef]
50. Mašek, T.; Starčević, K.; Mikulec, Ž. The influence of the addition of thymol, tannic acid or gallic acid to broiler diet on growth performance, serum malondialdehyde value and cecal fermentation. *Yeast* **2014**, *5*, 5.0. [CrossRef]
51. Kubena, L.; Byrd, J.; Young, C.; Corrier, D. Effects of tannic acid on cecal volatile fatty acids and susceptibility to Salmonella typhimurium colonization in broiler chicks. *Poult. Sci.* **2001**, *80*, 1293–1298. [CrossRef]
52. Ebrahim, R.; Liang, J.B.; Jahromi, M.F.; Shokryazdan, P.; Ebrahimi, M.; Li Chen, W.; Goh, Y.M. Effects of tannic acid on performance and fatty acid composition of breast muscle in broiler chickens under heat stress. *Ital. J. Anim. Sci.* **2015**, *14*, 3956. [CrossRef]
53. Mor-Mur, M.; Yuste, J. Emerging bacterial pathogens in meat and poultry: An overview. *Food Bioproc. Tech.* **2010**, *3*, 24. [CrossRef]
54. Adhikari, P.; Cosby, D.E.; Cox, N.A.; Franca, M.S.; Williams, S.M.; Gogal, R.M., Jr.; Ritz, C.W.; Kim, W.K. Effect of dietary fructooligosaccharide supplementation on internal organs Salmonella colonization, immune response, ileal morphology, and ileal immunohistochemistry in laying hens challenged with Salmonella enteritidis. *Poult. Sci.* **2018**, *97*, 2525–2533. [CrossRef]
55. Thung, T.; Mahyudin, N.; Basri, D.; Radzi, C.W.M.; Nakaguchi, Y.; Nishibuchi, M.; Radu, S. Prevalence and antibiotic resistance of Salmonella Enteritidis and Salmonella Typhimurium in raw chicken meat at retail markets in Malaysia. *Poult. Sci.* **2016**, *95*, 1888–1893. [CrossRef]
56. Wu-Yuan, C.; Chen, C.; Wu, R. Gallotannins inhibit growth, water-insoluble glucan synthesis, and aggregation of mutans streptococci. *J. Dent. Res.* **1988**, *67*, 51–55. [CrossRef]
57. Delimont, N.M.; Rosenkranz, S.K.; Haub, M.D.; Lindshield, B.L. Salivary proline-rich protein may reduce tannin-iron chelation: A systematic narrative review. *Nutr. Metab.* **2017**, *14*, 47. [CrossRef]
58. Karamać, M. Chelation of Cu (II), Zn (II), and Fe (II) by tannin constituents of selected edible nuts. *Int. J. Mol. Sci.* **2009**, *10*, 5485–5497. [CrossRef]
59. Tan, Z. Roles of Iron Uptake in the Survival, Colonization and Virulence of Salmonella in Animal Guts. Master's Thesis, University of Manitoba, Winnipeg, MB, Canada, 2019.
60. Mahadwar, G.; Chauhan, K.; Bhagavathy, G.; Murphy, C.; Smith, A.; Bhagwat, A. Swarm motility of S almonella enterica serovar T yphimurium is inhibited by compounds from fruit peel extracts. *Lett. Appl. Microbiol.* **2015**, *60*, 334–340. [CrossRef]
61. Sivasankar, C.; Jha, N.K.; Ghosh, R.; Shetty, P.H. Anti quorum sensing and anti virulence activity of tannic acid and it's potential to breach resistance in Salmonella enterica Typhi/Paratyphi A clinical isolates. *Microb. Pathog.* **2020**, *138*, 103813. [CrossRef]
62. Yang, Q.; Wang, L.; Gao, J.; Liu, X.; Feng, Y.; Wu, Q.; Baloch, A.B.; Cui, L.; Xia, X. Tannin-rich fraction from pomegranate rind inhibits quorum sensing in Chromobacterium violaceum and biofilm formation in Escherichia coli. *Foodborne Pathog. Dis.* **2016**, *13*, 28–35. [CrossRef]
63. Costabile, A.; Sanghi, S.; Martín-Pelaez, S.; Mueller-Harvey, I.; Gibson, G.R.; Rastall, R.A.; Klinder, A. Inhibition of Salmonella Typhimurium by tannins in vitro. *J. Food. Agric. Environ.* **2011**, *9*, 119–124.
64. Reyes, A.W.B.; Hong, T.G.; Hop, H.T.; Arayan, L.T.; Huy, T.X.N.; Min, W.; Lee, H.J.; Lee, K.S.; Kim, S. The in vitro and in vivo protective effects of tannin derivatives against Salmonella enterica serovar Typhimurium infection. *Microb. Pathog.* **2017**, *109*, 86–93. [CrossRef]
65. Daing, M.I.; Pathak, A.; Bhat, M.A.; Zargar, M.A. Antioxidant and antibacterial potential of condensed tannins containing tree leaves extract. *Vet. Pract.* **2017**, *18*, 118–121.
66. Maisetta, G.; Batoni, G.; Caboni, P.; Esin, S.; Rinaldi, A.C.; Zucca, P. Tannin profile, antioxidant properties, and antimicrobial activity of extracts from two Mediterranean species of parasitic plant Cytinus. *BMC Complement Altern. Med.* **2019**, *19*, 82. [CrossRef]

67. Elizondo, A.M.; Mercado, E.C.; Rabinovitz, B.C.; Fernandez-Miyakawa, M.E. Effect of tannins on the in vitro growth of Clostridium perfringens. *Vet. Microbiol.* **2010**, *145*, 308–314. [CrossRef]
68. Liu, H.; Li, K.; Zhao, J.; Deng, W. Effects of chestnut tannins on intestinal morphology, barrier function, pro-inflammatory cytokine expression, microflora and antioxidant capacity in heat-stressed broilers. *J. Anim. Physiol. Anim. Nutr.* **2018**, *102*, 717–726. [CrossRef]
69. Kamali Sangani, A.; Masoudi, A.A.; Hosseini, S.A. The effects of herbal plants on Mucin 2 gene expression and performance in ascetic broilers. *Iran. J. Vet. Med.* **2014**, *8*, 47–52. [CrossRef]
70. Moghaddam, H.S.; Moghaddam, H.N.; Kermanshahi, H.; Mosavi, A.H.; Raji, A. The effect of threonine on mucin2 gene expression, intestinal histology and performance of broiler chicken. *Ital. J. Anim. Sci.* **2011**, *10*, e14. [CrossRef]
71. Urdaneta, V.; Casadesús, J. Interactions between bacteria and bile salts in the gastrointestinal and hepatobiliary tracts. *Front. Med.* **2017**, *4*, 163. [CrossRef]
72. Li, X.; Bleumink-Pluym, N.M.; Luijkx, Y.M.; Wubbolts, R.W.; van Putten, J.P.; Strijbis, K. MUC1 is a receptor for the Salmonella SiiE adhesin that enables apical invasion into enterocytes. *PLoS Pathog.* **2019**, *15*, e1007566. [CrossRef]
73. Van Parys, A.; Boyen, F.; Dewulf, J.; Haesebrouck, F.; Pasmans, F. The use of tannins to control Salmonella typhimurium infections in pigs. *Zoonoses Public Health* **2010**, *57*, 423–428. [CrossRef]
74. Jamroz, D.; Wiliczkiewicz, A.; Skorupińska, J.; Orda, J.; Kuryszko, J.; Tschirch, H. Effect of sweet chestnut tannin (SCT) on the performance, microbial status of intestine and histological characteristics of intestine wall in chickens. *Br. Poult. Sci.* **2009**, *50*, 687–699. [CrossRef]
75. Adhikari, P.; Cosby, D.; Cox, N.; Kim, W. Colonization of mature laying hens with Salmonella Enteritidis by oral or intracloacal inoculation. *J. Appl. Poult. Res.* **2017**, *26*, 286–294. [CrossRef]
76. Chen, Y.; Zhang, H.; Cheng, Y.; Li, Y.; Wen, C.; Zhou, Y. Dietary l-threonine supplementation attenuates lipopolysaccharide-induced inflammatory responses and intestinal barrier damage of broiler chickens at an early age. *Br. J. Nutr.* **2018**, *119*, 1254–1262. [CrossRef]
77. Omonijo, F.A.; Liu, S.; Hui, Q.; Zhang, H.; Lahaye, L.; Bodin, J.-C.; Gong, J.; Nyachoti, M.; Yang, C. Thymol improves barrier function and attenuates inflammatory responses in porcine intestinal epithelial cells during lipopolysaccharide (LPS)-induced inflammation. *J. Agric. Food Chem.* **2018**, *67*, 615–624. [CrossRef]
78. van Ampting, M.T.; Schonewille, A.J.; Vink, C.; Brummer, R.J.M.; van der Meer, R.; Bovee-Oudenhoven, I.M. Damage to the intestinal epithelial barrier by antibiotic pretreatment of salmonella-infected rats is lessened by dietary calcium or tannic acid. *J. Nutr.* **2010**, *140*, 2167–2172. [CrossRef]
79. Adhikari, P.; Lee, C.; Cosby, D.; Cox, N.; Kim, W. Effect of probiotics on fecal excretion, colonization in internal organs and immune gene expression in the ileum of laying hens challenged with SalmonellaEnteritidis. *Poult. Sci.* **2019**, *98*, 1235–1242. [CrossRef]
80. Sieniawska, E.; Baj, T. Tannins. In *Pharmacognosy*; Elsevier: Amsterdam, The Netherlands, 2017; pp. 199–232.
81. Konishi, Y.; Kobayashi, S.; Shimizu, M. Transepithelial transport of p-coumaric acid and gallic acid in Caco-2 cell monolayers. *Biosci. Biotechnol. Biochem.* **2003**, *67*, 2317–2324. [CrossRef]
82. Grilli, G.; Borgonovo, F.; Tullo, E.; Fontana, I.; Guarino, M.; Ferrante, V. A pilot study to detect coccidiosis in poultry farms at early stage from air analysis. *Biosyst. Eng.* **2018**, *173*, 64–70. [CrossRef]
83. Chapman, H. Milestones in avian coccidiosis research: A review. *Poult. Sci.* **2014**, *93*, 501–511. [CrossRef]
84. Clark, E.L.; Macdonald, S.E.; Thenmozhi, V.; Kundu, K.; Garg, R.; Kumar, S.; Ayoade, S.; Fornace, K.M.; Jatau, I.D.; Moftah, A. Cryptic Eimeria genotypes are common across the southern but not northern hemisphere. *Int. J. Parasitol.* **2016**, *46*, 537–544. [CrossRef]
85. Li, C.; Yan, X.; Lillehoj, H.S.; Oh, S.; Liu, L.; Sun, Z.; Gu, C.; Lee, Y.; Xianyu, Z.; Zhao, H. Eimeria maxima-induced transcriptional changes in the cecal mucosa of broiler chickens. *Parasites Vectors* **2019**, *12*, 285. [CrossRef]
86. Yin, G.; Lin, Q.; Wei, W.; Qin, M.; Liu, X.; Suo, X.; Huang, Z. Protective immunity against Eimeria tenella infection in chickens induced by immunization with a recombinant C-terminal derivative of EtIMP1. *Vet. Immunol. Immunopathol.* **2014**, *162*, 117–121. [CrossRef]
87. Teng, P.-Y.; Yadav, S.; de Souza Castro, F.L.; Tompkins, Y.H.; Fuller, A.L.; Kim, W.K. Graded Eimeria challenge linearly regulated growth performance, dynamic change of gastrointestinal permeability, apparent ileal digestibility, intestinal morphology, and tight junctions of broiler chickens. *Poult. Sci.* **2020**, *99*, 4203–4216. [CrossRef]

88. Abbas, R.; Iqbal, Z.; Mansoor, M. Role of natural antioxidants for the control of coccidiosis in poultry. *Pak. Vet. J.* **2013**, *33*, 401.
89. Prescott, J.F.; Smyth, J.A.; Shojadoost, B.; Vince, A. Experimental reproduction of necrotic enteritis in chickens: A review. *Avian Dis.* **2016**, *45*, 317–322. [CrossRef]
90. Wu, S.-B.; Stanley, D.; Rodgers, N.; Swick, R.A.; Moore, R.J. Two necrotic enteritis predisposing factors, dietary fishmeal and Eimeria infection, induce large changes in the caecal microbiota of broiler chickens. *Vet. Microbiol.* **2014**, *169*, 188–197. [CrossRef]
91. Peek, H.; Landman, W. Coccidiosis in poultry: Anticoccidial products, vaccines and other prevention strategies. *Vet. Q.* **2011**, *31*, 143–161. [CrossRef]
92. Noack, S.; Chapman, H.D.; Selzer, P.M. Anticoccidial drugs of the livestock industry. *Parasitol. Res.* **2019**, *118*, 2009–2026. [CrossRef]
93. Levine, R.; Horst, G.; Tonda, R.; Lumpkins, B.; Mathis, G. Evaluation of the effects of feeding dried algae containing beta-1, 3-glucan on broilers challenged with Eimeria. *Poult. Sci.* **2018**, *97*, 3494–3500. [CrossRef]
94. Christaki, E.; Florou-Paneri, P.; Giannenas, I.; Papazahariadou, M.; Botsoglou, N.A.; Spais, A.B. Effect of a mixture of herbal extracts on broiler chickens infected with Eimeria tenella. *Anim. Res.* **2004**, *53*, 137–144. [CrossRef]
95. Abbas, R.Z.; Munawar, S.H.; Manzoor, Z.; Iqbal, Z.; Khan, M.N.; Saleemi, M.K.; Zia, M.A.; Yousaf, A. Anticoccidial effects of acetic acid on performance and pathogenic parameters in broiler chickens challenged with Eimeria tenella. *Pesqui. Vet. Bras.* **2011**, *31*, 99–103. [CrossRef]
96. Yang, C.; Kennes, Y.M.; Lepp, D.; Yin, X.; Wang, Q.; Yu, H.; Yang, C.; Gong, J.; Diarra, M.S. Effects of encapsulated cinnamaldehyde and citral on the performance and cecal microbiota of broilers vaccinated or not vaccinated against coccidiosis. *Poult. Sci.* **2020**, *99*, 936–948. [CrossRef]
97. Bortoluzzi, C.; Rothrock, M.J.; Vieira, B.S.; Mallo, J.J.; Puyalto, M.; Hofacre, C.; Applegate, T.J. Supplementation of protected sodium butyrate alone or in combination with essential oils modulated the cecal microbiota of broiler chickens challenged with coccidia and Clostridium perfringens. *Front. Sustain. Food Syst.* **2018**, *2*, 72. [CrossRef]
98. He, B.; Bortoluzzi, C.; King, W.; Graugnard, D.; Dawson, K.; Applegate, T. Zinc source influences the gene expression of zinc transporters in jejunum and cecal tonsils during broiler challenge with Eimeria maximaand Clostridium perfringens. *Poult. Sci.* **2019**, *98*, 1146–1152. [CrossRef]
99. Teng, P.-Y.; Fuller, A.L.; Kim, W.K. Evaluation of nitro compounds as feed additives in diets of Eimeria-challenged broilers in vitro and in vivo. *Poult. Sci.* **2020**, *99*, 1320–1325. [CrossRef]
100. Min, B.; Hart, S. Tannins for suppression of internal parasites. *J. Anim. Sci.* **2003**, *81*, E102–E109. [CrossRef]
101. Scalbert, A. Antimicrobial properties of tannins. *Phytochemistry* **1991**, *30*, 3875–3883. [CrossRef]
102. Chung, K.-T.; Lu, Z.; Chou, M. Mechanism of inhibition of tannic acid and related compounds on the growth of intestinal bacteria. *Food Chem. Toxicol.* **1998**, *36*, 1053–1060. [CrossRef]
103. Parisi, F.; Mancini, S.; Mazzei, M.; Forzan, M.; Turchi, B.; Perrucci, S.; Poli, A.; Paci, G. Effect of dietary supplementation of a mix of chestnut and quebracho tannins on intestinal morphology, bacterial load, eimeria spp oocyst excretion and immune response after vaccination in rabbits. *Am. J. Anim.Vet. Sci.* **2018**. [CrossRef]
104. Kaleem, Q.M.; Akhtar, M.; Awais, M.M.; Saleem, M.; Zafar, M.; Iqbal, Z.; Muhammad, F.; Anwar, M.I. Studies on Emblica officinalis derived tannins for their immunostimulatory and protective activities against coccidiosis in industrial broiler chickens. *Sci. World J.* **2014**, *2014*. [CrossRef]
105. Mishra, B.; Jha, R. Oxidative stress in the poultry gut: Potential challenges and interventions. *Front. Vet. Sci.* **2019**, *6*, 60. [CrossRef]
106. Wang, M.; Suo, X.; Gu, J.; Zhang, W.; Fang, Q.; Wang, X. Influence of grape seed proanthocyanidin extract in broiler chickens: Effect on chicken coccidiosis and antioxidant status. *Poult. Sci.* **2008**, *87*, 2273–2280. [CrossRef]
107. Mansoori, B.; Modirsanei, M. Effects of dietary tannic acid and vaccination on the course of coccidiosis in experimentally challenged broiler chicken. *Vet. Parasitol.* **2012**, *187*, 119–122. [CrossRef]
108. Attia, Y.A.; Al-Harthi, M.A.; El-Shafey, A.S.; Rehab, Y.A.; Kim, W.K. Enhancing tolerance of broiler chickens to heat stress by supplementation with vitamin E, vitamin C and/or probiotics. *Ann. Anim. Sci.* **2017**, *17*, 1155–1169. [CrossRef]

109. Quinteiro-Filho, W.; Ribeiro, A.; Ferraz-de-Paula, V.; Pinheiro, M.; Sakai, M.; Sá, L.; Ferreira, A.; Palermo-Neto, J. Heat stress impairs performance parameters, induces intestinal injury, and decreases macrophage activity in broiler chickens. *Poult. Sci.* **2010**, *89*, 1905–1914. [CrossRef]
110. Laudicina, D.C.; Marnett, L.J. Enhancement of hydroperoxide-dependent lipid peroxidation in rat liver microsomes by ascorbic acid. *Arch. Biochem. Biophys.* **1990**, *278*, 73–80. [CrossRef]
111. Choi, J.; Li, W.; Schindell, B.; Ni, L.; Liu, S.; Zhao, X.; Gong, J.; Nyachoti, M.; Yang, C. Molecular cloning, tissue distribution and the expression of cystine/glutamate exchanger (xCT, SLC7A11) in different tissues during development in broiler chickens. *Anim. Nutr.* **2020**, *6*, 107–114. [CrossRef]
112. Lin, H.; Decuypere, E.; Buyse, J. Acute heat stress induces oxidative stress in broiler chickens. *Comp. Biochem. Physiol. Part A Mol. Integr. Physiol.* **2006**, *144*, 11–17. [CrossRef]
113. Lee, S.; Kim, D.; Keum, M.; Han, E.; An, B.; Chang, H.; Choi, Y.; Moon, B.; Lee, K. Effects of fumonisin B1 and mycotoxin binders on growth performance, tibia characteristics, gut physiology, and stress indicators in broiler chickens raised in different stocking densities. *Poult. Sci.* **2018**, *97*, 845–854. [CrossRef]
114. Lauridsen, C. From oxidative stress to inflammation: Redox balance and immune system. *Poult. Sci.* **2019**, *98*, 4240–4246. [CrossRef]
115. Nawab, A.; Ibtisham, F.; Li, G.; Kieser, B.; Wu, J.; Liu, W.; Zhao, Y.; Nawab, Y.; Li, K.; Xiao, M. Heat stress in poultry production: Mitigation strategies to overcome the future challenges facing the global poultry industry. *J. Therm. Biol.* **2018**, *78*, 131–139. [CrossRef]
116. Yang, R.; Hui, Q.; Jiang, Q.; Liu, S.; Zhang, H.; Wu, J.; Lin, F.; Yang, C. Effect of Manitoba-grown red-osier dogwood extracts on recovering Caco-2 cells from H2O2-induced oxidative damage. *Antioxidants* **2019**, *8*, 250. [CrossRef]
117. Jiang, Q.; Zhang, H.; Yang, R.; Hui, Q.; Chen, Y.; Mats, L.; Tsao, R.; Yang, C. Red-Osier Dogwood Extracts Prevent Inflammatory Responses in Caco-2 Cells and a Caco-2 BBe1/EA. hy926 Cell Co-Culture Model. *Antioxidants* **2019**, *8*, 428. [CrossRef]
118. Terruzzi, I.; Montesano, A.; Senesi, P.; Villa, I.; Ferraretto, A.; Bottani, M.; Vacante, F.; Spinello, A.; Bolamperti, S.; Luzi, L. L-Carnitine reduces oxidative stress and promotes cells differentiation and bone matrix proteins expression in human osteoblast-like cells. *Biomed Res. Int.* **2019**, *2019*. [CrossRef]
119. Sahin, K.; Sahin, N.; Kucuk, O. Effects of chromium, and ascorbic acid supplementation on growth, carcass traits, serum metabolites, and antioxidant status of broiler chickens reared at a high ambient temperature (32 C). *Nutr. Res.* **2003**, *23*, 225–238. [CrossRef]
120. Yun, S.-H.; Moon, Y.-S.; SoHn, S.-H.; Jang, I.-S. Effects of cyclic heat stress or vitamin C supplementation during cyclic heat stress on HSP70, inflammatory cytokines, and the antioxidant defense system in Sprague Dawley rats. *Exp. Anim.* **2012**, *61*, 543–553. [CrossRef]
121. Kumbhar, S.; Khan, A.Z.; Parveen, F.; Nizamani, Z.A.; Siyal, F.A.; Abd El-Hack, M.E.; Gan, F.; Liu, Y.; Hamid, M.; Nido, S.A. Impacts of selenium and vitamin E supplementation on mRNA of heat shock proteins, selenoproteins and antioxidants in broilers exposed to high temperature. *Amb Express* **2018**, *8*, 1–10. [CrossRef] [PubMed]
122. Shakeri, M.; Cottrell, J.J.; Wilkinson, S.; Le, H.H.; Suleria, H.A.; Warner, R.D.; Dunshea, F.R. Growth performance and characterization of meat quality of broiler chickens supplemented with betaine and antioxidants under cyclic heat stress. *Antioxidants* **2019**, *8*, 336. [CrossRef]
123. Shi, J.; Yu, J.; Pohorly, J.E.; Kakuda, Y. Polyphenolics in grape seeds—biochemistry and functionality. *J. Med. Food.* **2003**, *6*, 291–299. [CrossRef]
124. Sahin, K.; Orhan, C.; Tuzcu, M.; Ali, S.; Sahin, N.; Hayirli, A. Epigallocatechin-3-gallate prevents lipid peroxidation and enhances antioxidant defense system via modulating hepatic nuclear transcription factors in heat-stressed quails. *Poult. Sci.* **2010**, *89*, 2251–2258. [CrossRef]
125. Ramnath, V.; Rekha, P. Brahma Rasayana enhances in vivo antioxidant status in cold-stressed chickens (Gallus gallus domesticus). *Indian J. Pharmacol.* **2009**, *41*, 115. [CrossRef]
126. Hosseini-Vashan, S.J.; Safdari-Rostamabad, M.; Piray, A.H.; Sarir, H. The growth performance, plasma biochemistry indices, immune system, antioxidant status, and intestinal morphology of heat-stressed broiler chickens fed grape (Vitis vinifera) pomace. *Anim. Feed Sci. Technol.* **2020**, *259*, 114343. [CrossRef]
127. Mujahid, A.; Yoshiki, Y.; Akiba, Y.; Toyomizu, M. Superoxide radical production in chicken skeletal muscle induced by acute heat stress. *Poult. Sci.* **2005**, *84*, 307–314. [CrossRef]

128. Nakagawa, T.; Yokozawa, T. Direct scavenging of nitric oxide and superoxide by green tea. *Food Chem. Toxicol.* **2002**, *40*, 1745–1750. [CrossRef]
129. Shepherd, E.; Fairchild, B. Footpad dermatitis in poultry. *Poult. Sci.* **2010**, *89*, 2043–2051. [CrossRef]
130. Abd El-Wahab, A.; Radko, D.; Kamphues, J. High dietary levels of biotin and zinc to improve health of foot pads in broilers exposed experimentally to litter with critical moisture content. *Poult. Sci.* **2013**, *92*, 1774–1782. [CrossRef]
131. Thøfner, I.C.N.; Poulsen, L.L.; Bisgaard, M.; Christensen, H.; Olsen, R.H.; Christensen, J.P. Correlation between footpad lesions and systemic bacterial infections in broiler breeders. *Vet. Res.* **2019**, *50*, 38. [CrossRef] [PubMed]
132. Heidemann Olsen, R.; Christensen, H.; Kabell, S.; Bisgaard, M. Characterization of prevalent bacterial pathogens associated with pododermatitis in table egg layers. *Avian Dis.* **2018**, *47*, 281–285. [CrossRef] [PubMed]
133. El-Wahab, A.A.; Visscher, C.; Beineke, A.; Beyerbach, M.; Kamphues, J. Experimental studies on the effects of different litter moisture contents and exposure time to wet litter on development and severity of foot pad dermatitis in young fattening turkeys. *Arch. Geflugelkd.* **2012**, *76*, 55–62.
134. Mayne, R.; Else, R.; Hocking, P. High litter moisture alone is sufficient to cause footpad dermatitis in growing turkeys. *Br. Poult. Sci.* **2007**, *48*, 538–545. [CrossRef] [PubMed]
135. Cengız, Ö.; Hess, J.; Bilgili, S. Feed enzyme supplementation does not ameliorate foot pad dermatitis in broiler chickens fed on a corn-soyabean diet. *Br. Poult. Sci.* **2012**, *53*, 401–407. [CrossRef]
136. Rezar, V.; Salobir, J. Effects of tannin-rich sweet chestnut (Castanea sativa mill.) wood extract supplementation on nutrient utilisation and excreta dry matter content in broiler chickens. *Eur. Poult. Sci.* **2014**, *78*, 1–10. [CrossRef]
137. Salmond, G.G. Supplementation of Chelated Zn Cu and Mn in Broiler Diets to Improve Performance Skin Integrity and the Incidence and Severity of Footpad Dermatitis. Master's Thesis, University of Pretoria, Pretoria, South Africa, 2015.
138. Mayne, R.; Powell, F.; Else, R.; Kaiser, P.; Hocking, P. Foot pad dermatitis in growing turkeys is associated with cytokine and cellular changes indicative of an inflammatory immune response. *Avian Dis.* **2007**, *36*, 453–459. [CrossRef]
139. Bostami, A.; Ahmed, S.; Islam, M.; Mun, H.; Ko, S.; Kim, S.; Yang, C. Growth performance, fecal noxious gas emission and economic efficacy in broilers fed fermented pomegranate byproducts as residue of fruit industry. *Int. J. Adv. Res* **2015**, *3*, 102–114.
140. Wang, S.-Y.; Huang, D.-J. Assessment of greenhouse gas emissions from poultry enteric fermentation. *Asian-Australas J. Anim. Sci.* **2005**, *18*, 873–878. [CrossRef]
141. Wang, Y.; Huang, M.; Meng, Q. Effects of atmospheric hydrogen sulfide concentration on growth and meat quality in broiler chickens. *Poult. Sci.* **2011**, *90*, 2409–2414. [CrossRef]
142. McCrory, D.; Hobbs, P. Additives to reduce ammonia and odor emissions from livestock wastes: A review. *J. Environ. Qual.* **2001**, *30*, 345–355. [CrossRef] [PubMed]
143. Aneja, V.P.; Schlesinger, W.H.; Erisman, J.W.; Behera, S.N.; Sharma, M.; Battye, W. Reactive nitrogen emissions from crop and livestock farming in India. *Atmos. Environ.* **2012**, *47*, 92–103. [CrossRef]
144. Xin, H.; Gates, R.S.; Green, A.R.; Mitloehner, F.M.; Moore Jr, P.A.; Wathes, C.M. Environmental impacts and sustainability of egg production systems. *Poult. Sci.* **2011**, *90*, 263–277. [CrossRef] [PubMed]
145. Malomo, G.A.; Bolu, S.A.; Madugu, A.S.; Usman, Z.S. Nitrogen Emissions and Mitigation Strategies in Chicken Production. *Anim. Husband. Nutrit* **2018**, *43*, 43–62.
146. Aboagye, I.A.; Oba, M.; Koenig, K.M.; Zhao, G.Y.; Beauchemin, K.A. Use of gallic acid and hydrolyzable tannins to reduce methane emission and nitrogen excretion in beef cattle fed a diet containing alfalfa silage. *J. Anim. Sci.* **2019**, *97*, 2230–2244. [CrossRef]
147. Ahmed, S.T.; Yang, C.-J. Effects of dietary Punica granatum L. by-products on performance, immunity, intestinal and fecal microbiology, and odorous gas emissions from excreta in broilers. *J. Poult. Sci.* **2017**. [CrossRef]
148. Ramah, A.; Yasuda, M.; Ohashi, Y.; Urakawa, M.; Kida, T.; Yanagita, T.; Uemura, R.; Bakry, H.H.; Abdelaleem, N.M.; El-Shewy, E.A. Different doses of tannin reflect a double-edged impact on broiler chicken immunity. *Vet. Immunol. Immunopathol.* **2020**, *220*, 109991. [CrossRef]

149. Karaffová, V.; Bobíková, K.; Levkut, M.; Revajová, V.; Ševčíková, Z.; Levkut, M. The influence of Farmatan®and Flimabend®on the mucosal immunity of broiler chicken. *Poult. Sci.* **2019**, *98*, 1161–1166. [CrossRef]
150. Erlejman, A.G.; Jaggers, G.; Fraga, C.G.; Oteiza, P.I. TNFα-induced NF-κB activation and cell oxidant production are modulated by hexameric procyanidins in Caco-2 cells. *Arch. Biochem. Biophys.* **2008**, *476*, 186–195. [CrossRef]
151. Diaz Carrasco, J.M.; Redondo, E.A.; Pin Viso, N.D.; Redondo, L.M.; Farber, M.D.; Fernandez Miyakawa, M.E. Tannins and bacitracin differentially modulate gut microbiota of broiler chickens. *Biomed. Res. Int.* **2018**, *2018*. [CrossRef]
152. Koo, B.; Nyachoti, C.M. Effects of thermally oxidized canola oil and tannic acid supplementation on nutrient digestibility and microbial metabolites in finishing pigs. *J. Anim. Sci.* **2019**, *97*, 2468–2478. [CrossRef] [PubMed]
153. Pacheco-Ordaz, R.; Wall-Medrano, A.; Goñi, M.G.; Ramos-Clamont-Montfort, G.; Ayala-Zavala, J.F.; González-Aguilar, G. Effect of phenolic compounds on the growth of selected probiotic and pathogenic bacteria. *Lett. Appl. Microbiol.* **2018**, *66*, 25–31. [CrossRef] [PubMed]
154. Viveros, A.; Chamorro, S.; Pizarro, M.; Arija, I.; Centeno, C.; Brenes, A. Effects of dietary polyphenol-rich grape products on intestinal microflora and gut morphology in broiler chicks. *Poult. Sci.* **2011**, *90*, 566–578. [CrossRef] [PubMed]
155. Shang, Y.; Kumar, S.; Oakley, B.; Kim, W.K. Chicken gut microbiota: Importance and detection technology. *Front. Vet. Sci.* **2018**, *5*, 254. [CrossRef]
156. Bilić-Šobot, D.; Kubale, V.; Škrlep, M.; Čandek-Potokar, M.; Prevolnik Povše, M.; Fazarinc, G.; Škorjanc, D. Effect of hydrolysable tannins on intestinal morphology, proliferation and apoptosis in entire male pigs. *Arch. Anim. Nutr.* **2016**, *70*, 378–388. [CrossRef]
157. Kim, T.; Silva, J.; Kim, M.; Jung, Y. Enhanced antioxidant capacity and antimicrobial activity of tannic acid by thermal processing. *Food Chem.* **2010**, *118*, 740–746. [CrossRef]
158. Adhikari, P.A.; Kim, W.K. Overview of prebiotics and probiotics: Focus on performance, gut health and immunity–a review. *Ann. Anim. Sci.* **2017**, *17*, 949–966. [CrossRef]
159. Huyghebaert, G.; Ducatelle, R.; Van Immerseel, F. An update on alternatives to antimicrobial growth promoters for broilers. *Vet. J.* **2011**, *187*, 182–188. [CrossRef]
160. Khalil, R.K. Influence of gallic acid and catechin polyphenols on probiotic properties of Streptococcus thermophilus CHCC 3534 strain. *World J. Microbiol. Biotechnol.* **2010**, *26*, 2069–2079. [CrossRef]
161. Jarquin, R.; Nava, G.; Wolfenden, A.; Donoghue, A.; Hanning, I.; Higgins, S.; Hargis, B. The evaluation of organic acids and probiotic cultures to reduce Salmonella enteriditis horizontal transmission and crop infection in broiler chickens. *Int. J. Poult. Sci* **2007**, *6*, 182–186. [CrossRef]
162. Al-Harbi, R.; Shaaban, M.; Al-Wegaisi, R.; Moharram, F.; El-Rahman, O.A.; El-Messery, S. Antimicrobial Activity and Molecular Docking of Tannins from Pimenta dioica. *Lett. Drug Des. Discov.* **2018**, *15*, 508–515. [CrossRef]
163. Khan, S.H.; Iqbal, J. Recent advances in the role of organic acids in poultry nutrition. *J. Appl. Anim. Res.* **2016**, *44*, 359–369. [CrossRef]
164. De Lange, C.; Pluske, J.; Gong, J.; Nyachoti, C. Strategic use of feed ingredients and feed additives to stimulate gut health and development in young pigs. *Livest. Sci.* **2010**, *134*, 124–134. [CrossRef]
165. Chen, J.; Li, Y.; Yu, B.; Chen, D.; Mao, X.; Zheng, P.; Luo, J.; He, J. Dietary chlorogenic acid improves growth performance of weaned pigs through maintaining antioxidant capacity and intestinal digestion and absorption function. *J. Anim. Sci.* **2018**, *96*, 1108–1118. [CrossRef] [PubMed]
166. Choi, J.; Wang, L.; Liu, S.; Lu, P.; Zhao, X.; Liu, H.; Lahaye, L.; Santin, E.; Liu, S.; Nyachoti, M. Effects of a microencapsulated formula of organic acids and essential oils on nutrient absorption, immunity, gut barrier function, and abundance of enterotoxigenic Escherichia coli F4 in weaned piglets challenged with E. coli F4. *J. Anim. Sci.* **2020**, *98*, skaa259. [CrossRef] [PubMed]
167. Gracia, L.G.; de Castro, M.L. Development and validation of a flow-injection method for the determination of albumin tannate, the active component of a pharmaceutical preparation. *J. Pharm. Biomed. Anal.* **1997**, *15*, 447–452. [CrossRef]

168. Choi, J.; Wang, L.; Ammeter, E.; Lahaye, L.; Liu, S.; Nyachoti, M.; Yang, C. Evaluation of lipid matrix microencapsulation for intestinal delivery of thymol in weaned pigs. *Transl. Anim. Sci.* **2020**, *4*, 411–422. [CrossRef]
169. Omonijo, F.A.; Kim, S.; Guo, T.; Wang, Q.; Gong, J.; Lahaye, L.; Bodin, J.-C.; Nyachoti, M.; Liu, S.; Yang, C. Development of novel microparticles for effective delivery of thymol and lauric acid to pig intestinal tract. *J. Agric. Food Chem.* **2018**, *66*, 9608–9615. [CrossRef]
170. Liu, H.; Gong, J.; Chabot, D.; Miller, S.S.; Cui, S.W.; Ma, J.; Zhong, F.; Wang, Q. Protection of heat-sensitive probiotic bacteria during spray-drying by sodium caseinate stabilized fat particles. *Food Hydrocoll.* **2015**, *51*, 459–467. [CrossRef]
171. Choi, J. Lipid Matrix Microencapsulation for Effective Delivery of Essential Oils and Organic Acids to Improve gut Health in Weaned Piglets. Master's Thesis, University of Manitoba, Winnieg, MB, Canada, 2019.
172. Tang, Z.; Huang, X.; Baxi, S.; Chambers, J.R.; Sabour, P.M.; Wang, Q. Whey protein improves survival and release characteristics of bacteriophage Felix O1 encapsulated in alginate microspheres. *Food Res. Int.* **2013**, *52*, 460–466. [CrossRef]
173. Jang, I.; Kwon, C.H.; Ha, D.M.; Jung, D.Y.; Kang, S.Y.; Park, M.J.; Han, J.H.; Park, B.-C.; Lee, C.Y. Effects of a lipid-encapsulated zinc oxide supplement on growth performance and intestinal morphology and digestive enzyme activities in weanling pigs. *J. Anim. Sci. Technol.* **2014**, *56*, 29. [CrossRef] [PubMed]
174. Liu, L.J.; Zhu, J.; Wang, B.; Cheng, C.; Du, Y.J.; Wang, M.Q. In vitro stability evaluation of coated lipase. *Asian-Australas J. Anim. Sci.* **2017**, *30*, 192. [CrossRef] [PubMed]
175. Zhang, Y.; Gong, J.; Yu, H.; Guo, Q.; Defelice, C.; Hernandez, M.; Yin, Y.; Wang, Q. Alginate-whey protein dry powder optimized for target delivery of essential oils to the intestine of chickens. *Poult. Sci.* **2014**, *93*, 2514–2525. [CrossRef] [PubMed]
176. Trabelsi, I.; Ktari, N.; Slima, S.B.; Bouchaala, K.; Salah, R.B. Effects of supplementation with L. plantarum TN8 encapsulated in alginate-chitosan in broiler chickens. *Int. J. Biol. Macromol.* **2016**, *89*, 677–681. [CrossRef]
177. Fernández, H.; Catanese, F.; Puthod, G.; Distel, R.A.; Villalba, J.J. Depression of rumen ammonia and blood urea by quebracho tannin-containing supplements fed after high-nitrogen diets with no evidence of self-regulation of tannin intake by sheep. *Small Rumin. Res.* **2012**, *105*, 126–134. [CrossRef]
178. Adejoro, F.A.; Hassen, A.; Akanmu, A.M. Effect of lipid-encapsulated acacia tannin extract on feed intake, nutrient digestibility and methane emission in sheep. *Animals* **2019**, *9*, 863. [CrossRef]
179. Wang, M.; Huang, H.; Hu, Y.; Huang, J.; Yang, H.; Wang, L.; Chen, S.; Chen, C.; He, S. Effects of dietary microencapsulated tannic acid supplementation on the growth performance, intestinal morphology, and intestinal microbiota in weaning piglets. *J. Anim. Sci.* **2020**, *98*, skaa112. [CrossRef]

3

Effects of Fennel Seed Powder Supplementation on Growth Performance, Carcass Characteristics, Meat Quality, and Economic Efficiency of Broilers under Thermoneutral and Chronic Heat Stress Conditions

Ahmed A. AL-Sagan [1], Shady Khalil [2], Elsayed O. S. Hussein [3] and Youssef A. Attia [4,*]

[1] King Abdulaziz City for Science & Technology, P.O. Box 6086, Riyadh 11442, Saudi Arabia; abdeen@kacst.edu.sa

[2] AlzChem Trostberg GmbH, 83308 Trostberg, Germany; shady.khalil@alzchem.com

[3] Department of Animal Production, College of Food and Agriculture Sciences, King Saud University, Riyadh 11451, Saudi Arabia; shessin@ksu.edu.sa

[4] Arid Land Agriculture Department, Faculty of Meteorology, Environment, and Arid Land Agriculture, King Abdulaziz University, Jeddah 21589, Saudi Arabia

* Correspondence: yaattia@kau.edu.sa

Simple Summary: Nowadays, great attention has been given to phytogenic products as a growth enhancer due to their safety and eco-friendly influences on animal nutrition. Thus, this investigation looked at the use of fennel seed powder at 0, 1.6, and 3.2% as a dietary additive on the performance, carcass traits, meat quality, and efficiency of the production of broiler chickens raised under thermoneutral and chronic heat stress conditions. In essence, 3.2% fennel seed powder in the diet of broilers enhanced the growth rate under chronic heat stress and decreased breast meat redness and temperature, suggesting that 3.2% fennel seed could be used as an agent for enhancing the broiler's tolerance during chronic heat stress (CHS) condition from 19 to 41 days of age.

Abstract: Nowadays, phytogenic products have received great attention as a growth promoter due to their safety and environmentally friendly effect as a replacement for classical growth promoters such as antibiotics in animal nutrition. Thus, this research seeks the possibility of using fennel seed powder (FSP) as a dietary additive from 19 to 41 days of age on productive performance, carcass traits, meat quality, and production efficiency of broiler chickens raised under thermoneutral and chronic heat stress conditions. Thus, 216 one-day-old Ross-308 broiler chicks were divided into two equal groups. The first group was placed in an independent temperature-controlled room at 23 ± 2 °C. The broiler chicks from the second group were placed in a heat-stressed room and exposed to chronic heat stress conditions (32 ± 2 °C) for seven hours per day from 8 a.m. to 3 p.m. The experimental design was 2 × 3 factorial including two environmental temperatures (thermoneutral vs chronic heat stress) and three experimental diets that contained 0, 1.6, and 3.2% FSP. The chickens were randomly assigned to 18-floor pens per room temperature, representing six replicates per treatment and six birds per replicate. The results showed that dietary fennel seed powder during days 19–41 of age enhanced the growth rate of broiler chickens and improved breast meat redness and reduced temperature under chronic heat stress. In conclusion, 3.2% of fennel seed powder could be used as an agent for enhancing the broiler's tolerance during chronic heat stress condition from 19 to 41 days of age. Moreover, it is necessary to study in further detail the nitrite and nitrate contents in FSP and their impacts on muscle redness (a*) as well as muscle temperature.

Keywords: broilers; heat stress; fennel seed; performance; carcass characteristics; meat quality

1. Introduction

In hot regions, the poultry industry is significantly affected by high ambient temperatures, especially when combined with high relative air moisture contents (humidity) [1–3]. A further increase of 0.6–2.5 °C is expected to occur due to global warming over the next 50 years [4]. Optimum temperatures for broiler performance are 34–32, 32–28, 28–26, 26–24, 18–24, and 18–24 °C for the first, second, third, fourth, fifth, and sixth weeks of age, respectively [5]. Therefore, it is necessary to reevaluate poultry management to minimize heat stress.

Signs of heat stress begin to appear when the environmental temperature increases above 30 °C [6]. High temperatures can cause serious physiological disturbances [7]. Since birds do not have sweat glands, they evaporate their excess sweat through panting [8]. It has been estimated that approximately 0.5 kcals of energy are lost per gram of water evaporated during panting [9]. Urinary output is another mechanism that accompanies the excreta, with an increase in water intake to reduce body heat. This deviation in normal physiology reduces broiler productivity and leads to economic losses [9]. Generally, heat stress is considered as an inducer of oxidative stress [3,10,11] and can negatively affect animal performance and meat quality [12,13]. Moreover, nutritional intervention, along with proper management, is required to relieve the negative effect of heat stress [1]. Nutritional management can include feed restriction and dietary antioxidant supplementation [14,15] as well as the maintenance of electrolyte balance [16].

Recently, there has been great interest in phytochemicals, which are effective stress-alleviating agents [17,18]. The use of phytochemicals from medicinal plants in poultry diets is more related to product quality and consumer health than to production efficiency [19]. Fennel (Foeniculum vulgare, Apiaceae) seeds contain health-promoting volatile essential oil compounds such as fenchone, anethole, myrcene, limonene, chavicol, cineole, anisic aldehyde, and pinene as well as amino acids, phenolic compounds, and flavonoids [20]. These seeds also have apoptotic, hepatoprotective, antithrombotic, antiviral, antimicrobial, antispasmodic, anti-inflammatory, antimutagenic, antipyretic, and antinociceptive properties [14,20]. These active ingredients are also known to have digestive, anti-flatulent, and carminative properties [21]. This medicinal plant is a potential source of natural phytochemicals, especially antioxidants [22]. Furthermore, it can be used as a therapeutic agent to alleviate heat stress and stress-related diseases [15]. Fennel seeds contain high levels of nitrites and nitrates (65 to 376.7 mg/100 g), which are known to play crucial roles in maintaining vascular and digestive functions [23]. In mammals, nitrates are reduced to nitrites and eventually to nitric oxide (NO), which plays an essential role in the maintenance of the cardiovascular system [23]. In mammals and poultry, NO plays an important role in the regulation of various physiological functions that are mediated by the hypothalamus such as thermoregulation, fever, water balance, and cardiovascular regulation [24]. It has been found that inducible nitric oxide synthase (iNOS) increases in cold stress, heat stress, and inflammatory response [25] and as a response to lipopolysaccharide injection [26]. Since NO is produced from arginine metabolism [27,28], it can be speculated that fennel seed powder (FSP) may supply some NO, helping to reduce arginine metabolism for NO production [29].

Therefore, the objective of this study was to evaluate the potential of different levels of fennel seed powder supplemented to broiler diet as a physiological stimulator for the growth and development of chicks through its impacts on production performance traits, carcass characteristics, digestive organs, and meat quality under thermoneutral and heat stress conditions.

2. Materials and Methods

2.1. Animals and Treatments

King Abdulaziz City for Science and Technology, Riyadh, Saudi Arabia approved the experimental procedures. It recommends animal rights, welfare, and minimal stress and did not cause any harm or suffering to animals according to the Royal Decree M59 in 14/9/1431H.

Two hundred and sixteen one-day-old Ross-308 broiler chicks were purchased from the Al-Wadi company (Saudi Arabia) and used in this experiment. The chicks were housed in thermostatically controlled rooms. They were fed with a commercial starter mash based on corn and soybean ration (crude protein (CP), 23.2%; metabolizable energy (ME), 12.97 MJ/kg) during day 1 to 18 of age, and were managed according to the Ross-308 broiler guide for husbandry and health care practice.

On day 19, the broilers were divided into two equal groups. The first group was placed in an independent temperature-controlled room at 23 ± 2 °C. The broiler chicks from the second group were placed in a heat-stressed room and exposed to chronic heat stress conditions (32 ± 2 °C) for seven hours per day from 8:00 a.m. to 3:00 p.m. The indoor relative humidity and temperature under thermoneutral conditions during the experimental period for the broiler was 51.2% and 23.1 °C, 57% and 23.9 °C, and 53% and 24.7 °C during days 19–26, 27–33, and 34–41 of age, respectively. Under chronic heat stress conditions, the indoor relative humidity and temperature was 44.9% and 31.5 °C, 47.8% and 31.8 °C, and 41.8% and 31.9 °C during days 19–26, 27–33, and 34–41 of age, respectively. The experimental diets were formulated according to [30], as shown in Table 1. The nitrate contents were calculated using the minimum average reported by [23].

Table 1. Ingredients and nutrient composition of finishers (days 19 to 41), diets as fed basis.

Ingredients and Composition, g/kg	Fennel, g/kg		
	0	16	32
Corn	621	611	602
Soybean meal, 48% CP	287	283	277
Gluten Meal	30.0	30.0	30.0
Fennel seed powder	0.0	16.0	32.0
Palm oil	24.4	22.8	21.8
Dicalcium phosphate	17.5	17.1	16.8
Limestone	7.0	6.8	6.8
Dl-methionine	0.6	0.7	0.8
Lysine-HCl	1.9	2.0	2.2
L-threonine	0.5	0.6	0.7
Premix [1]	5.0	5.0	5.0
Salt	4.6	4.5	4.4
Choline chloride, 60%	0.5	0.5	0.5
Calculated composition			
ME, MJ/kg [2]	13.18	13.18	13.18
Dry Matter, g/kg [3]	894	894	894
Crude protein, g/kg [2]	205	204	204
Digestible methionine, g/kg [2]	5.18	5.23	5.28
Digestible lysine, g/kg [2]	11.0	11.0	11.0
Digestible methionine + cysteine, g/kg [2]	8.4	8.4	8.4
Digestible threonine, g/kg	7.2	7.2	7.2
Digestible arginine, g/kg	11.6	11.4	11.2
Nitrate, mg/kg	19.5	54.6	89.7
Nitrite, mg/kg	NA	NA	NA
Ether extract, g/kg [3]	46.2	44.3	42.9
Crude fiber, g/kg [3]	51.6	52.4	53.8
Calcium, g/kg [2]	8.5	8.5	8.5
Available Phosphorus, g/kg [2]	4.2	4.2	4.2
Magnesium, g/kg	1.59	1.59	1.60
Potassium, g/kg	7.93	8.24	8.54
Na (Sodium), g/kg	2.0	2.0	2.0
Cl (Chloride), g/kg	3.78	3.73	3.69
Dietary electrolyte balance, mEq/kg	183	193	201

[1] Vitamin-mineral premix contains the following per kg: vitamin A, 2,400,000 IU; vitamin D, 1,000,000 IU; vitamin E, 16,000 IU; vitamin K, 800 mg; vitamin B1, 600 mg; vitamin B2, 1600 mg; vitamin B6, 1000 mg; vitamin B12, 6 mg; niacin, 8000 mg; folic acid, 400 mg; pantothenic acid, 3000 mg; biotin 40 mg; antioxidant, 3000 mg; cobalt, 80 mg; copper, 2000 mg; iodine, 400; iron, 1200 mg; manganese, 18,000 mg; selenium, 60 mg, and zinc, 14,000 mg; DL-methionine, 32 g; lysine, 25 g. NA, data not available in addition.

The experimental design was 2 × 3 factorial including two environmental temperatures (thermoneutral vs. chronic heat stress), and three experimental diets that contained 0, 1.6, and 3.2% FSP. The two rooms were equipped with ultrasonic humidifiers to supply constant relative humidity to each room. The chickens were randomly assigned to 18-floor pens per room temperature, representing six replicates per treatment. Each floor pen (0.75 × 0.75 m) had six birds. Fennel seeds were purchased from the local market (imported from India) and ground into a fine powder using an electric mill. The FSP was added to the experimental diet in a mash form at the top. To assure that the diets were consumed, we offered a per weighted amount of feed daily according to the expected feed intake of each pen. Before, next meal, we collect the residual and mixed well with the new amount. From our observation, birds consumed almost all feeds and excessed 97% daily. The provision of feed and water was ad libitum throughout the experimental period. The chicks were exposed to continuous light from 1–6 days of age, then to 23:1 light-dark cycle throughout the experimental period. Feeding on experimental diets was initiated at 19 days of age.

2.2. *Measurements*

2.2.1. Production Performance Measurements

Live body weight (LBW), feed intake (FI) as fed basis, and feed conversion ratio (FCR) were measured for each floor pen every week. Then, the data were averaged for each pen to calculate the FI, body weight gain (BWG), and the FCR for the entire experimental period (19–41 days of age), where the BWG and FCR were calculated as follows:

BWG = Final body weight at 41 days of age-initial body weight at 19 days of age.

FCR = Amount of feed consumed (g)/live weight of chicks (g).

Nitrate intake was calculated using the feed intake data and the average concentration (220.85 mg/100 g) of nitrate in fennel seeds [23] and the available literature data.

Dietary electrolyte balance = Na concentration × 10000/ molecular weights of Na+K concentration × 10000/molecular weights of K- chloride × 10000/ molecular weights Cl.

Nitrate intake = Amount of feed consumed × concentration of nitrates in feedstuffs.

European production index and economic efficiency were determined, according to [18].

2.2.2. Carcass and Meat Characteristics

On day 41, six chicks were randomly taken from each treatment, one per replicate to represent all treatment replicates, individually weighed, and slaughtered. The chicks were scalded at 54 °C for 2 min and then de-feathered, and finally, their heads were removed. After de-feathering, breast muscles and abdominal fat were immediately removed from the hot carcass and weighed. The internal organs were removed in a precise anatomical manner from the beginning of the esophagus to the end of the outlet [31].

The intestines were cleaned and weighed using a 0.1 g sensitive balance, and the data were expressed as a percentage of carcass weight. The clean carcass weight (dressing yield) was expressed as a percentage of live body weight. The dressing percentage was determined by dividing the clean carcass weight by the live body weight. Abdominal fat, leg weight, breast weight, intestinal weight, gizzard weight, liver weight, and heart weight were determined as a percentage of carcass weight, as indicated by [32].

2.2.3. pH, Temperature, and Meat Color

The pH of the boneless breast fillets was measured at 15 min post-mortem, using an electronic TPX-90i pH meter (Toko Chemical Laboratories Co. Ltd., Tokyo, Japan) with a needle-type electrode (CE201S-SR; Toko Chemical Laboratories Co. Ltd., Tokyo, Japan). Each sample was measured at three points, and their mean value was recorded. The breast muscle temperature was determined 15 min after slaughter, with a portable digital thermocouple (EcoScan Series, Temp JKT, Eutech Instruments,

Vernon Hills, IL, USA). Then, the CIE color parameters (L* (lightness), a* (redness), and b* (yellowness) values) were determined at three points of the breast muscle sample surface on the dorsal side using a CR-400 colorimeter (Konica Minolta Sensing Inc., Osaka, Japan).

2.2.4. Economic Efficiency

Economic efficiency was calculated according to [33]. The assessment was carried out using the income-cost divided by the cost and multiplied by 100. The income is equal to the selling price of body weight at the end of the experiment. The cost included the cost of day-old chicks, veterinary care and other husbandry costs, and cost of feeding. The difference between the income and cost is the net revenue, which is divided by total cost and multiplied by 100 to obtain the economic efficiency.

2.3. Statistical Analysis

The general linear model procedure of the SAS Software (Cary, NC, USA) [34] was used for the analysis of the obtained data. A 3 × 2 factorial experiment with six replicates was used in a completely randomized block design. The replicate was the experimental unit. Percentage data were transformed to arcsine before the analysis. Differences in treatment means were compared using the Bonferroni Test at $p < 0.05$.

3. Results

3.1. Growth Performance

Table 2 shows the effects of the inclusion of FSP in Ross-308 broiler diets and heat stress on BWG, FI, FCR, survival rate, and production index. The interaction effect (Temp × FSP) was observed on the BWG and production index. Under normal temperatures, no differences were noticed among the groups. However, under heat stress, the highest BWG and production index were recorded in the 3.2% FSP ($p < 0.05$). These results reflect the positive effects of feeding the chicks with a diet containing 3.2% FSP on the BWG and production index under heat stress conditions.

Table 2. Broiler body weight gain, feed intake, and feed conversion ratio following supplementation with fennel seed powder under thermoneutral or heat stress conditions.

Temp	Diet	BWG (g/bird)	FI (g)	FCR (g Feed/g Weight Gain)	Survival Rate (%)	European Production Index
Normal	Control	2149 ^{a}	3226	1.50	100	341 ^{a}
	FSP (1.6%)	2092 ^{a}	3137	1.50	100	332 ^{a}
	FSP (3.2%)	2141 ^{a}	3052	1.43	100	358 ^{a}
High	Control	1740 ^{c}	2832	1.63	100	255 ^{c}
	FSP (1.6%)	1751 ^{bc}	2854	1.63	100	256 ^{bc}
	FSP (3.2%)	1821 ^{b}	2992	1.64	97.2	264 ^{b}
SEM		31.2	29.9	0.017	0	11.6
ANOVA						
Temp * FSP		0.050	0.072	0.170	0.863	0.043
Temp		0.001	0.001	0.001	0.932	0.001
FSP		0.012	0.885	0.415	0.912	0.013

$^{a–c}$ Means with varying superscripts differ significantly ($p < 0.05$). BWG, body weight gain; FI, feed intake; FCR, feed conversion ratio. $^{a–c}$ Means with varying superscripts differ significantly ($p < 0.05$) BWG, body weight gain; FI, feed intake; FCR, feed conversion ratio.

Heat stress significantly decreased feed intake and impaired FCR compared to the thermoneutral group. However, the effect of FSP and the interaction between fennel and heat stress were not significant.

3.2. Carcass Characteristics

Table 3 presents the carcass characteristics affected by the inclusion of FSP in broiler diets under heat stress conditions. The results indicate that the percentages of dressing, abdominal fat, leg, gizzard, and liver were not significantly different among the heat stress groups and/or the dietary FSP groups. However, significant differences were observed in the percentages of breast muscle due to FSP diets, and in the intestine of broiler chicks due to the interaction effects (Table 3). The highest percentage of the intestine was observed in broilers fed with diets tested under thermoneutral conditions. Groups on CHS had significantly decreased heart percentage compared to the thermoneutral groups. The heart was significantly smaller in broilers fed the 1.6% fennel diet compared to the unsupplemented control. The breast was significantly greater in broilers fed the 1.6% fennel diet compared to the unsupplemented control.

Table 3. Carcass characteristics affected by supplementation with fennel seed powder under thermoneutral or heat stress conditions.

Temp.	Diet	Dressing (%)	Fat (%)	Leg (%)	Breast (%)	Intestine (%)	Gizzard (%)	Heart (%)	Liver (%)
Normal	Control	69.7	2.49	30.9	36.6	5.92 ^{b}	3.32	0.92	2.72
	FSP (1.6%)	69.0	2.09	28.7	39.4	8.17 ^{a}	3.34	0.84	2.91
	FSP (3.2%)	68.2	2.22	29.1	39.2	9.14 ^{a}	3.43	0.95	2.99
High	Control	68.8	1.99	29.7	36.6	8.08 ^{ab}	3.75	0.83	2.93
	FSP (1.6%)	68.7	1.98	29.8	38.5	7.70 ^{ab}	3.56	0.65	2.75
	FSP (3.2%)	69.2	2.27	28.5	38.0	7.67 ^{ab}	3.50	0.78	2.87
SEM		0.29	0.081	0.03	0.39	0.241	0.082	0.021	0.072
ANOVA									
Temp* FSP		0.385	0.361	0.173	0.791	0.003	0.652	0.640	0.510
Temp		0.894	0.248	0.658	0.355	0.856	0.136	0.001	0.867
FSP		0.713	0.489	0.069	0.030	0.03	0.902	0.021	0.801

^{a,b} Means with varying superscripts differ significantly ($p < 0.05$).

3.3. Meat Quality

The results of nitrate intake, pH, and L*, a*, and b* values as well as temperature are shown in Table 4. Nitrate intake was significantly affected by temperature and levels of FSP. Nitrate intake increased with increasing FSP within each temperature, and significantly decreased due to exposure to high temperature. CHS had a significant impact on broiler meat quality. Breast muscle pH was neither affected by CHS nor by dietary supplementation of FSP ($p > 0.05$). Breast muscle color of broilers subjected to heat stress, as indicated by lightness (L* values), redness (a* values), and yellowness (b* values), had variable responses compared to those under normal temperature conditions. An interaction effect was observed in the redness (a*) of breast muscles ($p < 0.05$). A significant difference between dietary treatments were found under normal temperature conditions (1.6 vs. 3.2 FSP), but non-significant differences under heat stress.

The dietary supplementation of FSP under both environmental conditions did not affect the yellowness (b* values) and lightness (L* values) of breast meat, but the effect of temperature on the b* and L* values was significant. The value of b* significantly decreased with CHS, while the L* value increased (Table 4).

Breast muscle temperatures were significantly influenced by heat stress and/or dietary FSP supplementation. The highest value (29.67 °C) was observed in broilers fed with the control diet under CHS. Moreover, broilers supplemented with 1.6 and 3.2% FSP had significantly lowered breast temperatures than the unsupplemented control.

Table 4. Broiler meat quality affected by supplementation with fennel seed powder under thermoneutral or heat stress conditions.

Temp	Diet	Nitrate Intake, mg/Chick	a*	b*	L*	Temp	pH
Normal	Control	62.8 [e]	4.04 [ab]	4.51	45.4	29.5 [a]	6.69
	FSP (1.6%)	171.4 [c]	5.31 [a]	4.31	47.2	28.2 [b]	6.74
	FSP (3.2%)	276.1 [a]	3.47 [bc]	4.88	45.6	27.9 [b]	6.71
High	Control	56.2 [f]	3.28 [bc]	3.49	47.1	29.7 [a]	6.73
	FSP (1.6%)	156.3 [d]	2.56 [bc]	2.89	47.9	29.3 [b]	6.61
	FSP (3.2%)	265.7 [b]	2.23 [c]	2.59	48.0	29.1 [b]	6.65
SEM		1.61	0.213	0.172	0.37	0.16	0.021
ANOVA							
Temp * FSP		0.115	0.029	0.098	0.612	0.007	0.509
Temp		0.005	0.0001	<0.0001	0.030	0.009	0.618
FSP		0.0001	0.016	0.596	0.311	0.139	0.932

[a–f] Means with varying superscripts differ significantly ($p < 0.05$).

3.4. Economic Efficiency

Table 5 shows the impact of temperature, funnel supplementation, and the interaction between them on economic efficiency. As expected, CHS had significant negative effects on the economic efficiency traits of broilers. On the other hand, funnel fortification had significant impacts on the total revenue, net revenue, and economic efficiency, showing a positive effect of 3.2% FSP. The interaction effect was significant for most of the traits, except for economic efficiency. The results indicate that feeding 3.2% FSP under CHS increased the feeding cost and total cost compared to other FSP levels. Total and net revenue was significantly decreased due to feeding 1.6% FSP under thermoneutral compared to other FSP levels. In addition, feeding 3.2% FSP under CHS increased the total and net revenue compared to 1.6% FSP.

Table 5. Economic efficiency of boilers affected by supplementation with fennel seed powder under thermoneutral or heat stress conditions.

Temp	Diet	Feeding Cost ($)	Total Cost ($)	Total Revenue ($)	Net Revenue ($)	Economic Efficiency (%)
Normal	Control	1.12 [a]	1.92 [a]	3.06 [a]	1.14 [a]	59.6
	FSP (1.6%)	1.10 [a]	1.90 [a]	2.98 [b]	1.081 [b]	57.0
	FSP (3.2%)	1.09 [a]	1.89 [a]	3.05 [a]	1.17 [a]	61.8
High	Control	1.00 [c]	1.79 [c]	2.48 [e]	0.681 [d]	37.9
	FSP (1.6%)	1.00 [c]	1.79 [c]	2.50 [d]	0.696 [d]	38.6
	FSP (3.2%)	1.05 [b]	1.85 [b]	2.60 [c]	0.749 [c]	40.6
SEM		0.032	0.030	0.056	0.0431	2.11
ANOVA						
Temp*FSP		0.022	0.022	0.001	0.023	0.134
Temp		0.001	0.001	0.001	0.001	0.001
FSP		0.270	0.270	0.001	0.002	0.013

[a–e] Means with varying superscripts differ significantly ($p < 0.05$). BWG, body weight gain; FI, feed intake; FCR, feed conversion ratio.

4. Discussion

The main effect of the temperature on broiler performance was significant and confirmed the success of heat stress induction. The adaptation to heat stress markedly influences broiler performance, production index, and economic efficiency [3,13,35]. Broilers at market weight generate approximately 5–10 kcals of energy per hour as a normal physiological process [9]. Under heat stress, broilers dissipate heat from their body to the surrounding environment, assuming that it is at a lower temperature than the body (41 °C) [9]. As body temperature rises above 41.5 °C, broilers are considered to be under heat stress, and their performance is more likely to be adversely affected. However, mortality usually occurs when the temperature is above 42 °C [9].

To lose the extra heat gained from the environment, birds must reduce the FI to decrease metabolic heat production from feed, and thus, their BWG decreases [16]. FSP intake promotes BWG and FCR. Likewise, dietary FSP supplementation significantly increased the growth of quail chicks [16]. In the present study, the best BWG, FCR, breast muscle, and liver percentage were obtained when broilers under normal condition were supplemented with 3.2% of FSP. It is worth noting that FI significantly decreased under CHS conditions among all treatments when compared to the control. The improved performance of broilers on FSP diets may be due to enhanced digestibility and enriched antioxidants status. Fennel also has strong antiviral, antimicrobial, and anti-inflammatory effects, which may improve gut health and eliminate pathogens. It was reported that the FSP is a rich source of essential oil (anethole, fenchone, methyl chavicol, limonene, phellandrene, camphene, pinine, anisic acid, and palmitic, oleic, linoleic, and petroselenic acids, volatile compounds, flavonoids, phenolics [19–21].

Our results showed significant alterations in broiler performance, production, and economic efficiency under CHS. This finding is in line with those reported by [13,16]. Dietary supplementation with 3.2% FSP improved broiler growth rate by 4.7% under CHS, while, under optimum temperatures, FSP-supplemented groups displayed similar growth, feed intake, FCR, production index, and economic efficiency. Moreover, our findings were in agreement with those of [20,36]. However, diets with fennel powder of 1, 2, and 3 g/kg resulted in significant improvements in broiler growth and FCR, while feed intake was unaffected [28]. Heat stress may cause a physiological dysfunction that triggers the body to utilize nutrients to synthesize critical proteins instead of using them for growth, which allows broilers to decrease against the oxidative damage of heat stress [2,3,37].

The results obtained on the carcass characteristics in the experimental conditions did not show differences in these characteristics, except for heart percentage, which was affected by heat stress. Supplementation of 1.6% and 3.2% FSP seemed to improve breast percentage. Treatments with FSP, particularly at 1.6%, tended to decrease heart percentages. The results of carcass characteristics in this study were in line with those obtained by [38], who reported that the addition of FSP at 1, 2, and 3 g/kg to the diet resulted in insignificant differences for all carcass characteristics. Similar results were also reported by [20]. The results indicate that FSP had no adverse effects on carcass traits. Moreover, phytogenic plants and their essential oils are used as preservation approaches to enhance the sensory attributes and prolong the shelf life of animal products [19,21,36,39].

The results showed that meat quality parameters such as meat color, which is an important visual criterion, were significantly affected by heat stress. The color of meat is correlated with myoglobin and hemoglobin concentrations and their status [40,41]. The current results are in agreement with the results reported by [3,42]. The highest value of a* was reported in the 1.6% FSP group under normal temperatures. However, it was only similar to the control group under the same conditions. It is worth noting that the a* value is extremely important to consumers.

Furthermore, a high redness (a*) value yields an undercooked appearance. The a* value can be affected by bird age, stress before slaughter, and consumption of dietary nitrates [43]. Nitrates have been reported to be at high concentrations (65 to 376.7 mg/100 g) in FSP [23,44–46], which may explain the obtained results and have both beneficial and adverse effects [47]. The change in a* value parallels the change in nitrate intake, which increases with increasing FSP concentration and decreased under CHS condition. The nitrate intakes ranged from 56.2 in the control groups to 276 mg/ bird during 19–41 days of age, with an average daily intake of nitrate ranging from 2.44 to 12 mg/day. The toxic level of nitrate for chickens ranges from 450–900 ppm and is 658 ppm for nitrite [30]. Thus, the intake of nitrate was far from the toxic level. The concentration of nitrate in animal feedstuffs is considered safe at 4, 44, and 22 mg/kg in soybean meal, oat grain, and maize grain, respectively [48,49]. Thus, nitrite poisoning is rare in poultry fed cereal/grain/oilseeds meal based diets [50–52].

Additionally, NO was known to react with a myoglobin increased a* value [27]. Nevertheless, the latter may act as pro- or anti-oxidant, according to its concentration. With higher NO concentration, it may rapidly react with superoxide radical forming proxy nitrite [47], which oxidizes myoglobin, resulting in low a* value [43]. On the other hand, low a* values may indicate more oxidized myoglobin in birds subjected to heat stress, as reported by [53].

Heat stress increases breast meat temperatures, which may accelerate oxidative rancidity and increase the peroxidation biomarker. Similarly, CHS was found to increase body temperature in a study conducted by [54]. However, in our study, the FSP-supplemented groups showed significant reductions in breast muscle temperature under both normal and heat stress conditions, revealing a beneficial effect of FSP. This finding, along with the increase in BWG in broilers under CHS and fed 3.2% FSP diet, can be explained on the basis that FSP contains NO, which increases blood flow to the body surface and upper respiratory tract to dissipate body temperature, as reported by [53], suggesting that FSP may be used as a CHS alleviating agent.

In the present investigation, no differences were observed in the pH of meat among different temperatures and/or fennel groups. The pH is one of the most important physical traits for the qualitative profile of meat and is commonly utilized as an assessment of sensory qualities of the technological properties of meat. Meat pH is related to the water-holding capacity [40,55], which is correlated positively with the texture, juiciness, and flavor of meat [56].

5. Conclusions

According to the findings of this study, the dietary inclusion of FSP, especially at 3.2%, resulted in a beneficial impact on the growth performance and carcass quality of broilers under heat stress conditions. Furthermore, CHS significantly decreased the production index and economic efficiency, but FSP did not affect both criteria.

Author Contributions: A.A.A.-S. conducted the experimental design, fieldwork, and first draft of the manuscript; S.K. carried out the data collection and contribution to the experimental set-up, E.O.S.H., lab work, and reading the manuscript; and Y.A.A. undertook the statistical analyses and proofreading of the manuscript. All authors have read and agreed to the published version of the manuscript.

Acknowledgments: The authors express their sincere gratitude and appreciation to King Abdulaziz City for Science and Technology (KACST) for allowing the use of its facilities to conduct this study.

References

1. Suganya, T.; Senthilkumar, S.; Deepa, K.; Amutha, R. Nutritional management to alleviate heat stress in broilers. *Int. J. Sci. Environ. Technol.* **2015**, *4*, 661–666.
2. Attia, Y.A.; Hassan, S.S. Broiler tolerance to heat stress at various dietary protein/energy levels. *Europ. Poult. Sci.* **2017**, *81*. [CrossRef]
3. Attia, Y.A.; Al-Harthi, M.A.; Elnaggar, A.S. Productive, physiological and immunological responses of two broiler strains fed different dietary regimens and exposed to heat stress. *Ita, J. Anim. Sci.* **2018**, *17*, 686–697. [CrossRef]
4. Yahav, S. Alleviating heat stress in domestic fowl: Different strategies. *World´s Poult. Sci. J.* **2009**, *65*, 719–732. [CrossRef]
5. Cassuce, D.C.; Tinôco, I.; de, F.F.; Baêta, F.C.; Zolnier, S.; Cecon, P.R.; Vieira, M.; de, F.A. Thermal comfort temperature update for broiler chickens up to 21 days of age. *Eng. Agrícola Jaboticabal* **2013**, *33*, 28–36. [CrossRef]
6. Yardibi, H.; Gülhan, T. The effects of vitamin E on the antioxidant system, egg production, and egg quality in heat-stressed laying hens. Turk. *J. Vet. Anim. Sci.* **2008** *32*, 319–325.

7. Whitehead, C.C.; Keller, T. An update on ascorbic acid in poultry. *World's Poult. Sci. J.* **2003**, *59*, 161–184. [CrossRef]
8. Vandana, G.D.; Sejian, V. Towards identifying climate resilient poultry birds. *JDVAR* **2018**, *7*, 84–85.
9. Leeson, S.; Summers, J.D. *Commercial Poultry Nutrition*, 3rd ed.; Nottingham University Press: Nottingham, UK, 2009; pp. 261–262.
10. Lin, H.; Eddy, D.; Johan, B. Acute heat stress induces oxidative stress in broiler chickens. *Comp. Biochem. Physiol. A Mol. Integr. Physiol.* **2006**, *144*, 11–17. [CrossRef]
11. Akbarian, A.; Joris, M.; Jeroen, D.; Maryam, M.; Abolghasem, G.; Stefaan, D.S. Association 0 -between heat stress and oxidative stress in poultry; mitochondrial dysfunction and dietary interventions with phytochemicals. *J. Anim. Sci. Biotechnol.* **2016**, *7*, 37. [CrossRef]
12. Akşit, M.; Yalçin, S.; Özkan, S.; Metin, K.; Özdemir, D. Effects of temperature during rearing and crating on stress parameters and meat quality of broilers. *Poult. Sci.* **2006**, *85*, 1867–1874. [CrossRef] [PubMed]
13. Attia, Y.A.; Hassan, R.A.; Tag El-Din, A.E. Abou- Shehema, B.M. Effect of ascorbic acid or increasing metabolizable energy level with or without supplementation of some essential amino acids on productive and physiological traits of slow-growing chicks exposed to chronic heat stress. *J. Anim. Physiol. Anim. Nutr.* **2011**, *95*, 744–755. [CrossRef] [PubMed]
14. Guimarães, R.; Barros, L.; Carvalho, A.M.; Ferreira, I.C.F.R. Infusions and decoctions of mixed herbs used in folk medicine: Synergism in antioxidant potential. *Phytotherapy Res.* **2011**, *25*, 1209–1214. [CrossRef]
15. Koppula, S.; Kumar, H. Foeniculum vulgare Mill (Umbelliferae) attenuates stress and improves memory in Wister rats. *Trop. J. Pharm. Res.* **2013**, *12*, 553–558. [CrossRef]
16. Gous, R.M.; Morris, T.R. Nutritional interventions in alleviating the effects of high temperatures in broiler production. *Proc. Nutr. Soc.* **2005**, *61*, 463–475. [CrossRef]
17. Sahin, K.; Orhan, C.; Smith, M.O.; Sahin, N. Molecular targets of dietary phytochemicals for the alleviation of heat stress in poultry. *World´s Poult. Sci. J.* **2013**, *69*, 113–123. [CrossRef]
18. Attia, Y.A.; Bakhashwain, A.A.; Bertu, N.K. Utilisation of thyme powder (Thyme Vulgaris, L.) as a growth promoter alternative to antibiotics for broiler chickens raised in a hot climate. *Europ. Poult. Sci.* **2018**, *82*. [CrossRef]
19. Gharaghani, H.; Shariatmadari, F.; Torshizi, M.A. Effect of fennel (Foeniculum vulgare Mill.) used as a feed additive on the egg quality of laying hens under heat stress. *Rev. Bras. Ciênc. Avícola* **2015**, *17*, 199–207. [CrossRef]
20. El-Deek, A.A.; Al-Harthi, M.A.; Attia, Y.A.; Hannfy, M.M. Effect of anise (Pimpinella anisum), ginger (Zingiber officinale roscoe) and fennel (Foeniculum vulgare) and their mixture on performance of broilers. *Arch. Geflugelkunde* **2003**, *67*, 92–96.
21. Badgujar, S.B.; Patel, V.V.; Bandivdekar, A.H. Foeniculum vulgare Mill: A Review of Its Botany, Phytochemistry, Pharmacology, Contemporary Application, and Toxicology. *BioMed Res. Int.* **2014**, *2014*, 842674. [CrossRef]
22. Rather, M.A.; Dar, B.A.; Sofi, S.N.; Bhat, B.A.; Qurishi, M.A. *Foeniculum vulgare mill:* A comprehensive review of its traditional use, phytochemistry, pharmacology, and safety. *Arab. J. Chem.* **2016**, *9*, 1574–1583. [CrossRef]
23. Swaminathan, A.; Sridhara, S.R.; Sinha, S.; Nagarajan, S.; Balaguru, U.M.; Siamwala, J.H.; Rajendran, S.; Saran, U.; Chatterjee, S. Nitrites derived from Foeniculum vulgare (fennel) seeds promotes vascular functions. *J. Food Sci.* **2012**, *77*, 273–279. [CrossRef] [PubMed]
24. Dunai, V.; Tzschentke, B. Impact of environmental thermal stimulation on activation of hypothalamic neuronal nitric oxide synthase during the prenatal ontogenesis in Muscovy ducks. *The Sci. World, J.* **2012**, *2012*, 416936. [CrossRef]
25. Zeng, T.; Li, J.J.; Wang, D.Q.; Li, G.Q.; Wang, G.L.; Lu, L.Z. Effects of heat stress on antioxidant defense system, inflammatory injury, and heat shock proteins of Muscovy and Pekin ducks: Evidence for differential thermal sensitivities. *Cell Stress Chaperones* **2014**, *19*, 895–901. [CrossRef] [PubMed]
26. Dantonio, V.; Batalhão, M.E.; Fernandes, M.H.; Komegae, E.N.; Buqui, G.A.; Lopes, N.P.; Bícego, K.C. Nitric oxide and fever: Immune-to-brain signaling vs. thermogenesis in chicks. *Am. J. Physiol. Regul. Integr. Comp. Physiol.* **2016**, *310*(10), R896–R905. [CrossRef] [PubMed]
27. Zhang, W.; Al-Hijazeen, M.; Himali, S.; Eun, J.L.; Dong, U.A. Breast meat quality of broiler chickens can be affected by managing the level of nitric oxide. *Poult. Sci.* **2013**, *92*, 3044–3049. [CrossRef]

28. Zhang, Z.Y.; Jia, G.Q.; Zuo, J.J.; Zhang, Y.; Lei, J.; Ren, L.; Feng, D.Y. Effects of constant and cyclic heat stress on muscle metabolism and meat quality of broiler breast fillet and thigh meat. *Poult. Sci.* **2012**, *91*, 2931–2937. [CrossRef]
29. Assreuy, J.; Cunha, F.Q.; Liew, F.Y.; Moncada, S. Feedback inhibition of nitric oxide synthase activity by nitric oxide. *Br. J. Pharmacol.* **1993**, *108*, 833–837. [CrossRef]
30. National Research Council. *Nutrient requirements of poultry*, 9th ed.; National Academy Press: Washington, DC, USA, 1991.
31. Fletcher, D.L. Poultry meat color in poultry meat science. *Poult. Sci. Symp. Ser.* **1999**, *25*, 159–175.
32. AL-Fayad, H.A.; Naji, S.A.; Abid, N.N. *Poultry products technology*, 2nd ed.; Ministry of Higher Education and Scientific Research, College of Agriculture, University of Baghdad: Baghdad, Iraq, 2011.
33. Attia, Y.A.; Hamed, R.S.; Abd El-Hamid, A.E.; Shahba, H.A.; Bovera, F. Effect of inulin and mannan-oligosaccharides compared with zinc-bacitracin on growing performance, nutrient digestibility and hematological profiles of growing rabbits. *Anim. Prod. Sci.* **2015**, *55*, 80–86. [CrossRef]
34. SAS Institute. *SAS/STAT User's Guide (Version 9.1)*; SAS Inst. Inc.: Cary, NC, USA, 2004.
35. Hassan, A.; Abd Elazeem, M.H.; Hussein, M.M.; Osman, M.M.; Abd El-wahed, Z.H. Effect of chronic heat stress on broiler chicks performance and immune system. *SCVMJ* **2007**, *12*, 55–68.
36. Gharaghani, H.; Shariatmadari, M.; Karimi Torshizi, A. Comparison of oxidative quality of meat of chickens fed corn or wheat based diets with fennel (Foeniculum vulgare mil.), antibiotic and probiotic as feed additive, under different storage conditions. *Arch. Geflugelkunde* **2013**, *77*, 199–205.
37. Butcher, G.D.; Miles, R.D. *Interrelationship of Nutrition and Immunity*; University of Florida Cooperative Extension Service, Institute of Food and Agricultural Sciences, EDIS: Gainesville, FL, USA, 2002; Volume VM139, pp. 1–8.
38. Mohammed, A.A.; Abbas, R.J. The effect of using fennel seeds (Foeniculum vulgare L.) on productive performance of broiler chickens. *Int. J. Poult. Sci.* **2009**, *8*, 642–644. [CrossRef]
39. Bozkurt, M.; Küçükyılmaz, K.; Pamukçu, M.; Çabuk, M.; Alçiçek, A.; Çatlı, U. Long-term effects of dietary supplementation with an essential oil mixture on the growth and laying performance of two layer strains. *Ita. J. Anim. Sci.* **2012**, *11*, 23–28. [CrossRef]
40. Brewer, S. Irradiation effects on meat color – a review. *Meat Sci.* **2004**, *68*, 1–17. [CrossRef]
41. Attia, Y.A.; El-Tahawy, W.S.; Abd El-Hamid, A.E.; Nizza, A.; El-Kelway, M.I.; Al-Harthi, M.A.; Bovera, F. Effect of feed form, pellet diameter and enzymes supplementation on carcass characteristics, meat quality, blood plasma constituents and stress indicators of broilers. *Arch. Tierz.* **2014**, *57*, 1–14. [CrossRef]
42. Yalçin, S.; Önenç, A.; Özkan, S.; Güler, H.C.; Siegel, P.B. Meat quality of heat-stressed broilers: Effects of thermal conditioning at pre- and postnatal stages. In *Book of Abstract. XVIIth European Symposium On the Quality of Poultry Meat, Doorwert, The Netherlands, 23–26 May 2005*; WPSA: Beekbergen, The Netherlands, 2005; p. 265.
43. Smith, D.P.; Northcutt, J.K. Red discoloration of fully cooked chicken products. *J. Appl. Poult. Res.* **2003**, *12*, 515–521. [CrossRef]
44. Koudela, M.; Petříková, K. Nutritional compositions and yield of sweet fennel cultivars - Foeniculum vulgare Mill. ssp. vulgare var. azoricum (Mill.). *Thell Hort. Sci.* **2008**, *35*, 1–6.
45. Bukhari, H.; Shehzad, A.; Saeed, K.; Sadiq, B.M.; Tanveer, S.; Iftikhar, T. Compositional Profiling of Fennel Seed. *Pak. J. Food Sci.* **2014**, *24*, 132–136.
46. Hord Norman, G.; Tang, Y.; Bryan Nathan, S. Food Sources of Nitrates and Nitrites: The Physiologic Context for Potential Health Benefits. *Am. J. Clin. Nut.* **2009**, *90*, 1–10. [CrossRef]
47. Møller Jens, K.S.; Skibsted Leif, H. Myoglobins - The link between discoloration and lipid oxidation in muscle and meat. *Quimica Nova* **2006**, *29*, 1270–1278.
48. Cockburn, A.; Brambilla, G.; Fernandez, M.L.; Arcella, D.; Bordajandi, L.R.; Cottrill, B.; van Peteghem, C.; Dorne, J.L. Nitrite in feed: From Animal health to human health. *Toxicol. Appl. Pharmacol.* **2013**, *270*, 209–217. [CrossRef] [PubMed]
49. Crowley, J.W. Effects of nitrate on livestock. American Society of Agricultural Engineers Paper Number 80-20026. Cited by Undersander, D., Combs, D., Shaver, R.,Thomas, D. Available online: http://www.uwex.edu/ces/forage/pubs/nitrate.htm1985 (accessed on 20 March 2017).

50. Attia, Y.A.; Abd El Hamid, A.E.; Ismaiel, A.M.; de Oliveira Maria, C.; Al-Harthi, M.A.; El- Naggar Asmaa, S.H.; Simon, G.A. Nitrate detoxification using antioxidants and probiotics in the water of rabbits. *Revista Colombiana De Ciencias Pecuarias (RCCP)* **2018**, *31*, 130–138. [CrossRef]
51. Attia, Y.A.; El Hamid, E.A.; Ismaiel, A.M.; El-Nagar, A. The detoxication of nitrate by two antioxidants or a probiotic and the effects on blood and seminal plasma profiles and reproductive function of NZW rabbit bucks. *Animal* **2013**, *7*, 591–601. [CrossRef]
52. Ologhobo, A.D.; Adegede, H.I.; Maduagiwu, E.N. Occurrence of nitrate, nitrite and volatile nitrosamines in certain feedstuffs and animal products. *Nutr Health.* **1996**, *11*, 109–114. [CrossRef] [PubMed]
53. Zhu, W.; Jiang, W.; Wu, L.Y. Dietary L-Arginine supplement alleviates hepatic heat stress and improves feed conversion ratio of Pekin ducks exposed to high environmental temperature. *J. Anim. Physiol. Anim. Nutr.* **2014**, *98*, 1124–1131. [CrossRef]
54. Vinoth, A.; Thirunalasundari, T.; Shanmugam, M.; Rajkumar, U. Effect of early age thermal conditioning on expression of heat shock proteins in liver tissue and biochemical stress indicators in colored broiler chicken. *Eur. J. Exp. Biol.* **2016**, *6*, 53–63.
55. Jung, S.; Choe, J.H.; Kim, B.; Yun, H.Z.; Kruk, A.; Jo, C. Effect of dietary mixture of gallic acid and linoleic acid on antioxidative potential and quality of chest meat from broilers. *Meat Sci.* **2010**, *86*, 520–526. [CrossRef]
56. Abdullah, A.Y.; Musallam, H.S. Effect of different levels of energy on carcass composition and meat quality of male black goat kids. *Livest Sci.* **2007**, *107*, 70–80. [CrossRef]

4

Fatty Acid Signatures in Different Tissues of Mediterranean Yellowtail, *Seriola dumerili* (Risso, 1810), Fed Diets Containing Different Levels of Vegetable and Fish Oils

Francesco Bordignon [1], Ana Tomás-Vidal [2], Angela Trocino [1,*], Maria C. Milián Sorribes [2], Miguel Jover-Cerdá [2] and Silvia Martínez-Llorens [2]

1 Department of Comparative Biomedicine and Food Science (BCA), University of Padova, Viale dell'Università 16, I-35020 Legnaro, Italy; francesco.bordignon.3@phd.unipd.it

2 Institute of Animal Science and Technology, Group of Aquaculture and Biodiversity, Universitat Politècnica de València, Camino de Vera, 14, 46071 València, Spain; atomasv@dca.upv.es (A.T.-V.); mamisor@etsia.upv.es (M.C.M.S.); mjover@dca.upv.es (M.J.-C.); silmarll@dca.upv.es (S.M.-L.)

* Correspondence: angela.trocino@unipd.it.

Simple Summary: Most of the studies performed to date mainly investigated on the effects of dietary substitution of fish oil with vegetable oils on growth and fatty acid composition of edible muscle tissues. On the other hand, a few assessed how dietary lipids are retained in other tissues, such as brain, liver, and adipose tissue, which would provide further insights into the fatty acid requirements of new farmed marine fish species such as *Seriola dumerili*. Thus, this study evaluated how the replacement of fish oil with different proportions of vegetable oils in diets affects the tissue-specific fatty acid composition (also known as signature) of brain, muscle, liver, and visceral fat of *S. dumerili*. The fatty acid composition of the diet had a strong effect on the fatty acid signature of muscle, liver, and visceral fat, whereas the brain signature was less sensitive to dietary changes. These new insights contribute to identify the essential fatty acids requirements of Mediterranean yellowtail and to define the conditions under which the physiological functions in these fish are preserved when they are fed diets with low fish oil levels to guarantee the sustainability of their production and welfare.

Abstract: The study aimed to evaluate how replacing different proportions of fish oil (FO) with vegetable oils (VO) in the diet of Mediterranean yellowtail, *Seriola dumerili* (Risso, 1810), affects the fatty acids (FA) signature, i.e., overall FA profile, in different tissues. A total of 225 Mediterranean yellowtail juveniles (initial live weight: 176 ± 3.62 g) were fed for 109 days with one of three diets: A control diet (FO 100), with FO as the only lipid source, or diets with 75% and 100% of FO replaced with a VO mixture. At the end of the feeding trial, the brains, muscles, livers, and visceral fat were sampled in four fish per tank (12 per treatment), and their fat were extracted and used for FA analysis. The FA signatures of red and white muscle, liver, and visceral fat tissues changed when the dietary FA source changed, whereas FA signatures in the brain were rather robust to such dietary changes. These new insights might help evaluate whether key physiological functions are preserved when fish are fed diets with low FO levels, as well as define the dietary FA requirements of Mediterranean yellowtail to improve the sustainability of the production and welfare of the fish.

Keywords: brain; muscle; liver; greater amberjack; EPA; DHA

1. Introduction

The Mediterranean yellowtail, *Seriola dumerili* (Risso, 1810), also called the greater amberjack, is a new, high-value candidate for production in marine aquaculture. This cosmopolitan fish is mainly produced in Japan [1], Spain [2], Italy [3], and recently in Vietnam, Korea, and China [4]. It has high growth rates (reaching 6 kg within 2.5 years of culture), and exceptionally high consumer acceptance [4,5].

Since the Mediterranean yellowtail is an emerging species in aquaculture, its nutritional requirements have been defined in terms of major nutrients (47%–50% crude protein (CP) and 12%–14% crude lipid (CL)) [6–10]; however, little information is available on its fatty acid requirements [11], which thus require further investigation. Only one previous study assessed the effects of dietary fatty acids on Mediterranean yellowtail body composition [11].

Fatty acids (FA) play key roles in fish health and nutrition. They maintain the structural and functional integrity of cell membranes, provide metabolic energy through their oxidative metabolism, contribute to visual and brain development, and are precursors of a group of paracrine hormones with relevant biological roles known as eicosanoids [12–14]. Fish health, growth, and reproduction are strongly dependent on n-3 and n-6 polyunsaturated fatty acids (PUFA), especially arachidonic acid (AA, C20:4 n-6), eicosapentaenoic acid (EPA, C20:5 n-3), and docosahexaenoic acid (DHA, C22:6 n-3) [13,15–17]. However, marine carnivorous fish such as *S. dumerili* have limited ability to bio-convert the essential precursors of PUFA, such as linoleic acid (LA, C18:2 n-6) and alpha-linolenic acid (ALA, C18:3 n-3), into these essential FA [18,19].

The overall fatty acid profile within a given tissue, also known as its FA signature [20], is strongly dependent on the physiological function(s) of the tissue itself [21]. The liver and muscles are the main sites of β-oxidation [22], whereas the brain stores n-3 LC-PUFA, mainly DHA, which perform neurological functions [23–25]. Nevertheless, in nature, the FA signatures of prey tissues can be transferred to their predators with little modification and in a predictable manner [20,26,27]. Thus, in recent decades, FA have been extensively used as biomarkers in riverine and marine ecosystems [20,28,29]. Additionally, under farming conditions [30–32], the available literature collectively supports the conclusion that there is a close association between the diet and fillet FA composition in farmed fish, as reviewed by Turchini et al. [19].

As in other species [33], understanding the FA distributions and signatures within Mediterranean yellowtail tissues might help us understand whether the physiological needs and essential fatty acid (EFA) requirements of fish are satisfied under farming conditions. This is particularly crucial when diets that contain low levels of fish oil (FO) are used in fish farming. In fact, to ensure the long-term sustainability of the aquaculture sector and to improve its competitiveness, FO has been increasingly replaced by vegetable oils (VO) in fish diets because FO has a high cost and is available in finite and limited amounts [19]. Compared to FO, plant seed oils are rich in C:18 PUFA [19], but lacking in n-3 highly unsaturated fatty acids (HUFA), which are essential for marine fish [13].

Most studies performed in farmed fish to date have focused mainly on the effects of dietary lipids on the FA composition of edible muscle tissues [19]. Only one study [33], which investigated FA signatures in gilthead seabream (*Sparus aurata* L.; 1758), assessed how dietary lipids were retained in different tissues (brain, liver, and mesenteric adipose tissue) to obtain further insights into the FA requirements of farmed marine fish.

Therefore, this study was performed to evaluate how the replacement of FO with different proportions of VO in the diet affects the tissue-specific FA signatures and their robustness in the brain, muscle, liver, and visceral fat tissues of Mediterranean yellowtail (*S. dumerili*).

2. Materials and Methods

A feeding trial was performed at the Fish Nutrition Laboratory (LAC) of the Institute of Animal Science and Technology (ICTA) of the Universitat Politècnica de València (Polytechnic University of Valencia, Spain). The experimental protocol was approved by the Committee of Ethics and Animal

Welfare of the Polytechnic University of Valencia, following the Spanish Royal Decree 53/2013 on the protection of animals used for scientific purposes.

The facility used a thermoregulated recirculating seawater system (65 m^3 capacity), with a rotary drum-type filter and a mechanical gravity biofilter with a volume of 2 m^3, equipped with 9 cylindrical fiberglass tanks with a capacity of 1750 L each with aeration.

2.1. Experimental Diets

Three isoproteic (59% CP, 50% digestible protein (DP)), and isolipidic (15% CL) extruded diets were formulated, with increasing levels of vegetable oils used to replace the fish oil in the diets as follows: 0% (control diet, FO 100), 75% (FO 25), and 100% (FO 0) replacement of FO with VO (Table 1). The mixture of vegetable oils used consisted of linseed oil, sunflower oil, and palm oil (4:3:3). Diets were prepared using a cooking-extrusion processor with a semi-industrial twin-screw extruder (CLEXTRAL BC-45; Firmity, St Etienne, France), at a screw speed of 100 rpm, temperature of 110 °C, and pressure of 40–50 atm, to obtain pellets 2–3 mm in diameter.

Table 1. Ingredients (g kg^{-1} as fed) and proximate composition (% dry matter; DM) of the experimental diets.

	Diet		
	FO 100	FO 25	FO 0
Fish meal	350	350	350
Wheat	100	100	100
Wheat gluten	140	140	140
Defatted soybean meal	185	185	185
Iberian meat meal	110	110	110
Fish oil	95	24	0
Linseed oil	-	28	38
Sunflower oil	-	21	28
Palm oil	-	22	29
Multivitamin and minerals mix	20	20	20
Proximate composition			
Dry matter, %	87.4	88.8	89.1
Crude protein, % DM	58.8	60.6	58.8
Crude lipid, % DM	15.9	15.1	16.6
Ash, %DM	8.4	10.3	8.3
Gross energy, MJ kg^{-1} DM	24.3	23.8	24.4

Fish oil (FO) 100 diet: Diet formulated with fish oil as lipid source; FO 25 diet: Diet in which fish oil was included at a content of 25%; FO 0 diet: Diet in which fish oil was totally substituted with vegetable oil. [1] Vitamins and mineral mixture (values are g kg^{-1}): Premix, 25; Hill, 10; DL-α-tocopherol, 5; ascorbic acid, 5; $(PO_4)_2Ca_3$, 5. Premix composition (values are IU kg^{-1}): Retinol acetate, 1,000,000; calcipherol, 500; DL-α-tocopherol, 10; menadione sodium bisulphite, 0.8; hidroclorhidrate thiamine, 2.3; riboflavin, 2.3; pyridoxine hydrochloride, 15; cyanocobalamin, 25; nicotinamide, 15; pantothenic acid, 6; folic acid, 0.65; biotin, 0.07; ascorbic acid, 75; inositol, 15; betaine, 100; polypeptides, 12.

The FA profile of the experimental diets changed depending on the relative proportions of the two lipid sources included in them. Diets containing more FO contained higher proportions of highly unsaturated fatty acids from the n-3 series, whereas diets containing more vegetable oils had higher proportions of LA and ALA (Table 2). In all diets, C16:0 accounted for the bulk of the saturated fatty acids (SFA) present, C18:1n-9 for most of the monounsaturated fatty acids (MUFA), and EPA and ARA for most of the n-3 long-chain polyunsaturated fatty acids (PUFA). The inclusion of VO as a lipid source increased the dietary proportion of C18:1 n-9 (27.1%, 29.6%, and 32.7%), C18:1 n-6 (12.7%, 13.4%, and 14.9%), and ALA (14.1%, 14.3%, and 15.4%), whereas it decreased the proportions of total SFA, EPA, and docosahexaenoic acid (DHA) in the diets (SFA: 27.8%, 27.1%, and 25.5% in the FO 100, FO 25, and FO 0 diets, respectively; EPA: 5.81%, 4.34%, and 2.77%; DHA: 13.0%, 7.6%, and 4.3%) (Table 2).

Table 2. Fatty acid composition (% of total fatty acid content) of the experimental diets.

	Diet		
	FO 100	FO 25	FO 0
14:0	3.28	2.38	1.65
16:0	18.88	19.58	18.93
17:0	0.53	0.24	0.16
18:0	5.07	4.88	4.71
Σ SFA	27.79	27.12	25.46
16:1 n-9	4.22	2.77	1.82
18:1 n-9	27.14	29.64	32.67
18:1 n-7	3.94	2.94	2.44
22:1 n-9	0.32	0.04	0.06
Σ MUFA	35.62	35.39	36.99
18:2 n-6	12.66	13.36	14.86
18:3 n-6	0.10	0.09	0.09
20:3 n-6	0.10	0.04	0.04
20:4 n-6	1.02	0.58	0.35
22:4 n-6	0.24	0.19	0.09
Σ n-6 PUFA	14.12	14.26	15.43
18:3 n-3	2.24	10.48	14.60
20:3 n-3	0.15	0.08	0.06
20:5 n-3	5.81	4.34	2.77
22:5 n-3	1.29	0.73	0.42
22:6 n-3	12.98	7.61	4.28
Σ n-3 PUFA	22.47	23.24	22.13
Σ PUFA	36.59	37.50	37.56
Σ n-6/Σ n-3	0.63	0.61	0.70
DHA/EPA	2.23	1.76	1.54

FO 100 diet: Diet formulated with fish oil as lipid source; FO 25 diet: Diet in which fish oil was included at a content of 25%; FO 0 diet: Diet in which fish oil was totally substituted with vegetable oil. Abbreviations: SFA: Saturated fatty acids; MUFA: Monounsaturated fatty acids; PUFA: Polyunsaturated fatty acids; DHA/EPA: 22:6 n-3/20:5 n-3.

2.2. *Fish and Experimental Design*

A total of 225 Mediterranean yellowtail juveniles obtained from a private hatchery (Futuna Blue S.A.; Cádiz, Spain) were transported to the Fish Nutrition Laboratory of the Universitat Politècnica de València for use in the feeding trial. After four weeks of acclimation, fish were individually weighed (initial weight: 176 ± 3.62 g), randomly distributed among nine tanks (25 fish per tank), and fed the experimental diets (FO 100, FO 25, or FO 0, with three tanks per diet) for 109 days.

During the trial, the water temperature averaged 21.5 ± 2.4 °C, the salinity was 31.5 ± 4.1 g L^{-1}, and the dissolved oxygen content was 8.0 ± 0.4 mg L^{-1}. The water pH ranged from 7.5 to 8.0, and the levels of nitrogenous compounds in the water were kept within the limits recommended for marine species.

2.3. *In Vivo Recordings*

Feed was distributed by hand, twice a day (at 09:00 and 16:00) for six days per week, until apparent satiation. Feed intake was recorded at each administration. Mortality was checked every day. At the beginning and at the end of the trial, fish were individually weighed after one day of feed deprivation and under light anaesthesia (10 mg L^{-1} clove oil containing 87% eugenol; Guinama®, Valencia, Spain). Fish health during weighing was assessed by direct observation.

After 109 days of feeding, four fish per tank (12 per treatment, 36 in total) were randomly sacrificed by lethal immersion in clove oil (150 mg L^{-1}). Fish were dissected to sample their brain, white and red muscle, liver, and visceral fat tissues, which were then frozen in liquid nitrogen and stored at −80 °C until subsequent analyses.

2.4. *Chemical Analyses of Diets and Fish Tissues*

Diets were ground up and analysed according to AOAC procedures [34] to determine their dry matter (by heating at 105 °C until a constant weight was attained), ash (by incinerating them at 550

°C for 5 h), crude protein (AOAC official method 990.03, by the DUMAS direct combustion method, using a LECO CN628 apparatus, LECO, ST. Joseph, MI, USA), and crude lipid content (by extraction with methyl ether using an ANKOMXT10 extractor, ANKOM Technology, Macedon, NY, USA). Gross energy content (GE) was calculated according to Brouwer [35], from the C (g) and N (g) balance in the feed (GE = $51.8 \times C - 19.4 \times N$). The carbon and nitrogen content were analysed based on the Dumas principle (TruSpec CN; Leco Corporation, St Joseph, MI, USA). All analyses were performed in triplicate.

After thawing, tissues were minced and homogenised. The crude fats in the brain, muscle, liver, and visceral fat tissue sampled were then extracted [36]. A direct method of fatty acid methyl ester (FAME) synthesis was used for this procedure [37]. The analysis of brain, muscle, liver, and visceral fat tissues was carried out using 10–30 mg of extracted crude fat from each tissue. First, 1 mL of tridecanoic acid (C13:0) was used as internal standard. Then, 0.7 mL of 10 N KOH and 5.3 mL of HPLC (high-performance liquid chromatography)-quality methanol were added. Tubes were incubated at 55 °C in a thermoblock for 1.5 h, and underwent vigorous shaking for 5 s every 20 min. After cooling at ambient temperature in a water bath, 1.5 mL of HPLC-quality hexane was added to the reaction tubes, which were then vortex-mixed and centrifuged at 1006× *g* for 5 min. After this, the hexane layer, containing the FAMEs, was placed into vials for analysis by gas chromatography. The vials were kept at −80 °C until gas chromatography was performed. The FAMEs were analysed in a Focus gas chromatograph (Thermo, Milan, Italy) equipped with a split/splitless injector and a flame ionisation detector. Separation of the methyl esters was performed in a SPTM 2560 fused silica capillary column (Supelco, PA, USA) (100 m × 0.25 mm × 0.2 µm film thickness). Helium was used as the carrier gas at a flow rate of 20 cm s^{-1}. Samples were injected with a split ratio of 1:100.

The initial oven temperature, set at 140 °C, was held constant for 5 min, and then increased by 4 °C min^{-1} to 240 °C, at which this temperature was then maintained for 30 min. The FA were identified by comparing their retention times with those of the standards supplied by Supelco. The content of each type of FA was expressed as the percentage of the total FA content.

2.5. Statistical Analyses

The fish live weight, tissue fat content, and FA composition data were compared by analysis of variance (ANOVA), with diet included as the main effect and tank included as a random effect. The PROC MIXED procedure of the Statistical Analysis System (SAS) software [38] was used for all analyses. Adjusted means were compared among treatment levels using Bonferroni's t-test. Differences between means with $p \leq 0.05$ were considered statistically significant.

3. Results

In the present study of *S. dumerili* juveniles (176 g of initial live weight), diets containing only VO as the lipid source had no effect on fish growth (Figure 1). In fact, fish fed FO 100, FO 25, and FO 0 diets reached a final weight of 423, 409, and 419 g, respectively ($p > 0.05$). The survival rate was lower in fish fed diets without fish oil compared with those in fish fed diets containing 100% (FO 100) or 25% (FO 25) fish oil (80.3% vs. 89.7% and 92.7%, respectively; $p < 0.05$) (Figure 1).

3.1. Brain

With regard to the brain, the dietary treatment did not affect the proportions of total SFA (32.9% of the total FA on average), MUFA (29.2%), PUFA (37.9%), n-3 PUFA (29.6%), or n-6 PUFA (8.0%) present in the tissue ($p > 0.05$) (Table 3). Nevertheless, the replacement of FO with VO decreased the relative content of EPA (3.01% vs. 2.81% and 2.85% in FO 100 vs. FO 25 and FO 0, respectively; $p < 0.001$) and DHA (23.8% vs. 21.4% in FO 100 and FO 25 vs. FO 0; $p < 0.01$), whereas it increased the content of ALA (0.81% vs. 1.63% vs. 2.66% in FO 100 vs. FO 25 vs. FO 0; $p < 0.001$).

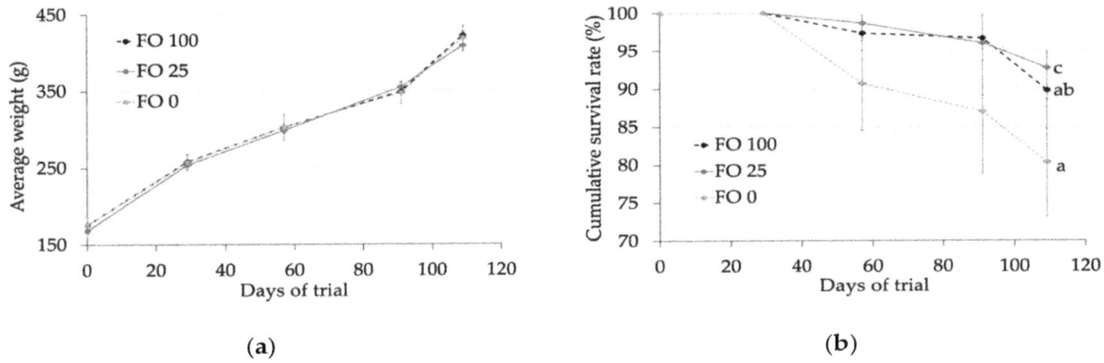

Figure 1. (**a**) Fish live weight (g) and (**b**) survival (%) of fish fed experimental diets during the 109 days of trial. Values are expressed as means ± standard error.

Table 3. Fat content and fatty acid composition (% of total fatty acid content) of the brain in Mediterranean yellowtail fed the experimental diets (n = 12 per diet). Values are expressed as least square (LS)means.

	Diet			*p*-Value	RSD
	FO 100	**FO 25**	**FO 0**		
Fat, % WW	3.22	4.23	3.52	0.999	0.951
Fatty acids, %					
10:0	0.42	0.46	0.57	0.083	0.138
14:0	0.76	0.64	0.68	0.134	0.141
15:0	0.12 ^{b}	0.10 ^{a}	0.09 ^{a}	<0.001	0.000
16:0	16.31	16.54	16.21	0.378	0.563
17:0	0.19 ^{b}	0.14 ^{a}	0.14 ^{a}	0.006	0.000
18:0	11.42	11.76	11.45	0.367	0.625
20:0	0.46	0.44	0.44	0.475	0.045
22:0	0.79	0.76	0.75	0.618	0.105
24:0	2.41	2.35	2.20	0.387	0.318
Σ SFA	32.90	33.17	32.51	0.278	0.932
16:1 n-9	2.67 ^{b}	2.40 ^{a}	2.34 ^{a}	<0.001	0.114
17:1 n-10	0.55	0.49	0.52	0.824	0.235
18:1 n-9	20.33 ^{ab}	19.93 ^{a}	21.49 ^{b}	0.009	1.050
18:1 n-7	2.80 ^{b}	2.58 ^{ab}	2.46 ^{a}	0.005	0.148
20:1 n-9	0.70 ^{b}	0.59 ^{a}	0.59 ^{a}	<0.001	0.063
22:1 n-9	0.32	0.32	0.28	0.271	0.045
24:1 n-9	2.16	2.07	1.98	0.378	0.261
Σ MUFA	29.47	28.44	29.66	0.139	1.336
18:2 n-6	4.25 ^{a}	4.39 ^{a}	5.55 ^{b}	0.016	1.080
20:2 n-6	0.26	0.23	0.25	0.298	0.055
20:4 n-6	2.65 ^{b}	2.51 ^{b}	2.27 ^{a}	<0.001	0.187
22:2 n-6	0.23 ^{b}	0.19 ^{ab}	0.18 ^{a}	0.028	0.045
22:4 n-6	0.22	0.22	0.18	0.150	0.055
Σ PUFA n-6	7.74	7.69	8.54	0.077	0.967
18:3 n-3	0.81 ^{a}	1.63 ^{b}	2.66 ^{c}	<0.001	0.644
20:3 n-3	0.10 ^{a}	0.16 ^{b}	0.19 ^{b}	<0.001	0.032
20:5 n-3	3.01 ^{b}	2.81 ^{a}	2.85 ^{a}	<0.001	0.118
22:5 n-3	2.13	2.26	2.14	0.483	0.219
22:6 n-3	23.83 ^{b}	23.83 ^{b}	21.43 ^{a}	0.006	1.668
Σ PUFA n-3	29.89	30.71	29.27	0.120	1.127
Σ PUFA	37.63	38.32	37.83	0.176	0.823
Σ n-6/Σ n-3	0.26	0.25	0.30	0.098	0.041
DHA/EPA	7.95 ^{ab}	8.50 ^{b}	7.54 ^{a}	0.023	0.780

FO 100 diet: Diet formulated with fish oil as lipid source; FO 25 diet: Diet in which fish oil was included at a content of 25%; FO 0 diet: Diet in which fish oil was totally substituted with vegetable oil. Abbreviations: WW: Wet weight; RSD: Residual standard deviation; SFA: Saturated fatty acids; MUFA: Monounsaturated fatty acids; PUFA: Polyunsaturated fatty acids; DHA/EPA: 22:6 n-3/20:5 n-3. The FA < 0.1% of the total FA (i.e., C12:0, C14:1, C18:3 n-6, and C30:3 n-6) are not given in the table. Different superscript letters indicate significant statistical differences among diets ($p \leq 0.05$).

3.2. Liver

In the liver, the replacement of FO with VO in the diet significantly decreased ($p < 0.001$) the proportion of the total SFA (24.3% and 23.3% vs. 20.2% of the total FA in FO 100 and FO 25 vs. FO 0), which was due to the lower proportion of C16:0 in VO than FO (13.9% and 13.4% vs. 11.6%) (Table 4).

Table 4. Fat content and fatty acid composition (% of total fatty acid content) of the liver in Mediterranean yellowtail fed the experimental diets ($n = 12$ per diet). Values are expressed as LS means.

	Diet			*p*-Value	RSD
	FO 100	FO 25	FO 0		
Fat, % WW	26.6	30.1	25.7	0.721	9.296
Fatty acids, %					
14:0	1.55 ^{c}	1.08 ^{b}	0.79 ^{a}	<0.001	0.129
15:0	0.26 ^{c}	0.15 ^{b}	0.10 ^{a}	<0.001	0.026
16:0	13.88 ^{b}	13.40 ^{b}	11.57 ^{a}	<0.001	0.885
17:0	0.49 ^{c}	0.32 ^{b}	0.24 ^{a}	<0.001	0.037
18:0	7.68 ^{b}	7.82 ^{b}	7.01 ^{a}	0.008	0.621
20:0	0.31 ^{c}	0.25 ^{b}	0.21 ^{a}	<0.001	0.014
24:0	0.06	0.07	0.05	0.369	0.020
Σ SFA	24.32 ^{b}	23.26 ^{b}	20.07 ^{a}	<0.001	1.317
16:1 n-9	3.26 ^{c}	2.23 ^{b}	1.64 ^{a}	<0.001	0.153
17:1 n-10	0.26	0.18	0.19	0.323	0.139
18:1 n-9	32.02 ^{a}	35.18 ^{b}	37.55 ^{c}	<0.001	1.371
18:1 n-7	7.30 ^{c}	5.77 ^{b}	4.95 ^{a}	<0.001	0.289
20:1 n-9	1.88 ^{b}	1.17 ^{a}	0.90 ^{a}	<0.001	0.314
22:1 n-9	0.37 ^{b}	0.24 ^{a}	0.16 ^{a}	<0.001	0.040
24:1 n-9	0.21 ^{b}	0.12 ^{a}	0.09 ^{a}	<0.001	0.063
Σ MUFA	45.29	44.90	45.47	0.613	1.431
18:2 n-6	15.23 ^{a}	15.21 ^{a}	16.96 ^{b}	0.002	1.10
18:3 n-6	0.12	0.11	0.11	0.482	0.026
20:2 n-6	1.23 ^{b}	0.95 ^{a}	0.88 ^{a}	<0.001	0.097
20:3 n-6	0.21 ^{c}	0.13 ^{b}	0.08 ^{a}	<0.001	0.021
20:4 n-6	0.89 ^{c}	0.53 ^{b}	0.39 ^{a}	<0.001	0.097
22:2 n-6	0.79 ^{c}	0.42 ^{b}	0.26 ^{a}	<0.001	0.052
22:4 n-6	0.26 ^{b}	0.08 ^{a}	0.05 ^{a}	<0.001	0.049
Σ PUFA n-6	18.73 ^{b}	17.43 ^{a}	18.74 ^{b}	0.036	1.207
18:3 n-3	2.48 ^{a}	8.60 ^{b}	11.27 ^{c}	<0.001	0.562
20:3 n-3	0.36 ^{a}	0.81 ^{b}	0.95 ^{c}	<0.001	0.117
20:5 n-3	2.21 ^{c}	1.45 ^{b}	1.10 ^{a}	<0.001	0.200
22:5 n-3	2.35 ^{c}	1.35 ^{b}	0.83 ^{a}	<0.001	0.199
22:6 n-3	4.26 ^{b}	2.29 ^{a}	1.57 ^{a}	<0.001	0.607
Σ PUFA n-3	11.66 ^{a}	14.45 ^{b}	15.72 ^{c}	<0.001	0.924
Σ PUFA	30.39 ^{a}	31.84 ^{a}	34.46 ^{b}	<0.001	1.703
Σ n-6/Σ n-3	1.62 ^{b}	1.20 ^{a}	1.20 ^{a}	<0.001	0.113
DHA/EPA	1.93 ^{b}	1.59 ^{ab}	1.36 ^{a}	<0.001	0.249

FO 100 diet: Diet formulated with fish oil as lipid source; FO 25 diet: Diet in which fish oil was included at a content of 25%; FO 0 diet: Diet in which fish oil was totally substituted with vegetable oil. Abbreviations: WW: Wet weight; RSD: Residual standard deviation; SFA: Saturated fatty acids; MUFA: Monounsaturated fatty acids; PUFA: Polyunsaturated fatty acids; DHA/EPA: 22:6 n-3/20:5 n-3. The FA < 0.1% of the total FA (i.e., C12:0, C14:1, C22:0, and C24:0) are not given in the table. Different superscript letters indicate significant statistical differences among diets ($p \leq 0.05$).

Moreover, the substitution of FO with VO in the diet decreased the relative content of C16:1 n-9 (3.26% vs. 2.23% vs. 1.64% in FO 100 vs. FO 25 vs. FO 0) and C17:1 n-7 (7.30% vs. 5.77% vs. 4.95%), whereas it increased that of C18:1 n-9 (32.0% vs. 35.2% vs. 37.6%) ($p < 0.001$). Thus, the proportion of the total MUFA (45.2% on average) was not affected by the dietary treatment ($p > 0.05$). With regard to n-3 PUFA, the inclusion of less FO in the diet decreased the liver tissue's relative content of EPA (2.21% vs. 1.45% vs. 1.10% in FO 100 vs. FO 25 vs. FO 0), C22:5 n-3 (2.35% vs. 1.35% vs. 0.83%), and DHA (4.26% vs. 2.29% and 1.57%), whereas it greatly increased ($p < 0.001$) its content of ALA (2.48% vs. 8.60% vs. 11.3%), and thus the total n-3 PUFA content. As for n-6 PUFA, the substitution of FO with VO decreased the AA proportion of liver tissue (0.89% vs. 0.53% vs. 0.39%; $p < 0.001$) and increased that of LA (15.2% and 15.2% vs. 17.0% in FO 100 and FO 25 vs. FO 0; $p < 0.01$).

3.3. Visceral Fat

In the visceral fat, the total substitution of FO with VO in the diet significantly decreased ($p < 0.001$) the proportion of the total SFA (23.6% and 23.8% vs. 22.5% of the total FA in FO 100 and FO 25 vs. FO 0) (Table 5). Moreover, the replacement of FO with VO decreased the C16:1 n-9 (4.07% vs. 2.98% vs. 2.48% in FO 100 vs. FO 25 vs. FO 0) and 18:1 n-7 (4.89% vs. 4.11% vs. 3.64%) proportions in the visceral fat, whereas it increased that of C18:1 n-9 (27.8% vs. 30.0% vs. 32.6%) ($p < 0.001$). With regard to n-3 PUFA, the inclusion of less FO in the diet decreased the EPA (3.68% vs. 2.76% vs. 2.07%), C22:5 n-3 (1.82% vs. 1.24% vs. 0.9%), and DHA (9.57% vs. 5.72% vs. 4.01%) proportions in the visceral fat, and increased that of ALA therein (3.76% vs. 9.44% vs. 12.0%) ($p < 0.001$), whereas the proportion of total n-3 PUFA (19.2% on average) was not affected by diet ($p > 0.05$). As for n-6 PUFA, the substitution of FO with VO increased the LA relative content (14.6% vs. 16.2% vs. 16.9%), which affected the total proportion of n-6 PUFA in this tissue (17.4% vs. 17.9% vs. 18.2%) ($p < 0.001$).

Table 5. Fat content and fatty acid composition (% of total fatty acid content) of the visceral fat in Mediterranean yellowtails fed the experimental diets (n = 12 per diet). Values are expressed as LS means.

	Diet			*p*-Value	RSD
	FO 100	FO 25	FO 0		
Fat, % WW	37.90	36.90	34.91	0.920	9.529
Fatty acids, %					
14:0	2.55 ^{c}	1.99 ^{b}	1.64	<0.001	0.094
15:0	0.32 ^{c}	0.22 ^{b}	0.16 ^{a}	<0.001	0.018
16:0	14.57 ^{a}	15.00 ^{b}	14.36 ^{a}	<0.001	0.293
17:0	0.44 ^{c}	0.32 ^{b}	0.26 ^{a}	<0.001	0.023
18:0	5.11 ^{a}	5.64 ^{b}	5.46^{b}	<0.001	0.207
20:0	0.36 ^{c}	0.32 ^{b}	0.29 ^{a}	<0.001	0.009
22:0	0.14 ^{a}	0.18 ^{b}	0.19 ^{b}	0.001	0.011
24:0	0.10	0.11	0.12	0.307	0.026
Σ SFA	23.63 ^{b}	23.80 ^{b}	22.53 ^{a}	<0.001	0.415
16:1 n-9	4.07 ^{c}	2.98 ^{b}	2.48 ^{a}	<0.001	0.184
17:1 n-10	0.37 ^{b}	0.20 ^{a}	0.16 ^{a}	<0.001	0.049
18:1 n-9	27.8 ^{a}	30.0 ^{b}	32.6 ^{c}	<0.001	0.547
18:1 n-7	4.89 ^{c}	4.11 ^{b}	3.64 ^{a}	<0.001	0.154
20:1 n-9	1.96 ^{c}	1.11 ^{b}	0.79 ^{a}	<0.001	0.116
22:1 n-9	0.40 ^{c}	0.22 ^{b}	0.15 ^{a}	<0.001	0.046
24:1 n-9	0.49 ^{c}	0.30 ^{b}	0.23 ^{a}	<0.001	0.038
Σ MUFA	39.95 ^{b}	38.87 ^{a}	40.07 ^{b}	<0.001	0.495
18:2 n-6c	14.64 ^{a}	16.17 ^{b}	16.89 ^{c}	<0.001	0.534
18:3 n-6	0.14 ^{b}	0.12 ^{a}	0.12 ^{a}	<0.001	0.003
20:2 n-6	0.92 ^{c}	0.64 ^{b}	0.48 ^{a}	<0.001	0.043
20:3 n-6	0.15 ^{b}	0.07 ^{ab}	0.06 ^{a}	0.023	0.075
20:4 n-6	0.67 ^{c}	0.43 ^{b}	0.32 ^{a}	<0.001	0.044
22:2 n-6	0.52 ^{c}	0.31 ^{b}	0.26 ^{a}	<0.001	0.044
22:4 n-6	0.34 ^{c}	0.18 ^{b}	0.10 ^{a}	<0.001	0.046
Σ PUFA n-6	17.37 ^{a}	17.92 ^{ab}	18.19 ^{b}	0.003	0.534
18:3 n-3	3.76 ^{a}	9.44 ^{b}	12.0 ^{c}	<0.001	0.843
20:3 n-3	0.22 ^{a}	0.26 ^{b}	0.28 ^{b}	<0.001	0.022
20:5 n-3	3.68 ^{c}	2.76 ^{b}	2.07 ^{a}	<0.001	0.150
22:5 n-3	1.82 ^{c}	1.24 ^{b}	0.90 ^{a}	<0.001	0.010
22:6 n-3	9.57 ^{c}	5.72 ^{b}	4.01 ^{a}	<0.001	0.634
Σ PUFA n-3	19.05	19.40	19.21	0.266	0.372
Σ PUFA	36.42 ^{a}	37.33 ^{b}	37.40 ^{b}	0.002	0.613
Σ n-6/Σ n-3	0.91	0.92	0.95	0.104	0.035
DHA/EPA	2.60 ^{b}	2.07 ^{a}	1.94 ^{a}	<0.001	0.153

FO 100 diet: Diet formulated with fish oil as lipid source; FO 25 diet: Diet in which fish oil was included at a content of 25%; FO 0 diet: Diet in which fish oil was totally substituted with vegetable oil. Abbreviations: WW: Wet weight; RSD: Residual standard deviation; SFA: Saturated fatty acids; MUFA: Monounsaturated fatty acids; PUFA: Polyunsaturated fatty acids; DHA/EPA: 22:6 n-3/20:5 n-3. The FA < 0.1% of the total FA (i.e., C10:0, C12:0, C14:1) are not given in the table. Different superscript letters indicate significant statistical differences among diets ($p \leq 0.05$).

3.4. Muscle

In the red muscle, the replacement of FO with VO did not affect the proportion of total SFA (24.0% of the total FA on average) (Table 6). Among MUFA, the replacement of FO with VO increased the relative content of C18:1 n-9 (27.2% vs. 29.0% vs. 31.8% in FO 100 vs. FO 25 vs. FO 0), whereas it decreased that of C18:1 n-7 (4.21% vs. 3.65% vs. 3.26% in FO 100 vs. FO 25 vs. FO 0) and C20:1 n-9 (1.91% vs. 1.05% vs. 0.74%) ($p < 0.001$). For n-3 PUFA, the inclusion of less FO in the diet decreased the proportions of EPA (3.22% vs. 2.60% vs. 2.14%), C22:5 n-3 (2.12% vs. 1.58% vs. 1.29%), and DHA (14.0% vs. 9.38% vs. 7.47%) in red muscle tissue, whereas it increased that of ALA (2.99% vs. 7.68% vs. 9.84%) ($p < 0.001$). In addition, the partial and total substitution of FO with VO in the diet increased the total n-6 PUFA content of red muscle (14.7% vs. 16.0% and 16.7%) due to changes in the proportions of LA ($p < 0.001$).

Table 6. Fat content and fatty acid composition (% of total fatty acid content) of the red muscle in Mediterranean yellowtails fed the experimental diets (n = 12 per diet). Values are expressed as LS means.

	Diet			p-Value	RSD
	FO 100	FO 25	FO 0		
Fat, % WW	4.37	4.89	4.49	0.500	0.816
Fatty acids, %					
14:0	1.81	1.81	1.26	0.105	0.602
16:0	14.41	15.01	14.40	0.125	0.455
17:0	0.44 ^{c}	0.30 ^{b}	0.24 ^{a}	<0.001	0.035
18:0	6.53 ^{a}	6.91 ^{b}	6.83 ^{b}	<0.001	0.184
20:0	0.38 ^{b}	0.33 ^{a}	0.32 ^{a}	<0.001	0.017
22:0	0.12 ^{a}	0.15 ^{b}	0.16 ^{b}	<0.001	0.015
Σ SFA	23.88	24.65	23.33	0.061	0.769
14:1 n-9	0.11	0.12	0.10	0.956	0.100
16:1 n-9	3.14 ^{b}	2.57 ^{ab}	2.05 ^{a}	0.005	0.125
17:1 n-10	0.31 ^{b}	0.21 ^{a}	0.16 ^{a}	<0.001	0.011
18:1 n-9	27.21 ^{a}	29.04 ^{b}	31.76 ^{c}	<0.001	0.509
18:1 n-7	4.21 ^{c}	3.65 ^{b}	3.26 ^{a}	<0.001	0.208
20:1 n-9	1.91 ^{c}	1.05 ^{b}	0.74 ^{a}	<0.001	0.134
22:1 n-9	0.36 ^{b}	0.18 ^{a}	0.13 ^{a}	<0.001	0.088
24:1 n-9	0.41 ^{c}	0.26 ^{b}	0.20 ^{a}	<0.001	0.038
Σ MUFA	37.62 ^{ab}	37.08 ^{a}	38.37 ^{b}	0.044	0.651
18:2 n-6c	13.05 ^{a}	14.83 ^{b}	15.76 ^{c}	<0.001	0.552
18:3 n-6	0.12	0.11	0.10	0.053	0.015
20:2 n-6	0.78 ^{b}	0.58 ^{a}	0.45 ^{a}	<0.001	0.027
20:4 n-6	0.91 ^{c}	0.65 ^{b}	0.54 ^{a}	<0.001	0.055
22:2 n-6	0.40 ^{c}	0.24 ^{b}	0.16 ^{a}	<0.001	0.031
22:4 n-6	0.51 ^{c}	0.29 ^{b}	0.21 ^{a}	<0.001	0.040
Σ PUFA n-6	14.70 ^{a}	15.95 ^{b}	16.68 ^{b}	<0.001	0.513
18:3 n-3	2.99 ^{a}	7.68 ^{b}	9.84 ^{c}	<0.001	0.705
20:3 n-3	0.20 ^{a}	0.25 ^{b}	0.27 ^{b}	<0.001	0.021
20:5 n-3	3.22 ^{c}	2.60 ^{b}	2.14 ^{a}	<0.001	0.116
22:5 n-3	2.12 ^{c}	1.58 ^{b}	1.29 ^{a}	<0.001	0.155
22:6 n-3	14.00 ^{c}	9.38 ^{b}	7.47 ^{a}	<0.001	0.946
Σ PUFA n-3	22.56 ^{b}	21.50 ^{ab}	21.01 ^{a}	0.036	0.884
Σ PUFA	38.50	38.27	38.30	0.925	0.881
Σ n-6/Σ n-3	1.54 ^{b}	1.35 ^{a}	1.26 ^{a}	<0.001	0.083
DHA/EPA	4.34 ^{b}	3.61 ^{a}	3.49 ^{a}	0.001	0.280

FO 100 diet: Diet formulated with fish oil as lipid source; FO 25 diet: Diet in which fish oil was included at a content of 25%; FO 0 diet: Diet in which fish oil was totally substituted with vegetable oil. Abbreviations: WW: Wet weight; RSD: Residual standard deviation; SFA: Saturated fatty acids; MUFA: Monounsaturated fatty acids; PUFA: Polyunsaturated fatty acids; DHA/EPA: 22:6 n-3/20:5 n-3. The FA < 0.1% of the total FA (i.e., C12:0, C15:0, C20:3 n-6, and C24:0) are not given in the table. Different superscript letters indicate significant statistical differences among diets ($p \leq 0.05$).

In the white muscle, the substitution of FO with VO in the diet decreased the proportion of total SFA (24.0% and 23.8% vs. 22.6% in FO 100 vs. FO 25 vs. FO 0) and increased that of C18:1 n-9 (27.6% vs. 29.0% vs. 31.7%); this also decreased the relative content of C16:1 n-9 (3.78% vs. 2.90% vs. 2.36%)

and C20:1 n-9 (1.76% vs. 1.09% vs. 0.76%) in white muscle tissue (Table 7). For n-3 PUFA, the inclusion of less FO in the diet decreased the EPA (3.41% vs. 2.87% vs. 2.26%), C22:5 n-3 (1.71% vs. 1.38% vs. 1.05%), and DHA (11.6% vs. 8.59% vs. 6.38%) content in white muscle tissue, whereas it increased the ALA content (3.76% vs. 9.44% vs. 12.0%) ($p < 0.001$). Nevertheless, the proportion of total n-3 PUFA (21.0% on average) was not affected ($p > 0.05$) by dietary treatment. Finally, the LA relative content in the white muscle was increased by this treatment (14.2% vs. 15.6% and 16.5%), which affected the total n-6 PUFA content (15.6% vs. 16.7% vs. 17.3%) ($p < 0.001$).

Table 7. Fat content and fatty acid composition (% of total fatty acid content) of the white muscle in Mediterranean yellowtails fed the experimental diets (n = 12 per diet). Values are expressed as LS means.

	Diet			*p*-Value	RSD
	FO 100	FO 25	FO 0		
Fat, % WW	5.84	5.94	6.03	0.941	0.770
Fatty acids, %					
14:0	2.28 ^{c}	1.80 ^{b}	1.47 ^{a}	<0.001	0.132
16:0	14.97 ^{b}	14.78 ^{b}	14.22 ^{a}	0.035	0.366
17:0	0.39 ^{c}	0.30 ^{b}	0.23 ^{a}	<0.001	0.037
18:0	5.89	6.21	6.13	0.050	0.171
20:0	0.36 ^{c}	0.33 ^{b}	0.32 ^{a}	<0.001	0.013
22:0	0.12 ^{a}	0.15 ^{b}	0.16 ^{b}	0.020	0.032
Σ SFA	23.94 ^{b}	23.80 ^{b}	22.63 ^{a}	0.009	0.494
14:1 n-9	0.28 ^{c}	0.22 ^{b}	0.15 ^{a}	<0.001	0.038
16:1 n-9	3.78 ^{c}	2.90 ^{b}	2.36 ^{a}	<0.001	0.200
17:1 n-10	0.33 ^{c}	0.22 ^{b}	0.17 ^{a}	<0.001	0.026
18:1 n-9	27.55 ^{a}	29.02 ^{b}	31.72 ^{c}	<0.001	0.759
18:1 n-7	4.15 ^{c}	3.56 ^{b}	3.20 ^{a}	<0.001	0.165
20:1 n-9	1.76 ^{c}	1.09 ^{b}	0.76 ^{a}	<0.001	0.175
22:1 n-9	0.29 ^{c}	0.19 ^{b}	0.12 ^{a}	<0.001	0.045
24:1 n-9	0.35 ^{c}	0.27 ^{b}	0.19 ^{a}	<0.001	0.048
Σ MUFA	38.63	37.50	38.63	0.050	0.615
18:2 n-6c	14.15 ^{a}	15.59 ^{b}	16.46 ^{b}	<0.001	0.552
18:3 n-6	0.14 ^{b}	0.13 ^{a}	0.12 ^{a}	0.001	0.009
20:2 n-6	0.86 ^{c}	0.63 ^{b}	0.49 ^{a}	<0.001	0.059
20:4 n-6	0.79 ^{c}	0.61 ^{b}	0.48 ^{a}	<0.001	0.076
22:2 n-6	0.42 ^{c}	0.28 ^{b}	0.19 ^{a}	<0.001	0.037
22:4 n-6	0.41 ^{c}	0.26 ^{b}	0.18 ^{a}	<0.001	0.052
Σ PUFA n-6	16.45 ^{a}	17.20 ^{ab}	17.80 ^{b}	0.001	0.493
18:3 n-3	3.93 ^{a}	8.25 ^{b}	10.84 ^{c}	<0.001	1.141
20:3 n-3	0.20 ^{a}	0.23 ^{b}	0.25 ^{c}	<0.001	0.019
20:5 n-3	3.41 ^{c}	2.87 ^{b}	2.26 ^{a}	<0.001	0.198
22:5 n-3	1.71 ^{c}	1.38 ^{b}	1.05 ^{a}	<0.001	0.136
22:6 n-3	11.63 ^{c}	8.59 ^{b}	6.38 ^{a}	<0.001	1.196
Σ PUFA n-3	20.60	21.21	20.77	0.830	0.773
Σ PUFA	37.48	38.70	38.74	0.500	0.825
Σ n-6/Σ n-3	0.80	0.81	0.86	0.086	0.069
DHA/EPA	3.43 ^{b}	2.95 ^{a}	2.82 ^{a}	0.019	0.234

FO 100 diet: Diet formulated with fish oil as lipid source; FO 25 diet: Diet in which fish oil was included at a content of 25%; FO 0 diet: Diet in which fish oil was totally substituted with vegetable oil. Abbreviations: WW: Wet weight; RSD: Residual standard deviation; SFA: Saturated fatty acids; MUFA: Monounsaturated fatty acids; PUFA: Polyunsaturated fatty acids; DHA/EPA: 22:6 n-3/20:5 n-3. The FA < 0.1% of the total FA (i.e., C12:0, C15:0, C20:3 n-6, and C24:0) are not given in the table. Different superscript letters indicate significant statistical differences among diets ($p \leq 0.05$).

4. Discussion

Successful growth performance may be achieved in fish fed with diets containing low levels of FO as long as their minimum EFA requirements are met [39]. In the larval stages, DHA plays a key role in promoting the development of neural membranes and eyes [40], improving growth, vitality, and survival [1,14,41], and preventing swimming disorders such as spinning and disorientation [41]. The accumulation of DHA in the brain during fish development has been measured in several marine species [14], which also show very low rates of DHA biosynthesis. Thus, any DHA deficiency during larval development will have serious consequences for the successful performance of fish larvae [14]. Moreover, DHA is essential for the development of schooling behaviour in the Mediterranean yellowtail [42–44]. Data in the published literature show that larvae (aged 3–7 days after hatching (dah)) of *Seriola dumerili* fed with rotifers [1] require a supply of DAH equal to 1.5% of the dry matter (DM) in their diet. Larvae (7 mm in length) of the related Japanese amberjack (*S. quinqueradiata* Temminck and Schlegel, 1845) fed with *Artemia* required a diet containing 3.9% of n-3 HUFA (2.5% and 1.3% DM of EPA and DHA, respectively) in a previous study [41]. Further, in *S. rivoliana* Valenciennes, 1833 larvae (aged 30 to 50 dah) had higher survival rates and better stress resistance when fed microdiets containing 3.2% DM of DHA [45].

In this experiment, the dietary treatments tested provided a minimum n-3 HUFA supply of 0.75% DM, which should be sufficient to meet the EFA requirements of most marine species (minimum value: 0.5% DM in juveniles and subadults of marine species [39]). Under different conditions, the total EFA requirements for juveniles (39 to 387 g of live weight in a 154-day feeding trial) of Mediterranean yellowtail have been estimated to be 1.2% DM [11]. In fact, juvenile fish, although they still need EFA, are likely to require lower levels of EPA and DHA in their diet than those needed in the larval stage. Nevertheless, dietary DHA deprivation in juveniles of pelagic marine species, such as carangids and tunids, can be particularly deleterious because of their fast growth rates [14], especially when they are fed diets containing low levels of FO.

During the grow-out phase, EFA deficiencies may reduce fish growth. In this regard, a substitution of 60% of the FO with VO in the diet is considered acceptable [46]. For carangid species, the growth performance of Japanese yellowtail (*S. quinqueradiata*) (average initial weight: 252 g; final weight 412 g) was not affected by the full replacement of FO with olive oil in a previous short-term feeding trial (40 days) [47]. On the contrary, yellowtail kingfish (*S. lalandi* Valenciennes, 1833) showed impaired growth rates when FO was totally substituted with canola or sunflower oil in their diet during a five-week-long feeding trial (with a weight change from 96 to 260 g) [47].

In the present study of *S. dumerili* (grown from 176 to 417 g of live weight, with 109 days of feeding), diets containing only VO as the lipid source had no detrimental impacts on the growth performance of these fish. Nevertheless, the survival rate was lower in fish fed diets without fish oil compared with those in fish fed diets containing 100% (FO 100) or 25% (FO 25) fish oil.

Indeed, Monge-Ortiz et al. [11] reported that diets including at least 525 g kg^{-1} of fish meal (FM) are likely to supply enough LC-PUFA to meet the needs of fish, even with the complete substitution of FO (with fish reaching a final weight of 397 g). In fact, FM usually contains up to 8%–15% DM of crude lipid, 30%–35% of which is composed of n-3 LC-PUFA [19]. In the present study, the experimental diets contained 350 g kg^{-1} of FM, which likely met the FA requirements of the studied fish during the grow-out phase.

The present study measured fat storage in different tissues, including the muscles, brain, liver, and visceral fat. Indeed, the locations of fat storage differ within and between fish species [15,48,49]. In a previous study [50], a high lipid content (53% DM) was found in the liver of farmed *S. dumerili*, which was consistent with that recorded in the present study (60% DM) and higher than the fat content of the muscle tissue (12% DM) [50].

The replacement of FO with VO in fish diets may lead to the accumulation of fat in the fish liver, generating a fatty liver syndrome, which may be the result of inefficient nutrient utilisation and increased rates of lipid peroxidation [51,52]. Nevertheless, we did not find differences in the liver fat content of the tested fish, even when they were fed with diets containing only VO as the lipid source, which was consistent with the findings of previous studies done on European seabass [53], gilthead sea bream [54], and turbot (*Psetta maxima* (L.; 1758)) [55].

Regardless of whether FO replacement affects fish growth and the lipid content of their tissues, its impact on fatty acid composition will vary depending on the dietary lipid content and source, as well as on the tissues considered [19]. The fatty acid signatures of fish tissues are closely related to their dietary FA composition [19]. Complete or partial FO replacement with VO blends is known to affect the FA compositions of muscles, fish organs, and fat storage tissues [19]. However, the magnitude of the changes in FA signatures varies among different fish species and tissues [56–58].

The FA signature of several wild and farmed fish was previously measured in muscle and liver tissues, but to our knowledge, few studies have investigated the FA signatures of other fatty tissues, such as those of the visceral fat and brain. Specifically, in wild and farmed Mediterranean yellowtail, the FA composition has been analysed in the muscles, liver, ovary [10,11,50,59,60], and eggs [61]. Other studies have been performed in other *Seriola* species, such as *S. lalandi* [47,62,63], *S. quinqueradiata* [64,65], *S. dorsalis* (Gill, 1863) [66,67], and *S. rivoliana* [45], which mainly focused on the FA composition of muscle tissues.

In the liver of Mediterranean yellowtail, we found a high proportion of MUFA and a low proportion of SFA, likely because their C16:0, C18:0, and C22:0 content was sufficient to maintain or even exceed the requirements of liver metabolic functions. In the brain and visceral fat, PUFA were highly represented (representing on average 37.9% and 37.1% of the total FA content in the brain and visceral fat, respectively). In fact, PUFA (especially C18 FA, such as LA and ALA) are generally stored in non-lipogenic tissues, especially the visceral fat, and are likely used as metabolic substrates for β-oxidation [58,68,69]. Moreover, the brain contained a high proportion of DHA (on average 23.0% of the total FA), which is known to regulate membrane fluidity, the blood-brain barrier, and the activities of certain enzymes, such as ionic channel proteins and nerve growth factors, both in mammals [70,71] and fish [1,14,45]. In salmonids, Atlantic cod, and flatfishes, DHA is also present at high levels in the brain [12,15,72], which agrees with our findings.

Consistent with other studies [56,73,74], the muscle, liver, and visceral fat FA signatures of Mediterranean yellowtail found in our study reflected the dietary FA profiles of these fish, with the FA signatures of these tissues thus showing a low robustness to changes in the dietary lipid sources from FO to VO. In contrast, the brain FA signature was less strongly affected by the dietary FA composition, whereas only ALA, EPA, and DHA changed. Nevertheless, in European seabass, brain lipids appeared to be sensitive to dietary lipid inputs [75]. On the contrary, a more recent study in juvenile European seabass fed n-3 LC-PUFA-deficient diets showed that polar lipids in their neural tissues had the highest capacity to regulate and preserve their DHA content [76]. In the present study, fish fed the diet with 100% VO showed a 10% lower DHA content in the brain than those fed diets with fish oil (FO 100 and FO 25) and a lower survival, which could have been related to some effects of EFA deficiency in them. In contrast, the administration of the FO 25 diet did not decrease the DHA proportion in the brain, which suggests that the substitution of 75% of the FO in the diet with VO could be suitable for use in diets for rearing Mediterranean yellowtail juveniles from 176 to 417 g of live weight. Further studies are necessary to confirm these results over longer feeding periods.

5. Conclusions

This study provided new findings on the FA signatures in different tissues of *Seriola dumerili* fed diets with decreasing levels of FO included in them. The FA composition of the diet had a strong effect on the FA compositions of muscle, liver, and visceral fat tissues. The brain had a FA signature that was more robust to dietary changes, whereas only some EFA (EPA and DHA) decreased when fish oil was totally replaced by vegetable oil.

Author Contributions: Conceptualization, S.M.-L.; A.T.-V.; and M.J.-C.; Formal analysis, F.B. and M.C.M.S.; Investigation, F.B. and M.C.M.S.; Data curation, A.T.-V.; F.B.; and A.T.; Writing—original draft preparation, F.B.; A.T.; A.T.-V.; and S.M.-L.; Writing—review and editing, F.B.; A.T.; A.T.-V.; S.M.-L.; and M.J.-C. All authors have read and agreed to the published version of the manuscript.

Acknowledgments: The authors gratefully acknowledge V. J. Moya Salvador and L. Ródenas for their technical support during chemical analyses of fats and fatty acids.

References

1. Matsunari, H.; Hashimoto, H.; Oda, K.; Masuda, Y.; Imaizumi, H.; Teruya, K.; Furuita, H.; Yamamoto, T.; Hamada, K.; Mushiake, K. Effects of docosahexaenoic acid on growth, survival and swim bladder inflation of larval amberjack (*Seriola dumerili*, Risso). *Aquac. Res.* **2013**, *44*, 1696–1705. [CrossRef]
2. Grau, A.; Riera, F.; Carbonell, E. Some protozoan and metazoan parasites of the amberjack from the Balearic Sea (western Mediterranean). *Aquac. Int.* **1999**, *7*, 307–317. [CrossRef]
3. Andaloro, F.; Pipitone, C. Food and feeding habits of the amberjack, *Seriola dumerili* in the Central Mediterranean Sea during the spawning season. *Cah. Biol. Mar.* **1997**, *38*, 91–96.
4. Sicuro, B.; Luzzana, U. The State of Seriola spp. other than yellowtail (*S. quinqueradiata*) farming in the World. *Rev. Fish. Sci. Aquac.* **2016**, *24*, 314–325. [CrossRef]
5. Mazzola, A.; Favaloro, E.; Sarà, G. Cultivation of the Mediterranean amberjack, *Seriola dumerili* (Risso, 1810), in submerged cages in the Western Mediterranean Sea. *Aquaculture* **2000**, *181*, 257–268. [CrossRef]
6. Jover, M.; García-Gómez, A.; Tomás, A.; De La Gándara, F.; Pérez, L. Growth of Mediterranean yellowtail (*Seriola dumerili*) fed extruded diets containing different levels of protein and lipid. *Aquaculture* **1999**, *179*, 25–33. [CrossRef]
7. Takakuwa, F.; Fukada, H.; Hosokawa, H.; Masumoto, T. Optimum digestible protein and energy levels and ratio for greater amberjack *Seriola dumerili* (Risso) fingerling. *Aquac. Res.* **2006**, *37*, 1532–1539. [CrossRef]
8. Vidal, A.T.; De La Gándara, F.; Gómez, A.G.; Cerdá, M.J. Effect of the protein/energy ratio on the growth of Mediterranean yellowtail (*Seriola dumerili*). *Aquac. Res.* **2008**, *39*, 1141–1148. [CrossRef]
9. Papadakis, I.E.; Chatzifotis, S.; Divanach, P.; Kentouri, M. Weaning of greater amberjack (*Seriola dumerili* Risso 1810) juveniles from moist to dry pellet. *Aquac. Int.* **2008**, *16*, 13–25. [CrossRef]
10. Haouas, W.G.; Zayene, N.; Guerbej, H.; Hammami, M.; Achour, L. Fatty acids distribution in different tissues of wild and reared *Seriola dumerili*. *Int. J. Food Sci. Technol.* **2010**, *45*, 1478–1485. [CrossRef]
11. Monge-Ortiz, R.; Tomás-Vidal, A.; Rodriguez-Barreto, D.; Martínez-Llorens, S.; Pérez, J.A.; Jover-Cerdá, M.; Lorenzo, A. Replacement of fish oil with vegetable oil blends in feeds for greater amberjack (*Seriola dumerili*) juveniles: Effect on growth performance, feed efficiency, tissue fatty acid composition and flesh nutritional value. *Aquac. Nutr.* **2018**, *24*, 605–615. [CrossRef]
12. Mourente, G.; Tocher, D.R.; Sargent, J.R. Specific accumulation of docosahexaenoic acid (22:6n-3) in brain lipids during development of juvenile turbot *Scophthalmus maximus* L. *Lipids* **1991**, *26*, 871–877. [CrossRef]
13. Sargent, J.; Bell, G.; McEvoy, L.; Tocher, D.; Estevez, A. Recent developments in the essential fatty acid nutrition of fish. *Aquaculture* **1999**, *77*, 191–199. [CrossRef]
14. Mourente, G. Accumulation of DHA (docosahexaenoic acid; 22:6n-3) in larval and juvenile fish brain. In Proceedings of the 26th Annual Larval Fish Conference, Os, Norway, 22–26 July 2002; Brownman, H., Skiftesvik, A.B., Eds.; Institute of Marine Research: Bergen, Norway, 2003; pp. 239–248.
15. Tocher, D.R.; Harvie, D.G. Fatty acid compositions of the major phosphoglycerides from fish neural tissues; (n-3) and (n-6) polyunsaturated fatty acids in rainbow trout (*Salmo gairdneri*) and cod (*Gadus morhua*) brains and retinas. *Fish. Physiol. Biochem.* **1988**, *5*, 229–239. [CrossRef] [PubMed]
16. Bell, J.G.; Castell, J.D.; Tocher, D.R.; MacDonald, F.M.; Sargent, J.R. Effects of different dietary arachidonic acid: Docosahexaenoic acid ratios on phospholipid fatty acid compositions and prostaglandin production in juvenile turbot (*Scophthalmus maximus*). *Fish. Physiol. Biochem.* **1995**, *14*, 139–151. [CrossRef] [PubMed]
17. Sargent, J.R.; McEvoy, L.A.; Bell, J.G. Requirements, presentation and sources of polyunsaturated fatty acids in marine fish larval feeds. *Aquaculture* **1997**, *155*, 117–127. [CrossRef]
18. Tocher, D.R.; Ghioni, C. Fatty acid metabolism in marine fish: Low activity of fatty acyl delta5 desaturation in gilthead sea bream (*Sparus aurata*) cells. *Lipids* **1999**, *34*, 433–440. [CrossRef]
19. Turchini, G.M.; Torstensen, B.E.; Ng, W.K. Fish oil replacement in finfish nutrition. *Rev. Aquac.* **2009**, *1*, 10–57. [CrossRef]
20. Iverson, S.J.; Field, C.; Bowen, W.D.; Blanchard, W. Quantitative fatty acid signature analysis: A new method. *Ecol. Monogr.* **2004**, *74*, 211–235. [CrossRef]
21. Sargent, J.R.; Bell, J.G.; Bell, M.V.; Henderson, R.J.; Tocher, D.R. The metabolism of phospholipids and polyunsaturated fatty acids in fish. In *Aquaculture: Fundamental and Applied Research*; Lahlou, B., Vitiello, P., Eds.; Coastal and Estuarine Studies; American Geophysical Union: Washington, DC, USA, 1993; Volume 43, pp. 103–124.

22. Henderson, J.R.; Tocher, D.R. The lipid composition and biochemistry of freshwater fish. *Prog. Lipid Res.* **1987**, *26*, 281–347. [CrossRef]
23. Furuita, H.; Takeuchi, T.; Uematsu, K. Effects of eicosapentaenoic and docosahexaenoic acids on growth, survival and brain development of larval Japanese flounder (*Paralichthys olivaceus*). *Aquaculture* **1998**, *161*, 269–279. [CrossRef]
24. Anderson, G.J.; Connoro, W.E.; Corliss, J.D. Docosahexaenoic acid is the preferred dietary n-3 fatty acid for the development of the brain and retina. *Pediatr. Res.* **1990**, *27*, 89–97. [CrossRef] [PubMed]
25. Bianconi, S.; Santillán, M.E.; del Rosario Solís, M.; Martini, A.C.; Ponzio, M.F.; Vincenti, L.M.; Schiöth, H.B.; Carlini, V.P.; Stutz, G. Effects of dietary omega-3 PUFAs on growth and development: Somatic, neurobiological and reproductive functions in a murine model. *J. Nutr. Biochem.* **2018**, *61*, 82–90. [CrossRef] [PubMed]
26. Bergé, J.P.; Barnathan, G. Fatty acids from lipids of marine organisms: Molecular biodiversity, roles as biomarkers, biologically active compounds, and economical aspects. *Adv. Biochem. Eng. Biotechnol.* **2005**, *96*, 49–125.
27. Thiemann, G.W. Using fatty acid signatures to study bear foraging: Technical considerations and future applications. *Ursus* **2008**, *19*, 59–72. [CrossRef]
28. Kaushik, S.J.; Corraze, G.; Radunz-Neto, J.; Larroquet, L.; Dumas, J. Fatty acid profiles of wild brown trout and Atlantic salmon juveniles in the Nivelle basin. *J. Fish. Biol.* **2006**, *68*, 1376–1387. [CrossRef]
29. Stowasser, G.; McAllen, R.; Pierce, G.J.; Collins, M.A.; Moffat, C.F.; Priede, I.G.; Pond, D. Trophic position of deep-sea fish-Assessment through fatty acid and stable isotope analyses. *Deep Res. Part I Oceanogr. Res. Pap.* **2009**, *56*, 812–826. [CrossRef]
30. Budge, S.M.; Penney, S.N.; Lall, S.P. Estimating diets of Atlantic salmon (*Salmo salar*) using fatty acid signature analyses; validation with controlled feeding studies. *Can. J. Fish. Aquat. Sci.* **2012**, *69*, 1033–1046. [CrossRef]
31. Magnone, L.; Bessonart, M.; Rocamora, M.; Gadea, J.; Salhi, M. Diet estimation of *Paralichthys orbignyanus* in a coastal lagoon via quantitative fatty acid signature analysis. *J. Exp. Mar. Biol. Ecol.* **2015**, *462*, 36–49. [CrossRef]
32. Happel, A.; Stratton, L.; Pattridge, R.; Rinchard, J.; Czesny, S. Fatty-acid profiles of juvenile lake trout reflect experimental diets consisting of natural prey. *Freshw. Biol.* **2016**, *61*, 1466–1476. [CrossRef]
33. Benedito-Palos, L.; Navarro, J.C.; Kaushik, S.; Pérez-Sánchez, J. Tissue-specific robustness of fatty acid signatures in cultured gilthead sea bream (*Sparus aurata* L.) fed practical diets with a combined high replacement of fish meal and fish oil. *J. Anim. Sci.* **2010**, *88*, 1759–1770. [CrossRef]
34. AOAC. *International Official Methods of Analysis of AOAC International*, 17th ed.; Association of Official Analytical Chemists: Arlington, VA, USA, 1990.
35. Brouwer, E. Report of Sub-Committee on Constants and Factors. In *Proceedings of the 3rd Symposium on Energy Metabolism*; Blaxter, K.L., Ed.; London Academic Press: London, UK, 1965; pp. 441–443.
36. Folch, J.; Lees, M.; Sloane Stanley, G.H. A simple method for the isolation and purification of total lipids from animal tissues. *J. Biol. Chem.* **1957**, *226*, 497–509.
37. O'Fallon, J.V.; Busboom, J.R.; Nelson, M.L.; Gaskins, C.T. A direct method for fatty acid methyl ester synthesis: Application to wet meat tissues, oils, and feedstuffs. *J. Anim. Sci.* **2007**, *85*, 1511–1521. [CrossRef]
38. SAS (Statistical Analysis System Institute, Inc.). *SAS/STAT(R) 9.3 User's Guide*, 3rd ed.; SAS Institute: Cary, NC, USA, 2013; Available online: https://support.sas.com/documentation/cdl/en/statug/63962/HTML/default/viewer.htm#titlepage.htm (accessed on 1 May 2019).
39. Tocher, D.R. Fatty acid requirements in ontogeny of marine and freshwater fish. *Aquac. Res.* **2010**, *41*, 717–732. [CrossRef]
40. Ishizaki, Y.; Masuda, R.; Uematsu, K.; Shimizu, K.; Arimoto, M.; Takeuchi, T. The effect of dietary docosahexaenoic acid on schooling behaviour and brain development in larval yellowtail. *J. Fish. Biol.* **2001**, *58*, 1691–1703. [CrossRef]
41. Furuita, H.; Takeuchi, T.; Watanabe, T.; Fujimoto, H.; Sekiya, S.; Imaizumi, K. Requirements of larval yellowtail for eicosapentanoic acid, docosahaenoic acid, and n-3 highly unsaturated fatty acid. *Fish. Sci.* **1996**, *62*, 372–379. [CrossRef]
42. Masuda, R.; Takeuchi, T.; Tsukamoto, K.; Sato, H.; Shimizu, K.; Imaizumi, K. Incorporation of dietary docosahexaenoic acid into the central nervous system of the yellowtail *Seriola quinqueradiata*. *Brain Behav. Evol.* **1999**, *53*, 173–179. [CrossRef]

43. Masuda, R.; Takeuchi, T.; Takeuchi, T.; Tsukamoto, K.; Tsukamoto, K.; Ishizaki, Y.; Kanematsu, M.; Imaizum, K. Critical involvement of dietary docosahexaenoic acid in the ontogeny of schooling behaviour in the yellowtail. *J. Fish. Biol.* **1998**, *53*, 471–484. [CrossRef]
44. Masuda, R.; Tsukamoto, K. School formation and concurrent developmental changes in carangid fish with reference to dietary conditions. *Environ. Biol. Fish.* **1999**, *56*, 243–252. [CrossRef]
45. Mesa-Rodriguez, A.; Hernández-Cruz, C.M.; Betancor, M.B.; Fernández-Palacios, H.; Izquierdo, M.S.; Roo, J. Effect of increasing docosahexaenoic acid content in weaning diets on survival, growth and skeletal anomalies of longfin yellowtail (*Seriola rivoliana*, Valenciennes 1833). *Aquac. Res.* **2018**, *49*, 1200–1209. [CrossRef]
46. Nasopoulou, C.; Zabetakis, I. Benefits of fish oil replacement by plant originated oils in compounded fish feeds. A review. *LWT-Food Sci. Technol.* **2012**, *47*, 217–224. [CrossRef]
47. Bowyer, J.N.; Qin, J.G.; Smullen, R.P.; Stone, D.A.J. Replacement of fish oil by poultry oil and canola oil in yellowtail kingfish (*Seriola lalandi*) at optimal and suboptimal temperatures. *Aquaculture* **2012**, *356–357*, 211–222. [CrossRef]
48. Bell, J.G.; Strachan, F.; Good, J.E.; Tocher, D.R. Effect of dietary echium oil on growth, fatty acid composition and metabolism, gill prostaglandin production and macrophage activity in Atlantic cod (*Gadus morhua* L.). *Aquac. Res.* **2006**, *37*, 606–617. [CrossRef]
49. Stoknes, I.S.; Økland, H.M.W.; Falch, E.; Synnes, M. Fatty acid and lipid class composition in eyes and brain from teleosts and elasmobranchs. *Comp. Biochem. Physiol.-B Biochem. Mol. Biol.* **2004**, *138*, 183–191. [CrossRef]
50. Rodríguez-Barreto, D.; Jerez, S.; Cejas, J.R.; Martin, M.V.; Acosta, N.G.; Bolaños, A.; Lorenzo, A. Comparative study of lipid and fatty acid composition in different tissues of wild and cultured female broodstock of greater amberjack (*Seriola dumerili*). *Aquaculture* **2012**, *360–361*, 1–9. [CrossRef]
51. Benedito-Palos, L.; Navarro, J.C.; Sitjà-Bobadilla, A.; Bell, G.J.; Kaushik, S.; Pérez-Sánchez, J. High levels of vegetable oils in plant protein-rich diets fed to gilthead sea bream (*Sparus aurata* L.): Growth performance, muscle fatty acid profiles and histological alterations of target tissues. *Br. J. Nutr.* **2008**, *100*, 992–1003. [CrossRef]
52. Piedecausa, M.A.; Mazón, M.J.; García García, B.; Hernández, M.D. Effects of total replacement of fish oil by vegetable oils in the diets of sharpsnout seabream (*Diplodus puntazzo*). *Aquaculture* **2007**, *263*, 211–219. [CrossRef]
53. Richard, N.; Mourente, G.; Kaushik, S.; Corraze, G. Replacement of a large portion of fish oil by vegetable oils does not affect lipogenesis, lipid transport and tissue lipid uptake in European seabass (*Dicentrarchus labrax* L.). *Aquaculture* **2006**, *261*, 1077–1087. [CrossRef]
54. Bouraoui, L.; Sánchez-Gurmaches, J.; Cruz-Garcia, L.; Gutiérrez, J.; Benedito-Palos, L.; Pérez-Sánchez, J.; Navarro, I. Effect of dietary fish meal and fish oil replacement on lipogenic and lipoprotein lipase activities and plasma insulin in gilthead sea bream (*Sparus aurata*). *Aquac. Nutr.* **2011**, *17*, 54–63. [CrossRef]
55. Regost, C.; Arzel, J.; Robin, J.; Rosenlund, G.; Kaushik, S.J. Total replacement of fish oil by soybean or linseed oil with a return to fish oil in turbot (*Psetta maxima*) 1. Growth performance, flesh fatty acid profile, and lipid metabolism. *Aquaculture* **2003**, *217*, 465–482. [CrossRef]
56. Bell, J.G.; McEvoy, J.; Tocher, D.R.; McGhee, F.; Campbell, P.J.; Sargent, J.R. Replacement of fish oil with rapeseed oil in diets of Atlantic Salmon (*Salmo salar*) affects tissue lipid compositions and hepatocyte fatty acid metabolism. *J. Nutr.* **2001**, *131*, 1535–1543. [CrossRef]
57. Bell, J.G.; Sargent, J.R. Arachidonic acid in aquaculture feeds: Current status and future opportunities. *Aquaculture* **2003**, *218*, 491–499. [CrossRef]
58. Torstensen, B.; Frøyland, L.; Lie, Ø. Replacing dietary fish oil with increasing levels of rapeseed oil and olive oil–effects on Atlantic salmon (*Salmo salar* L.) tissue and lipoprotein lipid composition and lipogenic enzyme activities. *Aquac. Nutr.* **2004**, *10*, 175–192. [CrossRef]
59. Saito, H. Lipid characteristics of two subtropical Seriola fishes, *Seriola dumerili* and *Seriola rivoliana*, with differences between cultured and wild varieties. *Food Chem.* **2012**, *135*, 1718–1729. [CrossRef]
60. Rodríguez-Barreto, D.; Jerez, S.; Cejas, J.R.; Martin, M.V.; Acosta, N.G.; Bolaños, A. Effect of different rearing conditions on body lipid composition of greater amberjack broodstock (*Seriola dumerili*). *Aquac. Res.* **2017**, *48*, 505–520. [CrossRef]

61. Rodríguez-Barreto, D.; Jerez, S.; Cejas, J.R.; Martin, M.V.; Acosta, N.G.; Bolaños, A.; Lorenzo, A. Ovary and egg fatty acid composition of greater amberjack broodstock (*Seriola dumerili*) fed different dietary fatty acids profiles. *Eur. J. Lipid Sci. Technol.* **2014**, *116*, 584–595. [CrossRef]
62. O'Neill, B.; Le Roux, A.; Hoffman, L.C. Comparative study of the nutritional composition of wild versus farmed yellowtail (*Seriola lalandi*). *Aquaculture* **2015**, *448*, 169–175. [CrossRef]
63. Rombenso, A.N.; Trushenski, J.T.; Drawbridge, M. Saturated lipids are more effective than others in juvenile California yellowtail feeds—Understanding and harnessing LC-PUFA sparing for fish oil replacement. *Aquaculture* **2018**, *493*, 192–203. [CrossRef]
64. Seno-O, A.; Takakuwa, F.; Hashiguchi, T.; Morioka, K.; Masumoto, T.; Fukada, H. Replacement of dietary fish oil with olive oil in young yellowtail *Seriola quinqueradiata*: Effects on growth, muscular fatty acid composition and prevention of dark muscle discoloration during refrigerated storage. *Fish. Sci.* **2008**, *74*, 1297–1306. [CrossRef]
65. Fukada, H.; Taniguchi, E.; Morioka, K.; Masumoto, T. Effects of replacing fish oil with canola oil on the growth performance, fatty acid composition and metabolic enzyme activity of juvenile yellowtail *Seriola quinqueradiata* (Temminck & Schlegel, 1845). *Aquac. Res.* **2017**, *48*, 5928–5939.
66. Bergman, A.M.; Trushenski, J.T.; Drawbridge, M. Replacing fish oil with hydrogenated soybean oils in feeds for yellowtail. *N. Am. J. Aquac.* **2018**, *80*, 141–152. [CrossRef]
67. Stuart, K.; Johnson, R.; Armbruster, L.; Drawbridge, M. Arachidonic acid in the diet of captive yellowtail and its effects on egg quality. *N. Am. J. Aquac.* **2018**, *80*, 97–106. [CrossRef]
68. Bell, J.G.; McGhee, F.; Campbell, P.J.; Sargent, J.R. Rapeseed oil as an alternative to marine fish oil in diets of post-smolt Atlantic salmon (*Salmo salar*): Changes in flesh fatty acid composition and effectiveness of subsequent fish oil "wash out". *Aquaculture* **2003**, *218*, 515–528. [CrossRef]
69. Stubhaug, I.; Lie, Ø.; Torstensen, B.E. Fatty acid productive value and β-oxidation capacity in Atlantic salmon (*Salmo salar* L.) fed on different lipid sources along the whole growth period. *Aquac. Nutr.* **2007**, *13*, 145–155. [CrossRef]
70. Ikemoto, A.; Nitta, A.; Furukawa, S.; Ohishi, M.; Nakamura, A.; Fujii, Y.; Okuyama, H. Dietary n-3 fatty acid deficiency decreases nerve growth factor content in rat hippocampus. *Neurosci. Lett.* **2000**, *285*, 99–102. [CrossRef]
71. Kitajka, K.; Puskas, L.G.; Zvara, A.; Hackler, L.; Barcelo-Coblijn, G.; Yeo, Y.K.; Farkas, T. The role of n-3 polyunsaturated fatty acids in brain: Modulation of rat brain gene expression by dietary n-3 fatty acids. *Proc. Natl. Acad. Sci. USA* **2002**, *99*, 2619–2624. [CrossRef]
72. Kreps, E.M.; Chebotarëva, M.A.; Akulin, V.N. Fatty acid composition of brain and body phospholipids of the anadromous salmon, *Oncorhynchus nerka*, from fresh-water and marine habitat. *Comp. Biochem. Physiol.* **1969**, *31*, 419–430. [CrossRef]
73. Caballero, M.J.; Obach, A.; Rosenlund, G.; Montero, D.; Gisvold, M.; Izquierdo, M.S. Impact of different dietary lipid sources on growth, lipid digestibility, tissue fatty acid composition and histology of rainbow trout, *Oncorhynchus mykiss*. *Aquaculture* **2002**, *214*, 253–271. [CrossRef]
74. Campos, I.; Matos, E.; Maia, M.R.G.; Marques, A.; Valente, L.M.P. Partial and total replacement of fish oil by poultry fat in diets for European seabass (*Dicentrarchus labrax*) juveniles: Effects on nutrient utilization, growth performance, tissue composition and lipid metabolism. *Aquaculture* **2019**, *502*, 107–120. [CrossRef]
75. Pagliarani, A.; Pirini, M.; Trigari, G.; Ventrella, V. Effect of diets containing different oils on brain fatty acid composition in sea bass (*Dicentrarchus labrax* L.). *Comp. Biochem. Physiol. Part B Biochem.* **1986**, *83*, 277–282. [CrossRef]
76. Skalli, A.; Robin, J.H.; Le Bayon, N.; Le Delliou, H.; Person-Le Ruyet, J. Impact of essential fatty acid deficiency and temperature on tissues' fatty acid composition of European sea bass (*Dicentrarchus labrax*). *Aquaculture* **2006**, *255*, 223–232. [CrossRef]

5

Comparison of Reconstituted, Acidified Reconstituted Milk or Acidified Fresh Milk on Growth Performance, Diarrhea Rate and Hematological Parameters in Preweaning Dairy Calves

Lingyan Li [1,†], Jiachen Qu [2,†], Xiaoyue Xin [1], Shuxin Yin [1] and Yongli Qu [1,*]

1 Key Laboratory of Efficient Utilization of Feed Resources and Nutrition Regulation in Cold Region of Heilongjiang Province, College of Animal Science and Veterinary Medicine, Heilongjiang Bayi Agricultural University, Daqing 163319, China; llytiger@163.com (L.L.); 15147012655@163.com (X.X.); SY20183040619@cau.edu.cn (S.Y)

2 College of Animal Science and Technology, Inner Mongolia Agricultural University, Hohhot 010018, China; qjc990313@163.com

* Correspondence: Ylqu007@126.com.

† These authors have contributed equally to this work.

Simple Summary: The preweaning phase is the period for the rapid growth and development of dairy calves. During this period, dairy calves receive their nutrients through milk. Feeding hygienic milk is of great benefit to optimum growth rate and health status of dairy calves. Previous studies focused on the effects of hygienic milk by acidification on dairy calves' health and growth. Reconstituted milk, as the common source of milk, is being used in dairy calves feeding. However, no previous studies reported the effects of feeding acidified reconstituted milk on dairy calves' health and growth. Our study will provide the evidence that the acidification of reconstituted milk had positive effects on growth performance and health status of preweaning dairy calves.

Abstract: The present experiment was carried out to assess the effects of reconstituted milk (RM), acidified reconstituted milk (ARM), and acidified fresh milk (AFM) on growth performance, diarrhea rate, and hematological parameters of preweaning dairy calves. For this purpose, a total of 27 Holstein female calves (one month of age) with initial body weight of (67.46 ± 4.08) kg were divided into three groups in such a way that each group contained nine calves. Calves were housed individually, and starter was offered ad libitum to each calf. The dietary treatments were RM, ARM, and AFM. The highest milk intake was observed in calves receiving AFM as compared to other treatments ($p < 0.01$). Calves fed AFM had more feed intake than those fed ARM and RM ($p < 0.01$). Feed efficiency was significantly lower for calves offered ARM than those offered RM and AFM ($p < 0.01$). A lower withers height growth was found for calves fed RM than those fed ARM and AFM ($p < 0.05$). Diarrhea rate and white blood cell (WBC) and lymphocytes (LYM) counts were greater for calves fed RM than those fed ARM and AFM ($p < 0.05$). These findings suggested that ARM and AFM had positive effects on growth performance and health status of the preweaning dairy calves.

Keywords: reconstituted milk; acidified reconstituted milk; acidified fresh milk; growth performance; diarrhea rate; hematological parameters; calves

1. Introduction

The preweaning phase is the period for the rapid growth and development of female dairy calves [1]. Providing calves with sufficient nutrients to achieve the optimum growth rate in preweaning phase is beneficial for promoting growth performance after adulthood, which positively associated with increased productivity in later life [2].

On large scale dairy farms, pre-weaning dairy calves are usually reared by milk or milk replacer [3–5]. Delay of feeding fresh milk to calves results in the multiplication of bacteria in the milk due to favorable conditions. Therefore, preservation of milk is recommended at large-scale dairy farms to overcome the consequences of bacterial load in fresh milk. It has been reported that acidification of milk to a pH value between 4.0 and 4.5 can inhibit pathogenic bacterial growth and enable milk to be preserved without refrigeration for a short time prior to feeding to calves [6,7]. It has also been reported that acidified milk protects calves from diseases caused by bacteria growth and reduces the incidence of diarrhea [8].

Reconstituted milk is resulted from the addition of water to dried or condensed milk powder in the amount necessary to reestablish the specified water:solids ratio. Reconstituted milk is also being used on the dairy farms for preweaning dairy calves [9]. However, to our knowledge, the acidification of reconstituted milk in preweaning growing calves has not been investigated. Previous studies only focused on acidified milk or acidified milk replacer in dairy calves [10–12]. Therefore, a comparative study was undertaken to compare the growth performance, diarrhea rate, and hematological parameters of preweaning dairy calves fed reconstituted milk, acidified reconstituted milk, or acidified fresh milk.

2. Materials and Methods

2.1. Calves, Feeds and Management

The animal care and experimental procedures were conducted by the Animal Welfare and Ethics Committee of Heilongjiang Bayi Agriculture University. Animal care and handling were followed the guidelines by the regulations for the Administration of Affairs Concerning Experimental Animals (The State Science and Technology Commission of China, 1988). Twenty-seven Holstein female calves procured from large dairy herd of same age with initial body weight of (67.46 ± 4.08) kg were selected and randomly divided into three groups in such a way that each group contained nine calves. Three types of milk were applied to different treatment groups—(1) reconstituted milk (RM), which was made by mixing milk powder with warm water (35–40 °C) in a weight ratio of 1:7; (2) acidified reconstituted milk (ARM), which was obtained by adding 30 mL formic acid (food grade) of 8.5% concentration in 1L reconstituted milk at temperature 5–10 °C; and (3) acidified milk made by using fresh milk (AFM), which was obtained by adding 30 mL formic acid (food grade) of 8.5% concentration in 1L fresh milk at temperature 5–10 °C. The fresh milk was procured from general and healthy herd and not pasteurized before feeding. Milk samples of 50 mL were collected every 10 days to detect the compositions. A milk analyzer (Foss Milkoscan 4000, Foss, Hillerod, Danmark) was used to determine the composition of milk, and milk composition are shown in Table 1. The pH value of acidified milk was determined by a pH meter (FE20, Mettler Toledo, Zurich, Switzerland). Milk samples of 50 mL were successively diluted (10^{-1} to 10^{-4}) using distilled water, and a volume of 1 mL of each dilution was inoculated onto the surface of plate count agar (PCA), violet red bile agar (VRBA), and MRS agar for the cultivation of total bacteria, lactobacillus, and *Escherichia coli*, respectively. Total bacteria and *Escherichia coli* were incubated at 37 °C for 48 and 24 h. Lactobacillus was incubated anaerobically at 37 °C for 72 h. After incubation, the numbers of total bacteria, lactobacillus and Escherichia coli colony-forming units (cfu) were calculated. Bacterial numbers of different types of milk are shown in Table 2.

Prior to onset of trial, all the calves were offered the same basal ration from day 1 to 21. Experimental treatments were applied from day 22 onward. Calves were given a nine-day adaptation period before the start of data collection. No abnormal behaviors were observed during the adaptation period, and the initial data were collected on day 30. Calves were housed individually in calf hutches, and 6 L milk were offered twice daily at 5:00 and 16:30. All calves had free-choice access to clean, fresh water and starter throughout the experiment. The chemical compositions of the starter diets are shown in Table 3.

The experiment period lasted for 40 days until weaning. Standard management and environmental conditions were ensured to avoid any stress as described in recent researches [13,14].

Table 1. Composition of different types of milk.

Items	ARM	RM	AFM
Milk lactose, %	4.08	4.48	4.63
Milk fat, %	4.30	4.55	4.74
Milk protein, %	2.86	2.94	3.13
Total solids, %	11.53	12.29	12.84
pH	4.26	6.87	4.32

ARM: acidified reconstituted milk. RM: reconstituted milk. AFM: acidified milk made by using fresh milk.

Table 2. Bacterial numbers of different types of milk.

Item	Diet			SEM	*p*-Value
	ARM	RM	AFM		
Total Bacterial Count (10^4 cfu/mL)	43.67 [a]	322.33 [b]	38.00 [a]	44.6	0.001
Lactobacillus (10^3 cfu/mL)	114.33	52.17	115.00	30.4	0.137
Escherichia coli (10^1 cfu/mL)	154.83 [a]	458.67 [b]	155.67 [a]	39.1	0.000

ARM: acidified reconstituted milk. RM: reconstituted milk. AFM: acidified milk made by using fresh milk. [a b] Mean values within a row with different superscript letter differ significantly ($p < 0.05$). SEM: Standard Error of Mean.

Table 3. Composition and nutrient levels of the starter (DM basis, %).

Ingredient	Content	Nutrient Levels	Content
Corn	33.00	NEL (Mcal/kg) [3]	1.69
Soybean meal	23.50	DM [4]	88.07
Expanded corn	10.00	CP [5]	20.95
DDGS [1]	8.00	EE [6]	3.31
Corn husk	5.07	NDF [7]	16.34
Extruded soybean	4.00	ADF [8]	5.48
Corn germ meal	9.00	Ash	7.40
Limestone	1.50	Ca [9]	0.95
MDCP [2]	0.15	P [10]	0.51
NaCl	0.77		
Beer yeast	2.00		
Elancoban	0.01		
Glucose	2.00		
Premix [11]	1.00		

[1] DDGS: Distillers Dried Grains with Solubles; [2] MDCP:mono-calcium and di-calcium phosphate; [3] NEL: net energy for lactation; [4] DM: dry matter; [5] CP: crude protein; [6] EE: ether extract; [7] NDF: neutral detergent fiber; [8] ADF: acid detergent fibre; [9] Ca: Calcium; [10] P: phosphorus; [11] The premix provided the following per kilogram of the diet: V_A 12,000 IU, V_D 4000 IU, V_E 2900IU, Cu 5000 mg, Fe 9000 mg, Mn 6000 mg, Se 67 mg, I 227 mg, Co 20 mg, Mg 9800 mg.

2.2. Growth Performance

Calves were weighed before the morning feeding for three consecutive days at the start and end of the experiment to calculate the initial body weight, final body weight, and average daily gain (ADG). Milk and starter consumption were recorded daily. Feed efficiency was determined from total feed consumption and ADG.

$$\text{Feed efficiency} = \frac{\text{Total feed intake}(\text{kg}/\text{day DM})}{\text{ADG}(\text{kg}/\text{day})} \tag{1}$$

In addition, body measurements of each calf including body length, withers height, heart girth, and shin circumference were taken at the beginning and end of the trial according to the study made by Li et al. (2014) [15].

2.3. Fecal Score and Diarrhea Rate

Fecal consistency was scored everyday on a scale of 1 to 4, where 1 = normal-firm but not hard, 2 = soft-does not hold form, 3 = runny-spreads easily, and 4 = devoid of solid matter. Fecal score of 3 or 4 were considered as diarrhea. The diarrhea rate was calculated according to the procedure described in the recent study [16] and formula is given below.

$$\text{Diarrhea rate} = \frac{\text{number of calves with diarrhea} \times \text{days of diarrhea}}{\text{total number of calves} \times \text{examined days}} \times 100 \tag{2}$$

2.4. Hematological Parameters

Blood samples (about 10 mL) were taken from the jugular vein in the morning, approximately 3h post feeding on the first and final day of the experiment. Blood was anticoagulated with EDTA for hematological analysis within three hours after collection. Hematological parameters values including white blood cell (WBC), lymphocytes (LYM), monocytes (MO), red blood cell (RBC), hemoglobin (Hb), and hematocrit (HCT) were detected by using automatic hematology analyzer (XN9000, Sysmex, Kobe, Japan).

2.5. Statistical Analysis

All the data were analyzed by PROC MIXED of SAS version 9.4 (SAS Institute Inc., Cary, NC, USA), the statistical model was

$$Y_{ij} = \mu + D_i + C_j + \varepsilon_{ij} \tag{3}$$

where Y_{ij} is the observed variable, μ is the overall mean, Di is the fixed effect of diet treatment, C_j is the random effect of calves, and ε_{ij} denotes the residual error. Significance was declared for $p < 0.05$, and trends were reported at $0.05 < p < 0.10$. When a significant effect of treatment was detected ($p < 0.05$), differences between the means were tested using Bonferroni multiple comparison test.

3. Results

3.1. Growth Performance

The results of growth performance of dairy calves are shown in Table 4. The initial BW, final BW, ADG, and starter intake were not different among treatments. Milk fed amount in liters were not different among different treatments. However dry matter intake from milk differed among treatments. Lowest dry matter intake from milk was observed in calves fed ARM, while the highest dry matter intake from milk was observed in calves fed AFM ($p < 0.01$). Calves fed AFM had increased total feed intake compared to those fed ARM and RM ($p < 0.01$). The value of feed efficiency was significantly lower for calves offered ARM than those offered RM and AFM ($p < 0.01$), which suggested that the feed efficiency was better for calves fed ARM than RM and AFM.

Table 4. Growth performance of dairy calves fed different types of milk.

Item	Diet			SEM	p-Value
	ARM	RM	AFM		
Initial BW (kg)	67.60	67.60	67.20	2.33	0.513
Final BW (kg)	104.40	102.20	101.80	3.79	0.766
ADG (kg/day)	0.92	0.88	0.96	0.25	0.952
Milk intake (L/day)	11.28	11.17	11.41	0.25	0.637
Milk intake (kg/day DM)	1.34 ^{a}	1.41 ^{b}	1.51 ^{c}	0.03	0.002
Starter intake (kg/day DM)	0.43	0.40	0.42	0.08	0.994
Total feed intake (kg/day DM)	1.76 ^{a}	1.81 ^{a}	1.92 ^{b}	0.03	0.002
Feed efficiency (kg intake/kg gain)	1.85 ^{a}	2.05 ^{b}	2.00 ^{b}	0.03	0.000

ARM: acidified reconstituted milk. RM: reconstituted milk. AFM: acidified milk made by using fresh milk. $^{a\,b\,c}$ Mean values within a row with different superscript letter differ significantly ($p < 0.05$). SEM: Standard Error of Mean.

Body measurements data of different treatments are presented in Table 5. Body measurements were not different at the beginning of the experiment. Smaller heart girth was observed in the end of the experiment for calves fed AFM than those fed ARM and RM ($p < 0.05$). Compared with calves fed ARM and AFM, a lower withers height growth could be found for calves fed RM ($p < 0.05$). Similarly, final withers height was lower for calves fed RM as compare to ARM and AFM ($p < 0.05$).

Table 5. Body measurements of dairy calves fed different types of milk.

Item	Diet			SEM	p-Value
	ARM	RM	AFM		
Initial					
Body length (cm)	83.80	83.60	82.60	0.76	0.275
Withers height (cm)	83.60	83.80	83.00	0.87	0.634
Heart girth (cm)	94.00	93.60	92.20	1.37	0.411
Shin circumference (cm)	12.00	11.80	11.60	0.26	0.335
Final					
Body length (cm)	94.60	94.80	93.80	1.17	0.671
Withers height (cm)	95.20 ^{a}	91.00 ^{b}	94.60 ^{a}	1.44	0.026
Heart girth (cm)	107.20 ^{a}	108.00 ^{a}	103.00 ^{b}	1.78	0.034
Shin circumference (cm)	12.70	12.80	12.40	0.44	0.649
Growth					
Body length (cm)	10.8	11.2	11.2	1.22	0.931
Withers height (cm)	11.6 ^{a}	7.20 ^{b}	11.6 ^{a}	1.45	0.015
Heart girth (cm)	13.2	14.4	10.80	1.76	0.157
Shin circumference (cm)	0.70	1.00	0.80	0.29	0.597

ARM: acidified reconstituted milk. RM: reconstituted milk. AFM: acidified milk made by using fresh milk. $^{a\,b}$ Mean values within a row with different superscript letter differ significantly ($p < 0.05$). SEM: Standard Error of Mean.

3.2. Fecal Score and Diarrhea Rate

Fecal score and diarrhea rate of dairy calves were presented in Table 6. Average fecal consistency score and diarrhea rate during 30–50 day, 51–60 day, and 61–70 day were greater for calves fed RM than those fed ARM and AFM ($p < 0.01$).

Table 6. Fecal score and diarrhea rate of dairy calves fed different types of milk.

Item	Diet			SEM	*p*-Value
	ARM	RM	AFM		
Fecal consistency score					
30–50 day	1.12 [a]	1.48 [b]	1.21 [a]	0.10	0.003
51–60 day	1.20	1.45	1.30	0.13	0.207
61–70 day	1.29 [a]	1.82 [b]	1.63 [b]	0.15	0.006
Diarrhea rate (%)					
30–50 day	0.72 [a]	11.07 [b]	0.74 [a]	0.03	0.000
51–60 day	0.17 [a]	10.87 [b]	0.23 [a]	0.20	0.000
61–70 day	2.90 [a]	9.40 [b]	2.48 [a]	0.41	0.000

ARM: acidified reconstituted milk. RM: reconstituted milk. AFM: acidified milk made by using fresh milk. [a b] Mean values within a row with different superscript letter differ significantly ($p < 0.05$). SEM: Standard Error of Mean.

3.3. Hematological Parameters

Hematological parameters of different treatments are presented in Table 7. Hematological parameters values of WBC, LYM, MO, RBC, HB, and HCT were not different among the treatments at the start of the experiment. At the end of the trial, higher white blood cell and lymphocytes counts were found in calves offered RM ($p < 0.05$).

Table 7. Hematological parameters of dairy calves fed different types of milk.

Item	Diet			SEM	*p*-Value
	ARM	RM	AFM		
Initial					
WBC [1] (10^9/L)	10.57	11.06	10.81	0.98	0.887
LYM [2] (10^9/L)	3.29	5.04	3.38	1.62	0.496
MO [3] (10^9/L)	1.69	1.70	2.57	0.51	0.185
RBC [4] (10^{12}/L)	4.49	4.69	4.57	0.48	0.921
Hb [5] (g/L)	130.60	135.60	131.40	5.21	0.601
HCT [6] (L/L)	0.22	0.23	0.22	0.01	0.383
Final					
WBC (10^9/L)	12.17 [a]	28.55 [b]	12.52 [a]	4.88	0.008
LYM (10^9/L)	3.21 [a]	4.83 [b]	3.00 [a]	0.71	0.049
MO (10^9/L)	1.28	1.36	1.26	0.31	0.947
RBC (10^{12}/L)	5.52	6.11	5.78	0.40	0.365
Hb (g/L)	116.20	121.40	118.00	4.61	0.537
HCT (L/L)	0.23	0.26	0.25	0.02	0.279

ARM: acidified reconstituted milk. RM: reconstituted milk. AFM: acidified milk made by using fresh milk. [a b] Mean values within a row with different superscript letter differ significantly ($p < 0.05$). [1] WBC: white blood cell; [2] LYM: lymphocytes; [3] MO: monocytes; [4] RBC: red blood cell; [5] Hb: hemoglobin; [6] HCT: hematocrit. SEM: Standard Error of Mean.

4. Discussion

4.1. Growth Performance

Preweaning growth of female dairy calves is considered an important factor of milk production in their later stage of life. Feed intake, ADG, and feed efficiency are important indicators for growth performance of female dairy calves. Feed intake affects the level of nutrient intake, thus affecting ADG and final BW. ADG and final BW play an important role in assessing the feed efficiency and overall health status of dairy female calves [17]. Akins reported that female diary calves should gain 0.75 kg per day to reach optimum weight of weaning at eight weeks of age [18]. In the present study, ADG of calves in different treatment groups exceeded 0.75 kg/day. Although ADG were in favor of calves fed ARM and AFM compared to those fed RM (0.92, 0.96 vs. 0.88 kg/day), however the differences were not significant. Eren found that the calves offered acidified milk had increased ADG than those offered whole milk [19]. Zhang et al. reported that acidified milk replacers contribute to the improvement of growth rate of calves [12]. Dry matter intakes (DMI) from milk was the highest for calves fed AFM. This was expected as the highest total solid concentration of AFM which resulted in the higher consumption of DM from milk by calves. DMI from starter were not different for calves fed ARM, RM, and AFM, which were 0.43, 0.40, and 0.42 kg/day DM, respectively. The results are consistent with other research [10–12,20]. However, Sun et al. found a reduction in the intake of starter for the calves fed acidified milk with butyric acid addition [16]. Interestingly, in the current study, starter feed DMI was similar among all the group, which could be explained by different acidifier and experimental treatments as reported in the previous studies. Overall, ARM, RM, and AFM behave similarly on the intake of starter feed in the current study.

Body measurements usually reflect body growth and development, which can be used to estimate liveweight [21]. Calves assigned to the ARM and AFM treatment groups exhibited greater withers height gain in the body measurements than the RM calves. The results suggested that calves reared using ARM and AFM had better skeletal growth. Todd et al. reported the calves fed acidified milk replacer were associated with greater preweaning structural growth [5]. However, some studies found that body measurements of calves were not affected by feeding acidified milk or acidified milk replacer [3,10,16].

4.2. Fecal Score and Diarrhea Rate

Diarrhea is the most important cause of disease of calves, with high morbidity and mortality, which threatens the health and growth performance of calves and is a major cause of economic loss for cattle raisers [22]. Results in the present study showed that calves fed ARM and AFM exhibited lower fecal consistency scores and diarrhea rate. It has been reported that lower pH values inhibit the growth of total bacteria and *Escherichia coli* [23], and the current study provided the evidence that ARM and AFM contained lower number of total bacteria and *Escherichia coli* (Table 2). The lower number of total bacteria and *Escherichia coli* reduces the growth and reproduction of saprophytic bacteria that harm the health of gastro-intestinal tract and cause diarrhea. On the other hand, feeding acidified milk can increase the acidity in the intestinal tract of calves, as reported by Woodford et al. [24]. Relatively higher acidity of intestinal tract is likely to exhibit a bacterial influence, therefore reducing the incidence of diarrhea among calves. Therefore, the lower incidence of diarrhea in ARM and AFM could be explained by the lower pH of intestine caused by ARM and AFM and by the lower number of total bacteria and *Escherichia coli*.

4.3. Hematological Parameters

Hematological parameters such as WBC, LYM, and RBC counts and Hb concentration are considered as important clinical indicators that are widely used to reflect health and disease status [25]. Diseases of the preweaning calves is one of the major cause of economic loss in dairy production [26]. Thus, the measurement of hematological parameters could be used to evaluate the health and physiological status of the calves [27]. WBC are the major part of the immune system that helps fight infection and defend the body against other foreign materials. Different types of WBC are involved in recognizing intruders, killing harmful bacteria, and creating antibodies to protect the body against future exposure to some bacteria and viruses [28,29]. LYM are one of several different types of white blood cells that are also important in the immune system, with T cells being responsible for directly killing many foreign invaders and B cells producing antibodies to guard the body against infection [30]. Some research reported that mean value of WBC and LYM were 8.08–12.75 (10^9/L) and 2.67–4.77 (10^9/L) of weaning Hanwoo and Holstein calves [31–33]. In the present study, mean value of WBC was 28.55 (10^9/L) and LYM was 4.83 (10^9/L) for calves fed RM, which were much higher than that of the research mentioned above. However, mean value of WBC and LYM for calves offered ARM and AFM were very close to the values of the research we have reviewed. Therefore, it can be concluded that the calves offered RM were susceptibility to infection, which may result in detrimental effects on calf health.

5. Conclusions

Results from this research indicated that calves fed acidified reconstituted milk or acidified fresh milk had greater average daily gain and withers height growth, lower diarrhea rate, and white blood cell and lymphocytes counts. We concluded that calves offered reconstituted milk were susceptible to infection, which may have a negative effect on calf health. Acidification of reconstituted milk as well as fresh milk had positive effects on growth performance and health status of the preweaning calves.

Author Contributions: Conceptualization, L.L., J.Q., and Y.Q.; methodology, X.X. and Y.Q.; formal analysis, L.L. and J.Q.; investigation, X.X. and S.Y.; data curation, L.L. and S.Y.; writing—original draft preparation, L.L., J.Q., X.X., and S.Y.; writing—review and editing, L.L. and Y.Q.; visualization, L.L.; project administration, L.L. and Y.Q.; funding acquisition, Y.Q.

Acknowledgments: The authors would like to acknowledge Guang Shao, the dairy farm manager, and the workers of Yuda Animal Husbandry Co., Ltd, for their help on collecting experimental data.

References

1. Diao, Q.Y.; Zhang, R.; Yan, T. Current research progresses on calf rearing and nutrition in China. *J. Integr. Agr.* **2017**, *16*, 2805–2814. [CrossRef]
2. Drackley, J.K. Calf nutrition from birth to breeding. *Vet. Clin. North Am. Food Anim. Pract.* **2008**, *24*, 55–86. [CrossRef] [PubMed]
3. Bayram, B.; Yanar, M.; Güler, O.; Metin, J. Growth performance, health and behavioural characteristics of Brown Swiss calves fed a limited amount of acidified whole milk. *Ital. J. Anim. Sci.* **2007**, *6*, 273–279. [CrossRef]
4. Khan, M.; Weary, D.; Von Keyserlingk, M. Invited review: Effects of milk ration on solid feed intake, weaning, and performance in dairy heifers. *J. Dairy Sci.* **2011**, *94*, 1071–1081. [CrossRef] [PubMed]
5. Todd, C.; Leslie, K.; Millman, S.; Bielmann, V.; Anderson, N.; Sargeant, J.; DeVries, T. Clinical trial on the effects of a free-access acidified milk replacer feeding program on the health and growth of dairy replacement heifers and veal calves. *J. Dairy Sci.* **2017**, *100*, 713–725. [CrossRef] [PubMed]
6. Collings, L.; Proudfoot, K.; Veira, D. The effects of feeding untreated and formic acid-treated colostrum ad libitum on intake and immunoglobulin levels in dairy calves. *Can. J. Anim. Sci.* **2011**, *91*, 55–59. [CrossRef]
7. Todd, C.; Millman, S.; Leslie, K.; Anderson, N.; Sargeant, J.; DeVries, T. Effects of milk replacer acidification and free-access feeding on early life feeding, oral, and lying behavior of dairy calves. *J. Dairy Sci.* **2018**, *101*, 8236–8247. [CrossRef]
8. Güler, O.; Yanar, M.; Bayrum, B.; Metin, J. Performance and health of dairy calves fed limited amounts of acidified milk replacer. *S. Afr. J. Anim. Sci.* **2006**, *36*, 149–154. [CrossRef]
9. Jenkins, K.; Bona, A. Performance of calves fed combinations of whole milk and reconstituted skim milk powder. *J. Dairy Sci.* **1987**, *70*, 2091–2094. [CrossRef]
10. Yanar, M.; Güler, O.; Bayram, B.; METİN, J. Effects of feeding acidified milk replacer on the growth, health and behavioural characteristics of Holstein Friesian calves. *Turk. J. Vet. Anim. Sci.* **2006**, *30*, 235–241.
11. Hill, T.; Bateman II, H.; Aldrich, J.; Quigley, J.; Schlotterbeck, R. Evaluation of ad libitum acidified milk replacer programs for dairy calves. *J. Dairy Sci.* **2013**, *96*, 3153–3162. [CrossRef] [PubMed]
12. Zhang, R.; Diao, Q.-Y.; Zhou, Y.; Yun, Q.; Deng, K.-D.; Qi, D.; Tu, Y. Decreasing the pH of milk replacer containing soy flour affects nutrient digestibility, digesta pH, and gastrointestinal development of preweaned calves. *J. Dairy Sci.* **2017**, *100*, 236–243. [CrossRef] [PubMed]
13. ur Rahman, M.A.; Chuanqi, X.; Huawei, S.; Binghai, C. Effects of hay grass level and its physical form (full length vs. chopped) on standing time, drinking time, and social behavior of calves. *J. Vet Behav.* **2017**, *21*, 7–12. [CrossRef]

14. ur Rahman, M.A.; Qi, X.C.; Binghai, C. Nutrient intake, feeding patterns and abnormal behavior of growing bulls fed different concentrate levels and a single fiber source (corn stover silage). *J. Vet Behav.* **2019**, *33*, 46–53. [CrossRef]
15. Li, L.; Zhu, Y.; Wang, X.; He, Y.; Cao, B. Effects of different dietary energy and protein levels and sex on growth performance, carcass characteristics and meat quality of F1 Angus× Chinese Xiangxi yellow cattle. *J. Anim. Sci. Biotechnol.* **2014**, *5*, 21. [CrossRef] [PubMed]
16. Sun, Y.; Li, J.; Meng, Q.; Wu, D.; Xu, M. Effects of butyric acid supplementation of acidified milk on digestive function and weaning stress of cattle calves. *Livest Sci.* **2019**, *225*, 78–84. [CrossRef]
17. Dingwell, R.; Wallace, M.; McLaren, C.; Leslie, C.; Leslie, K. An evaluation of two indirect methods of estimating body weight in Holstein calves and heifers. *J. Dairy Sci.* **2006**, *89*, 3992–3998. [CrossRef]
18. Akins, M.S. Dairy heifer development and nutrition management. *Vet. Clin. North Am. Food Anim. Pract.* **2016**, *32*, 303–317. [CrossRef] [PubMed]
19. Eren, V. The effect of acidified milk on body weight gain, some blood parameters and health in calves. *Yüzüncü Yıl Üniversitesi Veteriner Fakültesi Dergisi* **2009**, *20*, 17–21.
20. Jaster, E.; McCoy, G.; Tomkins, T.; Davis, C. Feeding acidified or sweet milk replacer to dairy calves. *J. Dairy Sci.* **1990**, *73*, 3563–3566. [CrossRef]
21. Lukuyu, M.N.; Gibson, J.P.; Savage, D.; Duncan, A.J.; Mujibi, F.; Okeyo, A. Use of body linear measurements to estimate liveweight of crossbred dairy cattle in smallholder farms in Kenya. *Springer Plus.* **2016**, *5*, 63. [CrossRef] [PubMed]
22. Gomez, D.E.; Weese, J.S. Viral enteritis in calves. *Can. Vet. J.* **2017**, *58*, 1267–1274. [PubMed]
23. Stewart, S.; Godden, S.; Bey, R.; Rapnicki, P.; Fetrow, J.; Farnsworth, R.; Scanlon, M.; Arnold, Y.; Clow, L.; Mueller, K. Preventing bacterial contamination and proliferation during the harvest, storage, and feeding of fresh bovine colostrum. *J. Dairy Sci.* **2005**, *88*, 2571–2578. [CrossRef]
24. Woodford, S.; Whetstone, H.; Murphy, M.; Davis, C. Abomasal pH, nutrient digestibility, and growth of Holstein bull calves fed acidified milk replacer. *J. Dairy Sci.* **1987**, *70*, 888–891. [CrossRef]
25. Kelada, S.N.; Aylor, D.L.; Peck, B.C.; Ryan, J.F.; Tavarez, U.; Buus, R.J.; Miller, D.R.; Chesler, E.J.; Threadgill, D.W.; Churchill, G.A. Genetic analysis of hematological parameters in incipient lines of the collaborative cross. *G3 Genes Genomes Genet.* **2012**, *2*, 157–165. [CrossRef] [PubMed]
26. Mohri, M.; Sharifi, K.; Eidi, S. Hematology and serum biochemistry of Holstein dairy calves: Age related changes and comparison with blood composition in adults. *Res. Vet. Sci.* **2007**, *83*, 30–39. [CrossRef] [PubMed]
27. Roland, L.; Drillich, M.; Iwersen, M. Hematology as a diagnostic tool in bovine medicine. *J. Vet. Diagn. Invest.* **2014**, *26*, 592–598. [CrossRef] [PubMed]
28. Hoffman, W.; Lakkis, F.G.; Chalasani, G. B cells, antibodies, and more. *Clin. J. Am. Soc. Nephrol.* **2016**, *11*, 137–154. [CrossRef] [PubMed]
29. Karlmark, K.; Tacke, F.; Dunay, I. Monocytes in health and disease—Minireview. *Eur. J. Microbiol. Immunol.* **2012**, *2*, 97–102. [CrossRef] [PubMed]
30. Görgens, A.; Radtke, S.; Horn, P.; Giebel, B. New relationships of human hematopoietic lineages facilitate detection of multipotent hematopoietic stem and progenitor cells. *Cell Cycle* **2013**, *12*, 3478–3482. [CrossRef]
31. Roodposhti, P.M.; Dabiri, N. Effects of probiotic and prebiotic on average daily gain, fecal shedding of Escherichia coli, and immune system status in newborn female calves. *Asian-Australas. J. Anim. Sci.* **2012**, *25*, 1255. [CrossRef] [PubMed]
32. Kim, M.-H.; Yang, J.-Y.; Upadhaya, S.D.; Lee, H.-J.; Yun, C.-H.; Ha, J.K. The stress of weaning influences serum levels of acute-phase proteins, iron-binding proteins, inflammatory cytokines, cortisol, and leukocyte subsets in Holstein calves. *J. Vet. Sci.* **2011**, *12*, 151–157. [CrossRef] [PubMed]
33. Sarker, M.; Ko, S.; Lee, S.; Kim, G.; Choi, J.; Yang, C. Effect of different feed additives on growth performance and blood profiles of Korean Hanwoo calves. *Asian-Australas. J. Anim. Sci.* **2009**, *23*, 52–60. [CrossRef]

6

Effects of Partial Replacment of Dietary Forage using Kelp Powder (*Thallus laminariae*) on Ruminal Fermentation and Lactation Performances of Dairy Cows

Fuguang Xue [1,†], Fuyu Sun [1,2,†], Linshu Jiang [3], Dengke Hua [1], Yue Wang [1], Xuemei Nan [1], Yiguang Zhao [1,*] and Benhai Xiong [1,*]

1 State Key Laboratory of Animal Nutrition, Institute of Animal Science, Chinese Academy of Agricultural Sciences, Beijing 100193, China; xuefuguang123@163.com (F.X.); sfycaas@163.com (F.S.); dengke_h@163.com (D.H.); wangyue9313@163.com (Y.W.); xuemeinan@126.com (X.N.)

2 College of Animal Science and Technology, China Agricultural University, Beijing 100193, China

3 Beijing Key Laboratory for Dairy Cow Nutrition, Beijing University of Agriculture, Beijing 102206, China; jls@bua.edu.cn

* Correspondence: zhaoyiguang@caas.cn (Y.Z.); xiongbenhai@caas.cn (B.X.)

† These authors contributed equally to this work.

Simple Summary: Kelp powder has been widely used as feed ingredient or additives in monogastric animal production and presented positive effects on production performance and intestinal microbiota because of its abundance in biomass output, novel oligosaccharides, and iodine. However, little information is available for the nutritional effects of kelp on ruminants. Therefore, the objective of the present study was to investigate the effects of kelp partial replacing dietary forage on rumen fermentation and production performance of dairy cows. Results indicated that, a substitution of 5% kelp powder for forage in the diet reduced the fiber content in the feeding diet and then regulated the proliferation of ruminal microbiota, which lead to a significant reduction of NH_3-N while increasing acetate, propionate, and total volatile fatty acids (TVFA) concentrations in rumen, and finally partly increasing milk production and milk fat content. The present study may provide more comprehensive cognition of the nutritional value of kelp powder as a dietary raw material. It may further promote the utilization of kelp powder and provide a useful dietary raw material to the husbandry of dairy cows.

Abstract: Background: Kelp powder, which was rich in novel oligosaccharides and iodine might be utilized by the rumen microbiome, promoted the ruminal fermentation and finally enhanced the lactation performance of dairy cows. Therefore, the purpose of this study was to investigate the effects of kelp powder partially replacing dietary forage on rumen fermentation and lactation performance of dairy cows. (2) Methods: In the present study, 20 Chinese Holstein dairy cows were randomly divided into two treatments, a control diet (CON) and a kelp powder replacing diet (Kelp) for a 35-d long trial. Dry matter intake (DMI), milk production, milk quality, ruminal fermentable parameters, and rumen microbiota were measured to investigate the effects of kelp powder feeding on dairy cows. (3) Results: On the lactation performance, kelp significantly increased milk iodine content and effectively enhanced milk production and milk fat content. On the fermentable aspects, kelp significantly raised TVFA while reducing the ammonia-N content. On the rumen microbial aspect, kelp feeding significantly promoted the proliferation of *Firmicutes and Proteobacteria* while suppressing *Bacteroidetes*. (4) Conclusion: kelp powder as an ingredient of feedstuff might promote the rumen fermentation ability and effectively increase milk fat and iodine content, and consequently improve the milk nutritional value.

Keywords: dairy cows; metagenomic sequencing; kelp powder; rumen fermentation

1. Introduction

Kelp powder (*Thallus laminariae*), attracts more attentions in recent years in the husbandry because of its high-yielding biomass and abundant nutritional content such as crude protein, amino acids, polyunsaturated fatty acids, and oligosaccharides [1,2]. Kelp has been used as the feed ingredient or feed additive in the production of monogastric animals such as layers, broilers, and pigs, and expressed positive effects on animal performance, such as reducing the incidence of diseases and promoting production performance [3–6]. Whereas little information is available on kelp nutritional effects on ruminants, while it has been found that that kelp may potentially lower somatic cell count and improve body condition of dairy cattle [7].

Ruminants have high digestibility of feeding ingredients for the ruminal microbial ecosystem comprised of an immense variety of microbiota which execute the complex anaerobic metabolism [8]. Bacteria in the rumen contribute more than 95% of the whole biomass and perform the main function in rumen fermentation [9,10]. For this purpose, oligosaccharides which were not degraded in monogastric creatures could be degraded in the rumen and hydrolyzed into mono- or di- saccharides and produced formate, acetate, and lactate as the end products [11]. Moreover, ruminal microbiota might be altered for the increasing ruminal end products and so might the fermentation procedure. Finally, the production performance might be influenced.

Therefore, we hypothesized that oligosaccharides in kelp will be digested in rumen, increase the volatile fatty acids (VFAs) content, and finally promote the lactation performance of ruminants. The objective of the present study was to investigate the effects of partially replacing dietary forage using kelp powder on ruminal microbiota, ruminal fermentation parameters, and the lactation performance of dairy cows.

2. Materials and Methods

All experimental protocols performed in this study were approved by the Animal Ethics Committee of the Chinese Academy of Agricultural Sciences (Beijing, China). The experimental procedures used in this study were in accordance with the recommendations of the academy's guidelines for animal research.

2.1. Kelp Powder

The kelp powder (brown algae, purity 100%, Wanfang Biological Co., Ltd., Xi'an, China) used in the current study was harvested in Weihai city, Shandong province, China (37°16′ N, 122°41′ E) in July 2017. It was then dried and milled as a gray-green powder. The chemical composition of the kelp powder is shown in Table S1.

2.2. Animals and Experimental Design

A completely randomized design was applied in the present study. Twenty Chinese Holstein lactating dairy cows (589 ± 19.9 kg BW; 160 ± 18 DIM, milk yield (22 ± 2.3 kg/day)) were randomly divided into two treatments. Treatments included a control diet (CON) and a kelp powder replacing diet (Kelp). In the kelp treatment, kelp was used partly replace corn silage, oat hay and fatty acid calcium which resulted in kelp content came to 5% (dry matter (DM) basis). The diets were formulated according to NRC (2001) to meet or exceed the energy requirements of Holstein dairy cows yielding 20 kg of milk/day with 3.5% milk fat and 3.0% true protein. Details of ingredient analysis and chemical composition of diets were shown in Table 1.

Table 1. Ingredients and chemical composition of the diets (DM basis) [A,B].

Items	Kelp	Control
Ingredients (%)		
Corn silage	27.0	29.1
Oat hay	20.4	22.8
Kelp powder	5.0	—
Alfalfa haylage	7.2	7.2
Ground corn	3.6	3.6
Extruded soybeans	2.6	2.6
Soybean meal	9.9	9.9
Rapeseed meal	4.5	4.5
Cottonseed meal	2.6	2.6
Pressed corn	2.2	2.2
Cottonseed	2.7	2.7
Corn hull	9.6	9.6
Fat powder	1.2	1.2
Fatty acid calcium	1.2	1.7
Vitamin/mineral premix	0.3	0.3
Nutrient composition (%, unless otherwise stated)		
NE_L (Mcal/kg)	1.40	1.41
CP	12.81	12.96
EE	3.42	3.57
NDF	36.78	38.14
ADF	20.69	22.15

[A] Nutrient composition of the experimental diets were calculated according to NRC (2001); 0% Kelp = no kelp powder in the diet; 5% Kelp = 5% kelp powder in the diet; EE = ether extract; NE_L = net energy for lactation and calculated according to NRC (2001). [B] Vitamin/mineral premix contained (DM basis): 15.7% of Ca, 4.1% of P, 1600 mg/kg of Fe, 700 mg/kg of Cu, 3500 mg/kg of Mn, 7500 mg/kg of Zn, 80 mg/kg of Se, 400 mg/kg of I, 50 mg/kg of Co, 190,000 IU/kg of vitamin A, 55,000 IU/kg of vitamin D and 1900 IU/kg of vitamin E. NDF = neutral detergent fiber, ADF = acid detergent fiber.

2.3. *Feeding Management*

All cows were raised in individual stalls of a 35 d-long period feeding procedure. The cows were fed three times a day (07:00, 13:00 and 18:00 h, respectively) ad libitum in order to ensure that the total mixed ration (TMR) were fresh and available for the cows for at least 20 h a day. Cows were milked three times a day (09:00, 15:00, and 20:00 h, respectively) and were managed with natural lighting. All cows were dewormed before the commencement of this study.

2.4. *Sampling and Parameters Measurement*

During the experimental period, automatic feeding equipment (made by Institute of Animal Science Chinese Academy of Agricultural Sciences, Beijing, China and NanShang Husbandry Science and Technology Ltd. Henan, China) was used to record dry matter intake. Milking facilities (90 Side-by-Side Parallel Stall Construction, Afimilk, Israel) were applied to record milk production of each cow.

On the last day of the experiment, a gastric rumen sampler was used to collect rumen fluid samples through the esophagus at 3 h after the morning feeding. Collected samples were strained through four layers of cheesecloth to obtain rumen fluid. Rumen fluid was then divided into two parts. One part was processed to analyze the pH value, rumen volatile fatty acid (VFAs) and ammonia-N (NH_3-N). The other part was put into the liquid nitrogen immediately after adding stabilizer and then stored at −80 °C for DNA extraction. Rumen contents were strained through four layers of cheesecloth with a mesh size of 250 μm. The pH of each rumen fluid sample was measured immediately using a portable pH meter (Testo 205, Testo AG, Lenzkirch, Germany). Individual and total VFAs (TVFA) in the aliquots of ruminal fluid were determined by gas chromatograph (GC-2010, Shimadzu, Kyoto,

Japan). Concentration of NH_3-N was determined by indophenol method and the absorbance value was measured through UV-2600 ultraviolet spectrophotometer (Tianmei Ltd., China).

Milk samples were collected from individual cows during the last four consecutive days in 100-mL vials in each milking period. Samples were preserved with 2-bromo-2-nitropropan-1,3-diol and stored at 4 °C, before sent to the Milk and Dairy Products Quality Supervision and Testing Center, Ministry of Agriculture (Beijing, China) for analyses of milk protein, fat, lactose, somatic cell count and milk iodine content by a mid-infrared spectroscopy (Fossomatic 4000, Foss Electric A/S, Hillerød, Denmark).

2.5. *DNA Extraction and Sequencing Process*

DNA for metagenomics sequencing was extracted from the rumen fluid samples by using the QIAamp DNA Stool Mini Kit (Qiagen, Hilden, Germany) according to manufacturer's protocols. The DNA concentration and purity were quantified with TBS-380 and NanoDrop2000, respectively. DNA quality was examined with the 1% agarose gels electrophoresis system.

The bacteria 16S rRNA gene was amplified using the barcoded universal primers 338F (5′-barcorde-ACTCCTRCGGGAGGCAGCAG-3) and 806R (5′-GGACTACCVGGGTATCTAAT-3′) spanning the V3–V4 hyper variable region. The 16s RNA sequencing procedure was conducted through Illumina HiSeq 4000 platform (Illumina Inc., San Diego, CA, USA). Quality filtering on the raw tags were performed under specific filtering conditions to obtain the high-quality clean tags according to the QIIME (V1.7.0) quality-controlled process. The tags were compared with the reference database using UCHIME algorithm to detect chimera sequences, and then the chimera sequences were removed. Then the effective tags were finally obtained. Sequences analysis were performed by Uparse software (Uparse v7.0.1001, Tiburon, California, USA). Sequences with >97% similarity were assigned to the same OTUs. A representative sequence for each OTU was screened for further annotation. For each representative sequence, the GreenGene Database was used based on RDP classifier algorithm to annotate taxonomic information. Abundance information of OTUs was normalized using a standard of sequence number corresponding to the sample with the least sequences. Subsequent analysis of alpha diversity and beta diversity were all performed basing on this output normalized data. Alpha diversity is applied in analyzing complexity of species diversity for a sample through six indices, including, Chao1, Shannon, Simpson, ACE. All indices in our samples were calculated with QIIME (Version 1.7.0) and displayed with R software (Version 3.3.1, R Core Team, Vienna, Austria)). Beta diversity analysis was used to evaluate differences of samples in species complexity. Beta diversity on unweighted unifrac were calculated by QIIME software (Version 1.7.0, San Diego, CA, USA). Cluster analysis was preceded by principal coordinates analysis (PCoA), which was applied to reduce the dimension of the original variables. PCoA analysis was displayed by WGCNA package, stat packages, and ggplot2 package in R software. Unweighted Pair-Group Method with Arithmetic Means (UPGMA) Clustering was performed as a type of hierarchical clustering method to interpret the distance matrix using average linkage and was conducted by QIIME.

2.6. *Statistical Analysis*

Results were presented as means ± SEM. A normal distribution test of ruminal pH, VFAs, and NH_3-N were first conducted using SAS procedure "proc univariate data = test_normal" and then Student's *t*-test of SAS 9.2 was applied to analyze the differences of parameters between CON and Kelp treatments. p-value < 0.05 was considered to be significance and $0.05 \leq p < 0.10$ was considered as a tendency. Barplot, principal coordinate analysis (PCoA), hierarchical clustering analysis (HCA) for different rumen bacteria were conducted using R package. Spearman correlations between bacteria communities and ruminal fermentation variables were assessed using the PROC CORR procedure of SAS 9.2 (SAS Institute, Inc., Cary, NC, USA). Relative abundance of all phyla of bacteria were chosen to conduct the correlation analysis. A correlation matrix was created and visualized in a heatmap format using R package version 3.3.1.

3. Results

3.1. Animal Production Performance and Rumen Fermentation Parameters

The animal production performance was the most concerned in the dairy cattle production. In the present study, as shown in Table 2, the DMI, milk production, milk fat increased after kelp feeding, however, not significantly. Particularly, the milk iodine content significantly increased in the kelp treatment compared with CON treatment ($p < 0.05$). On the aspect of fermentation parameters, the kelp significantly raised ruminal total VFAs content while reduced the ammonia-N content ($p < 0.05$). For the individual VFA, kelp feeding significantly increased ruminal acetate, propionate and isobutyrate content.

Table 2. Effects of kelp powder on animal productivity and ruminal fermentation.

Items	CON	Kelp	SEM	*p*-Value
DMI (kg/day)	16.2	16.6	0.468	0.114
Milk production (kg/day)	21.1	22.7	1.894	0.235
Milk fat (%)	3.8	3.9	0.126	0.146
Milk protein(%)	3.41	3.42	0.061	0.106
Milk iodine(mg/L)	0.07 [b]	0.12 [a]	0.020	0.029
Ruminal pH	6.45	6.38	0.194	0.216
Ammonia-N (mg/dL)	17.8 [a]	13.1 [b]	1.933	<0.001
Acetate (mmol/L)	65.7 [b]	67.1 [a]	0.773	0.038
Propionate (mmol/L)	17.2 [b]	19.2 [a]	0.632	0.027
A/P	3.83	3.57	0.098	0.001
Butyrate (mmol/L)	16.3	16.1	0.137	0.356
Isobutyrate (mmol/L)	1.6 [b]	3.9 [a]	0.115	0.041
Valerate(mmol/L)	2.21	2.14	0.064	0.241
Isovalerate (mmol/L)	2.31	2.50	0.160	0.216
TVFA (mmol/L)	104.5 [b]	110.8 [a]	1.836	0.016

[a,b] means within a row with different letters differed significantly ($p < 0.05$); SEM = standard error of the mean. A/P = acetate/propionate.

3.2. Sequencing Information

The bacteria 16S rRNA gene was amplified using the barcoded universal primers 338F and 806R spanning the V3–V4 hyper variable region. The sequencing reads number of each samples were between 30,000–45,000 and the mean length of each reads were more than 430 nt. All sequencing information has been shown in Additional File 2. Sequences with > 97% similarity were assigned to the same OTUs and then aligned to GreenGene Database based on RDP classifier algorithm to annotate taxonomic information. Totally, 1338 OTUs, 17 phyla and more than 220 genera were identified in the present study, and all the taxonomy information is displayed in Additional File 3.

3.3. Effects of Kelp Replacing Diet on Ruminal Bacteria

3.3.1. α-Diversity

All identified bacteria were chosen for further analysis to investigate the effects of kelp replacing diet on the community of ruminal bacteria. Alpha diversity is applied in analyzing complexity of species diversity for a sample through Chao1, Shannon, Simpson, ACE. All these indices of α-diversity were displayed in Table 3. Based on the results, no significant changes were found in the kelp treatment compared with CON.

Table 3. Effects of kelp powder on α-abundance of ruminal microbiota.

Items	CON	Kelp	FC	Log 2 FC	SE	*p*-Value
ACE	1182.5	1170.2	0.99	−0.02	0.204	0.853
Chao1	1192.4	1189.9	1.00	0.00	0.076	0.97
Shannon	5.42	5.31	0.98	−0.03	0.08	0.398
Simpson	0.014	0.018	1.29	0.36	0.009	0.216
SOBs	1035.3	1016	0.98	−0.03	0.12	0.759

Note, FC = fold change, SE = standard error of the mean.

3.3.2. β-Diversity

Beta diversity analysis was then conducted to evaluate differences of samples in species complexity. PCoA based on unweighted UniFrac distance metrics was conducted to compare bacterial profile among the three treatments. As shown in Figure 1, PCoA axes 1 and 2 accounted for 29.77% and 11.52% of the total variation, respectively. Based on the results, bacteria community in kelp treatment could be separated from those in the CON by PCo1 except Kelp3.

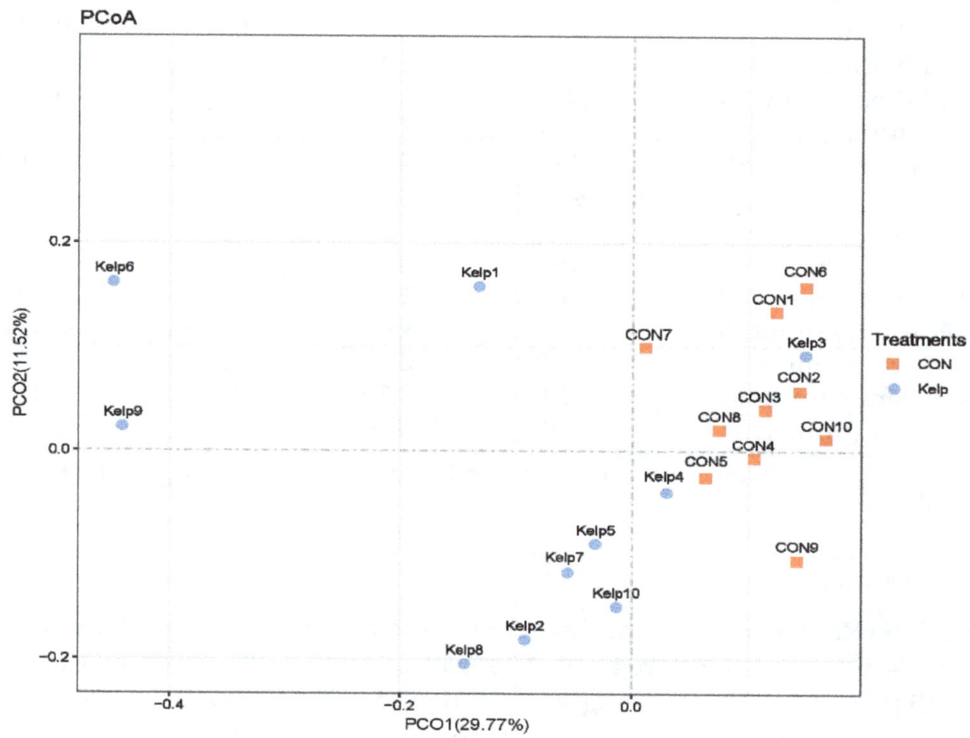

Figure 1. Principal coordinate analysis (PCoA) of bacteria community structures in CON and kelp treatment. CON = control diet; Kelp = kelp powder replacing diet.

Since the kelp made a difference to the whole bacteria community, a differential analysis on ruminal bacteria in different levels was then conducted to investigate the effects of kelp on the abundances of ruminal bacteria. Results are shown in Tables 4 and 5, in the level of phyla and genera, respectively. Based on the Table 4, *Bacteroidetes*, *Firmicutes*, and *Proteobacteria* contribute the three most abundant phyla of ruminal bacteria and the kelp feeding significantly promoted the proliferation of *Firmicutes* and *Proteobacteria* while suppressing *Bacteroidetes*. No significant effects were found of kelp feeding on other phyla. At the genera level, as shown in Table 5, *Prevotella*, *Ruminococcaceae*, *Ruminococcus*, *Bacteroidales*, *Prevotellaceae*, *Succinivibrionaceae*, *Lachnospiraceae*, and *Succiniclasticum* were the most abundant genera in the present study. *Prevotella*, *Bacteroidale*, *Rikenellacea*, *Selenomonas*, and *Saccharofermentans* were significantly decreased while *Ruminococcaceae*, *Ruminococcus*, *Lachnospiraceae*, and *Sharpea* were significantly increased after kelp feeding. There were no significant changes of other genera.

Table 4. Effects of kelp powder on the abundances of ruminal microbiota in phyla level.

Items	CON	Kelp	FC	log2 FC	SE	*p*-Value
p__Bacteroidetes	14570 [a]	10110 [b]	0.69	−0.53	0.10	<0.001
p__Firmicutes	7007 [b]	12098 [a]	1.73	0.79	0.13	0.008
p__Fibrobacteres	91.4	36.8	0.40	−1.31	0.46	0.088
p__Proteobacteria	424.6 [b]	1787.3 [a]	4.21	2.07	0.53	0.024
p__Spirochaetae	179.3	242.1	1.35	0.43	0.32	0.408
p__Saccharibacteria	133.5	143	1.07	0.099	0.32	0.765
p__Tenericutes	264	214.4	0.81	−0.30	0.52	0.462
p__Verrucomicrobia	33.8	19	0.56	−0.83	0.31	0.277
others	39.4	54.2	1.38	0.46	0.27	0.286

[a,b] means within a row, different letters differed significantly ($p < 0.05$); SE = standard error.

Table 5. Effects of kelp powder on the abundances of ruminal microbiota in genera level.

Items	CON	Kelp	FC	Log2 FC	SE	*p*-Value
g__Prevotella	9897.2 [a]	6376.8 [b]	0.64	−0.63	0.08	<0.001
g__Ruminococcaceae	1470.3 [b]	3111.1 [a]	2.12	1.08	0.17	0.021
g__Ruminococcus	539.8 [b]	2603.8 [a]	4.82	2.27	0.25	0.004
g__Bacteroidales	1969.7 [a]	1064.1 [b]	0.54	−0.89	0.15	0.002
g__Prevotellaceae	1472	1199.2	0.81	−0.30	0.15	0.185
g__Succinivibrionaceae	636.3	1238.8	1.95	0.96	0.56	0.247
g__Lachnospiraceae	566.4 [b]	1069.2 [a]	1.89	0.92	0.15	0.008
g__Succiniclasticum	626.6	653.8	1.04	0.06	0.33	0.876
g__Eubacterium	502.7	521.8	1.04	0.05	0.27	0.767
g__Rikenellaceae	600.5 [a]	315.6 [b]	0.53	−0.93	0.13	0.016
g__Christensenellaceae	430.8	428.6	0.99	−0.01	0.28	0.986
g__Selenomonas	437.1	240.3	0.55	−0.86	0.28	0.012
g__Acetitomaculum	171.7	446.2	2.60	1.38	0.18	0.121
g__Butyrivibrio	177.1	255.1	1.44	0.53	0.31	0.615
g__Treponema	285.8[a]	124 [b]	0.43	−1.20	0.42	0.022
g__Erysipelotrichaceae	140.8	204.2	1.45	0.54	0.51	0.539
g__Anaeroplasma	214.9 [a]	113.3 [b]	0.53	−0.92	0.35	0.003
g__Lachnobacterium	170.3	143.4	0.84	−0.25	0.18	0.453
g__Candidatus Saccharimonas	169.6	111	0.65	−0.61	0.22	0.05
g__Saccharofermentans	146.9 [a]	81.2 [b]	0.55	−0.86	0.31	0.016
g__Sharpea	31.2 [b]	189.7 [a]	6.08	2.60	0.22	0.016
g__Roseburia	84.6	123.2	1.46	0.54	0.6	0.535
g__Ruminiclostridium	50.5	150.9	2.99	1.58	0.49	0.159
g__Schwartzia	93.5	77.5	0.83	−0.27	0.33	0.642
g__Lachnospiraceae	79.8	81.7	1.02	0.03	0.41	0.868
g__Fibrobacter	77.3	52.9	0.68	−0.55	0.14	0.654
g__Mogibacterium	58	64.9	1.12	0.16	0.53	0.768
g__Succinivibrio	60	34	0.57	−0.82	0.28	0.338
others	3665.1	3447.5	0.94	−0.088	0.48	0.471

[a,b] means within a row, different letters differed significantly ($p < 0.05$); SE = standard error.

Hierarchical clustering analysis (HCA) and heat map analysis was conducted to further understand the effects of kelp feeding on ruminal bacteria profile. Results of HCA analysis are displayed in Figure 2. Bacteria which belongs to CON and kelp treatments could separate into two big clusters clearly except CON7. Ruminal bacteria in genera level gathered into two big clusters. The genera in upper half part were increased after kelp feeding which consists of *g__Eubacterium*, *g__Christensenellaceae*, *g__Ruminococcaceae*, *g__Acetitomaculum*, *g__Butyrivibrio*, and *g__Mogibacterium*. Lower half part consists of those genera which were decreased after kelp feeding, such as *g__Succinivibrionaceae*, *g__Fibrobacter*, *g__Anaeroplasma*, *g__Rikenellaceae*, and *g__Saccharofermentans*.

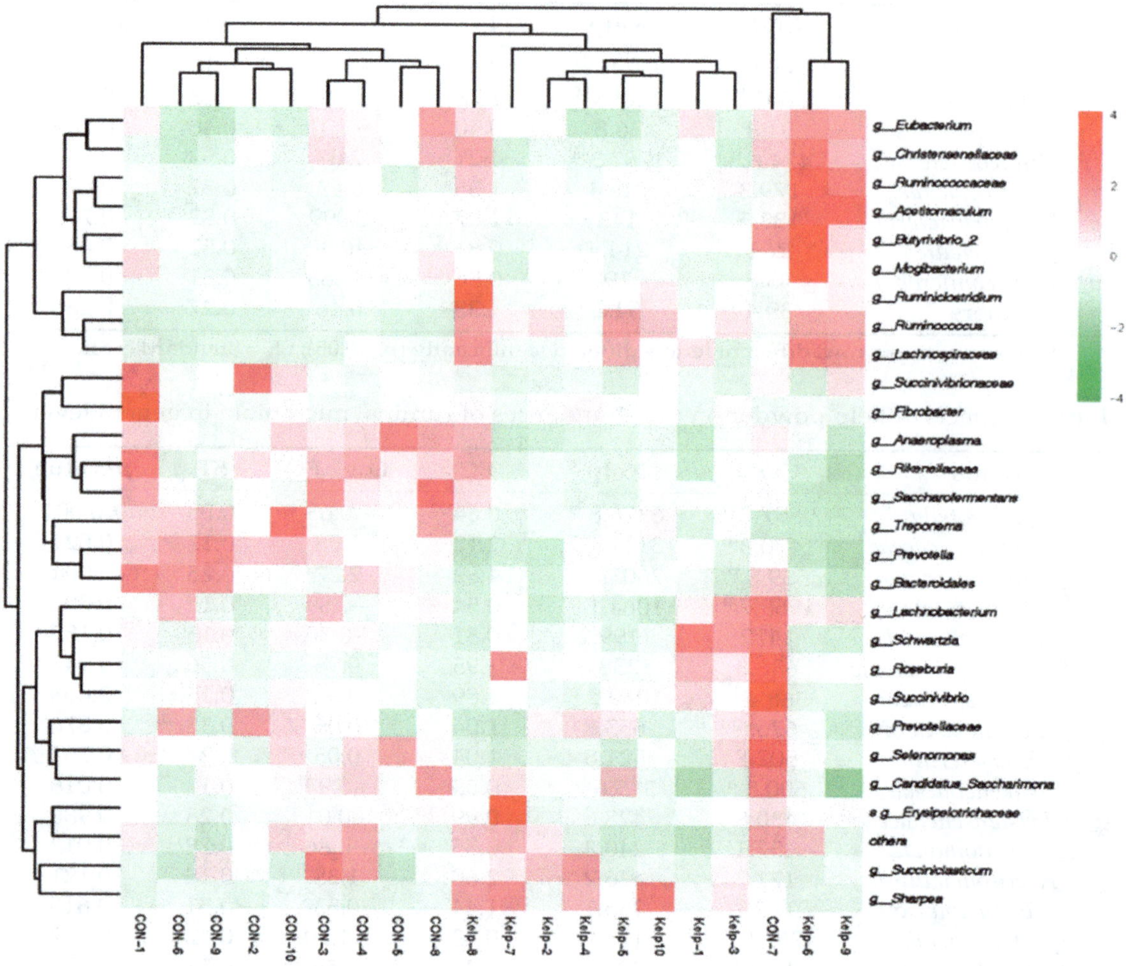

Figure 2. Hierarchical clustering analysis (HCA) and heat map analysis on abundances of ruminal bacteria content between CON and kelp treatment on the level of genera. Rows represent genera of bacteria and columns represent samples. Cells were colored based on the relative abundance of bacteria measured in rumen, red represents high rumen levels while green represents low signal intensity and white cells showing the intermediate level.

3.4. Correlations between Bacteria Communities and Ruminal Fermentation Parameters

At last, all phyla and most abundant genera were selected to conduct the correlation analysis between bacteria and fermentation parameters for the purpose to investigate the effects of ruminal bacteria on ruminal fermentation. Results are shown in Figures 3 and 4, respectively. All phyla could be separated into two parts based on the correlations with ruminal fermentation parameters. One was positively correlated with ruminal pH and VFA content, while negatively correlated with NH_3-N, this part was mainly comprised by *Firmicutes* and *Proteobacteria*.

The other part was just conversed, which was positive with NH_3-N while negative correlated with ruminal VFAs content. This part mainly consisted of *Bacteroidetes*, *Fibrobacteres*, and *Saccharibacteria*. As to the genera level, bacteria could also separate into two big clusters. The first part, which consists of *Ruminiclostridium*, *Acetitomaculum*, *Butyrivibrio*, *Ruminococcaceae*, *Ruminococcus*, *Lachnospiraceae*, and *Succiniclasticum* was positively correlated with pH, TVFAs, acetate, propionate, and Isobutyrate while being negatively correlated with NH_3-N. The other part was mainly comprised by *Prevotella*,

Saccharimonas, Saccharofermentans, Bacteroidales, Prevotellaceae, Succinivibrionaceae, and *Selenomonas* which performed conversed correlation with the fermentation parameters compared with the first part.

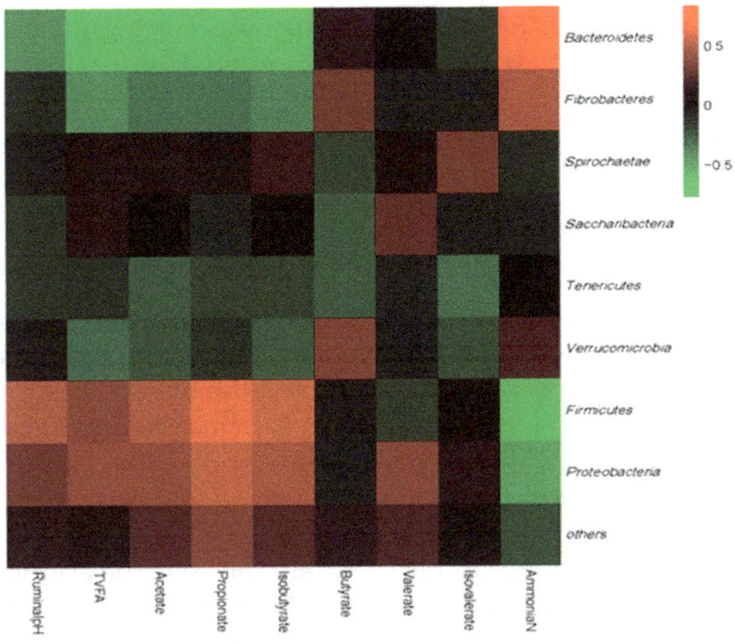

Figure 3. Correlation analyses between abundances of ruminal bacteria and DMI, milk quality parameters, and ruminal fermentation parameters on the level of phyla. The red color represents a positive correlation while the green color represents a negative correlation.

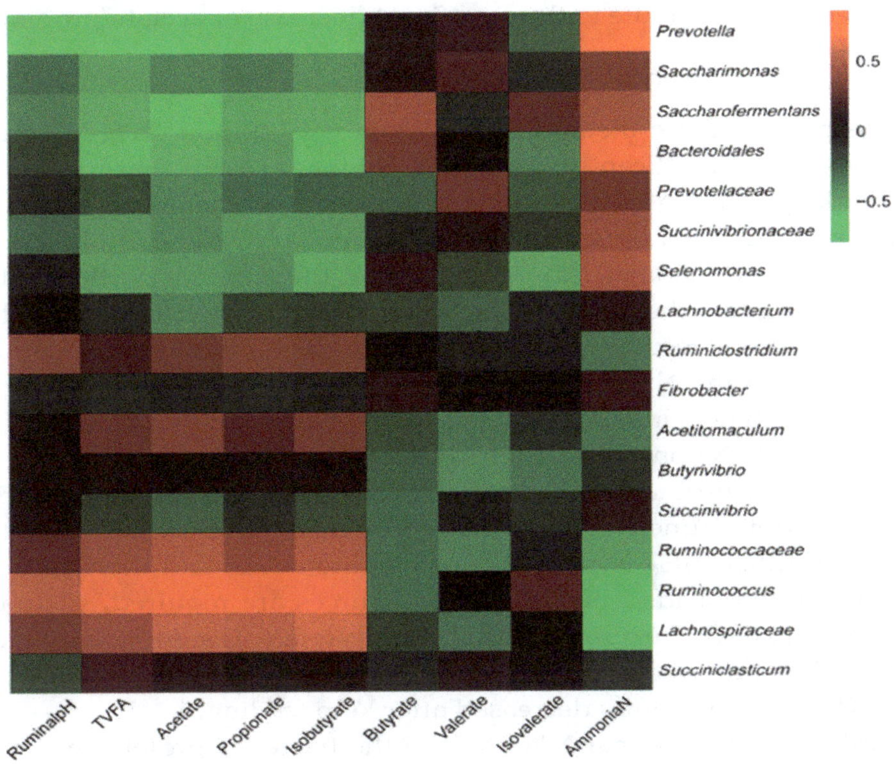

Figure 4. Correlation analyses between the most abundant genera of ruminal bacteria and DMI, milk quality parameters, and ruminal fermentation parameters. The red color represents a positive correlation while the green color represents a negative correlation.

4. Discussion

4.1. Effects of Kelp Feeding on Ruminal Carbohydrate Metabolism

In the present study, ruminal TVFA, acetate and propionate significantly increased after kelp powder diet feeding, which indicated that kelp promoted ruminal carbohydrates metabolism. In ruminal conditions, carbohydrates were always degraded into mono- or di-saccharides, and then hydrolyzed into VFAs [12]. The kelp was rich in oligosaccharides, which are always undegradable in monogastrics, which could be fermented by ruminal microbiota [11]. Because of the lower molecular weight and the simpler structure [13], oligosaccharides might be easier digestible than the polysaccharides such as cellulose. When replacing dietary forage with kelp powder, more carbohydrates were digested in the kelp feeding diets, and therefore, more VFAs were synthesized.

Kelp treatment significantly decreased the ratio of acetate/propionate. In ruminal conditions, when dietary high-fermentable content increased, it always decreased the ratio of acetate/propionate [14]. In the present study, kelp feeding diets partially decreased the fiber content, which might lead to the reduction of the mainly fiber-degrading bacteria such as *Bacteroidetes* and *Fibrobacteres* [15,16]. Besides, the kelp was added through powder form, which significantly reduced the particle size compared with their substitution, decreasing the retention time in rumen and further reducing the reaction time between fiber-degrading bacteria and dietary fiber [17].

Inversely, the easier digestible oligosaccharides help increase the abundances of *Firmicutes* and *Proteobacteria* which mainly degrade the highly-fermentable carbohydrates in ruminal conditions [18,19]. For the reasons given above, kelp feeding diets contain more highly-fermentable carbohydrate and might lead to accumulation of glucose. In ruminal conditions, glucose digested into pyruvate through glycolytic pathway [20], and then hydrolyzed into acetyl-CoA, which was the precursor of acetate with the help of pyruvate formate-lyase (PFL) [21]. When provided more high-fermentable carbohydrate, pyruvate accumulated and then turn to form more lactate, oxaloacetic acid [22], and further developed into propionate. This might contribute to the significant decrease of acetate/propionate.

4.2. Effects of Kelp Feeding on Ruminal Nitrogen Metabolism

Apart from the carbohydrate metabolism, the average rumen NH_3-N concentration in the kelp feeding treatment was significantly decreased compared with that in the control treatment, which indicated that kelp might affect the ruminal nitrogen(N) metabolism. The ruminal NH_3-N concentration could be influenced by absorption of rumen wall, ruminal outflow rate, and rumen microbial community [23]. Previous study reported that the ruminal outflow rate was negatively associated with forage length in the diet [24], the longer forage particles reduced the contact surface area with the microorganisms, resulting in a lower synthesis efficiency of rumen microbial protein and consequently increased the NH_3-N concentration [25]. In the present study, kelp was added through powder form which help increase attached area for ruminal microbiota, thereafter leading to the decrease of ruminal NH_3-N concentration in kelp feeding.

Pisulewski et al. (2010) obtained best results of ruminal NH_3-N utilization content with 88–133 mg/L [26], which opportunely included the content in kelp treatment. In ruminal conditions, ruminal bacteria could utilize NH_3-N to synthesize amino acids, and then form bacteria proteins. For synthesis of some amino acids, specific carbon sources are required, such as isobutyrate for valine [27]. In the present study, we observed that isobutyrate was significantly increased after kelp feeding. The increased isobutyrate might enhance the activity of ruminal bacteria to utilize more NH_3-N. Therefore, NH_3-N significantly decreased after kelp feeding.

In addition, utilization of ruminal NH_3-N into the bacterial protein in the rumen is energy dependent, providing adequate ruminal available energy is requirement for ruminal NH_3-N utilization [28]. In ruminal conditions, available energy was mainly provided by VFAs [29], which were found to significantly increase in kelp treatment. Therefore, the activity of rumen microorganisms will be enhanced by the provided energy, and more NH_3-N could be incorpoarated into bacteria protein.

4.3. *Effect of Kelp Feeding on Milk Performance*

Although not significantly, the milk production and milk fat percentage were increased after kelp feeding to some extent in the present study. Milk fat was a necessary parameter of milk quality and benefits to the human health [30]. In the current study, kelp feeding treatment could effectively promote the milk fat percentage through improving ruminal acetate content [31]. Increasing the replacement percentage of dietary forage with kelp might lead to increasing milk fat, consequenty improving the benefits of milk to human health.

Milk iodine content significantly increased in the present study, which was in line with Kaufmann [32]. For ruminants, iodine deficiency would result in low thyroid hormone concentrations and prolonged gestation [33]. The abundant iodine content in kelp might contribute to the partial increase of milk production.

5. Conclusions

In summary, a substitution of 5% kelp powder for forage in the diet reduced the fiber content in the feeding diet and then regulated the proliferation of ruminal microbiota, which lead to a significantly reduced NH_3-N content while increasing acetate and TVFA concentrations in the rumen. This result indicated that kelp powder as a dietary raw material might promoted the rumen fermentation ability and then benefit the lactation performance of dairy cows. Moreover, it could effectively increase milk fat and iodine content and consequently improve milk's nutritional value.

Author Contributions: Designed the study, F.X., Y.Z., and B.X.; Supervison and revised the manuscript, F.X., L.J., and B.X.; Conducted the experiment, F.X., F.S.; Analyzed the data, F.X., Y.W., D.H., and X.N.; Wrote the manuscript, F.X.; English editing, Y.Z.

Acknowledgments: This work was funded by Beijing Dairy Industry Innovation Team (bjcystx-ny-1). The authors thank the staff in the dairy cattle breeding center of Sanyuan Group (Beijing, China) for their assistance in the animal management.

References

1. Overton, T.R.; Yasui, T. Practical applications of trace minerals for dairy cattle. *J. Anim. Sci.* **2014**, *92*, 416–426. [CrossRef] [PubMed]
2. Yu, L.L.; Browning, J.F.; Burdette, C.Q.; Caceres, G.C.; Chieh, K.D.; Davis, W.C.; Kassim, B.L.; Long, S.E.; Murphy, K.E.; Oflaz, R.; et al. Development of a kelp powder (*Thallus laminariae*) Standard Reference Material. *Anal. Bioanal. Chem.* **2018**, *410*, 1265–1278. [CrossRef] [PubMed]
3. Abudabos, A.M.; Okab, A.B.; Aljumaah, R.S.; Samara, E.M.; Abdoun, K.A.; Alhaidary, A.A. Nutritional Value of Green Seaweed (Ulva Lactuca) for Broiler Chickens. *Ital. J. Anim. Sci.* **2013**, *12*, 612–620. [CrossRef]
4. Pereira, M.; Tala, F.; Fernandez, M.; Subida, M.D. Effects of kelp phenolic compounds on the feeding-associated mobility of the herbivore snail Tegula tridentata. *Mar. Environ. Res.* **2015**, *112*, 40–47. [CrossRef] [PubMed]
5. Garima, K.; Bruce, R.; Glenn, S.; Nikhil, T.; Franklin, E.; Alan, C.; Jeff, H.; Balakrishnan, P. Feed supplementation with red seaweeds, Chondrus crispus and Sarcodiotheca gaudichaudii, affects performance, egg quality, and gut microbiota of layer hens. *Poult. Sci.* **2014**, *93*, 2991–3001.
6. Heins, B.J.; Chester-Jones, H. Effect of feeding kelp on growth and profitability of group-fed calves in an organic production system. *Prof. Ani. Sci.* **2015**, *31*, 368–374. [CrossRef]
7. Antaya, N.T.; Soder, K.J.; Kraft, J.; Whitehouse, N.L.; Guindon, N.E.; Erickson, P.S.; Conroy, A.B.; Brito, A.F. Incremental amounts of Ascophyllum nodosum meal do not improve animal performance but do increase milk iodine output in early lactation dairy cows fed high-forage diets. *J. Dairy Sci.* **2015**, *98*, 1991–2004. [CrossRef]
8. Pitta, D.W.; Indugu, N.; Kumar, S.; Vecchiarelli, B.; Sinha, R.; Baker, L.D.; Bhukya, B.; Ferguson, J.D. Metagenomic assessment of the functional potential of the rumen microbiome in Holstein dairy cows. *Anaerobe* **2016**, *38*, 50–60. [CrossRef]

9. Hess, M.; Sczyrba, A.; Egan, R.; Kim, T.W.; Chokhawala, H.; Schroth, G.; Luo, S.; Clark, D.S.; Chen, F.; Zhang, T. Metagenomic Discovery of Biomass-Degrading Genes and Genomes from Cow Rumen. *Science* **2011**, *331*, 463–467. [CrossRef]
10. Puniya, A.K.; Singh, R.; Kamra, D.N. *Rumen Microbiology: From Evolution to Revolution*; Springer: New Delhi, India, 2015.
11. An, D.; Cai, S.; Dong, X. Actinomyces ruminicola sp. nov., isolated from cattle rumen. *Int. J. Syst. Evol. Microbiol.* **2006**, *56*, 2043–2048. [CrossRef]
12. Hall, M.B.; Larson, C.C.; Wilcox, C.J. Carbohydrate source and protein degradability alter lactation, ruminal, and blood measures. *J. Dairy Sci.* **2010**, *93*, 311–322. [CrossRef] [PubMed]
13. Li, Z.; Bai, H.; Zheng, L.; Jiang, H.; Cui, H.; Cao, Y.; Yao, J. Bioactive polysaccharides and oligosaccharides as possible feed additives to manipulate rumen fermentation in rusitec fermenters. *Int. J. Biol. Macromol.* **2018**, *109*, 1088–1094. [CrossRef] [PubMed]
14. Vyas, D.; Beauchemin, K.A.; Koenig, K.M. Using organic acids to control subacute ruminal acidosis and fermentation in feedlot cattle fed a high-grain diet. *J. Anim. Sci.* **2015**, *93*, 3950. [CrossRef]
15. Varki, A. Biological roles of oligosaccharides: All of the theories are correct. *Glycobiology* **1993**, *3*, 97–130. [CrossRef] [PubMed]
16. Ransom-Jones, E.; Jones, D.L.; McCarthy, A.J.; McDonald, J.E. Thefibrobacteres: An important phylum of cellulose-degrading bacteria. *Microb. Ecol.* **2012**, *63*, 267–281. [CrossRef] [PubMed]
17. Rodríguez-Prado, M.; Calsamiglia, S.; Ferret, A. Effects of Fiber Content and Particle Size of Forage on the Flow of Microbial Amino Acids from Continuous Culture Fermenters. *J. Anim. Sci.* **2004**, *87*, 1413–1424. [CrossRef]
18. Naas, A.E.; Mackenzie, A.K.; Mravec, J.; Schuckel, J.; Willats, W.G.T.; Eijsink, V.G.H.; Pope, P.B. Do rumen bacteroidetes utilize an alternative mechanism for cellulose degradation? *MBIO* **2014**, *5*, 1401–1414. [CrossRef]
19. Ludwig, W.; Schleifer, K.H.; Whitman, W.B. Revised Road Map to the Phylum Firmicutes. In *Bergey's Manual® of Systematic Bacteriology*; Springer: New York, NY, USA, 2009; pp. 1–13.
20. Pan, X.H.; Yang, L.; Xue, F.G.; Xin, H.R.; Jiang, L.S.; Xiong, B.H.; Beckers, Y. Relationship between thiamine and subacute ruminal acidosis induced by a high-grain diet in dairy cows. *J. Anim. Sci.* **2016**, *99*, 8790–8801. [CrossRef]
21. Xue, F.; Pan, X.; Jiang, L.; Guo, Y.; Xiong, B.H. GC–MS analysis of the ruminal metabolome response to thiamine supplementation during high grain feeding in dairy cows. *Metabolomics* **2018**, *14*, 67. [CrossRef]
22. Pan, X.; Nan, X.; Yang, L.; Jiang, L.; Xiong, B. Thiamine status, metabolism and application in dairy cows: A review. *Br. J. Nutr.* **2018**, *120*, 491–499. [CrossRef]
23. Whitelaw, F.G.; Milne, J.S.; Chen, X.B. The effect of a rumen microbial fermentation on urea and nitrogen metabolism of sheep nourished by intragastric infusion. *Exp. Physiol.* **1991**, *76*, 91–101. [CrossRef] [PubMed]
24. Owens, F.N.; Zinn, R.A.; Kim, Y.K. Limits to starch digestion in the ruminant small intestine. *J. Anim. Sci.* **1986**, *63*, 1634. [CrossRef] [PubMed]
25. Devant, M.; Ferret, A.; Gasa, J.; Calsamiglia, S.; Casals, R. Effects of protein concentration and degradability on performance, ruminal fermentation, and nitrogen metabolism in rapidly growing heifers fed high-concentrate diets from 100 to 230 kg body weight. *J. Anim. Sci.* **2000**, *78*, 1667–1676. [CrossRef] [PubMed]
26. Pisulewski, P.M.; Okorie, A.U.; Buttery, P.J.; Haresign, W.; Lewis, D. Ammonia concentration and protein synthesis in the rumen. *Proc. Nutr. Soc.* **2010**, *32*, 759–766. [CrossRef]
27. Allison, M.J. Biosynthesis of amino acids by ruminal microorganisms. *J. Anim. Sci.* **1969**, *29*, 797–807. [CrossRef]
28. Wallace, R.J. Ruminal microbiology, biotechnology, and ruminant nutrition: Progress and problems. *J. Anim. Sci.* **1994**, *72*, 2992–3003. [CrossRef]
29. Ryle, M.; Ørskov, E.R. *Energy Nutrition of Rumen Micro-Organisms*; Springer: Dordrecht, The Netherlands, 1990; pp. 10–27.
30. Lock, A.L.; Bauman, D.E. Modifying milk fat composition of dairy cows to enhance fatty acids beneficial to human health. *Lipids* **2004**, *39*, 1197–1206. [CrossRef]

31. Storry, J.E.; Rook, J.A.F. The relationship in the cow between milk-fat secretion and ruminal volatile fatty acids. *Br. J. Nutr.* **1966**, *20*, 217–228. [CrossRef]
32. Kaufmann, S.; Rambeck, W.A. Iodine supplementation in chicken, pig and cow feed. *J. Anim. Physiol. Anim. Nutr.* **2010**, *80*, 147–152.
33. Davis, S.R.; Collier, R.J.; McNamara, J.P.; Head, H.H.; Sussman, W. Effects of thyroxine and growth hormone treatment of dairy cows on milk yield, cardiac output and mammary blood flow. *J. Anim. Sci.* **1988**, *66*, 70. [CrossRef]

7

Effect of Mineral Supplementation on the Macromineral Concentration in Blood in Pre- and Postpartum Blackbelly Sheep

Juan Carlos Moyano Tapia [1,2,*], Simon Alexander Leib [1], Pablo Roberto Marini [2,3,4] and Maria Laura Fischman [2,5,6]

1. Facultad de Ciencias de la Vida/Centro de Investigación, Posgrado y Conservación Amazónica, Universidad Estatal Amazónica, Puyo 160150, Ecuador; simonleib@gmx.de
2. Centro Latinoamericano de Estudios de Problemáticas Lecheras (CLEPL), Casilda 2170, Argentina; pmarini@unr.edu.ar (P.R.M.); fischman@fvet.uba.ar (M.L.F.)
3. Facultad de Ciencias Veterinarias, Universidad Nacional de Rosario, Casilda 2170, Argentina
4. Consejo de Investigaciones (CIC-UNR), Rosario 2000, Argentina
5. Facultad de Ciencias Veterinarias, Universidad de Buenos Aires, CABA 1427, Argentina
6. Instituto de Investigación y Tecnología en Reproducción Animal (INITRA), Universidad de Buenos Aires, CABA 1427, Argentina

* Correspondence: jmoyano@uea.edu.ec

Simple Summary: The objective of this study was to determine the effect of mineral supplementation on the serum concentration of calcium, phosphorus, and magnesium in pre- and postpartum Blackbelly sheep throughout three successive lambing periods under free grazing conditions in the Ecuadorian Amazon Region. The field work was carried out between January 2015 and February 2018 using 20 Blackbelly females. The flock was randomly divided into two groups: Group 1 (G1), who were fed forage plus a supplementation (Pecutrin® Mineral supplement plus vitamins A, D3, and E. Bayer HealthCare); and Group 2 (G2), who were fed forage only without mineral supplementation. In this study, we showed that Blackbelly sheep raised under free grazing conditions (G2) had very low serum calcium values, and supplementation was unable to improve them. Meanwhile, phosphorus and magnesium values were below the required levels, but after supplementation (G1), they exceeded the minimum threshold. Mineral supplementation in the rearing of sheep in grazing systems is necessary during the entire production cycle, but it must be done taking into account the soil–plant–animal relationship, specifically for the Amazonian Region systems.

Abstract: The objective of this study was to determine the effect of mineral supplementation on the serum concentration of calcium, phosphorus, and magnesium in pre- and postpartum Blackbelly sheep throughout three successive lambing periods under free grazing conditions in the Ecuadorian Amazon Region. The field work was carried out between January 2015 and February 2018 using 20 Blackbelly sheep belonging to the Centre for Research, Postgraduate Studies and Conservation of Amazon Biodiversity, Ecuador. The flock was randomly divided into two groups: Group 1 (G1) was fed with forage plus a supplementation (Pecutrin® Mineral supplement plus vitamins A, D3, and E. Bayer HealthCare) and Group 2 (G2) was fed only with forage without mineral supplementation. Three blood samples from the coccygeal vein were taken from each sheep 30 days before lambing, 30 days after, and 60 days after lambing. Concerning the average of calcium, significant differences were found at different times inside each group and also between them ($p < 0.0001$ in both cases). As for the phosphorus, significant differences were found between the means of the groups for all times from 30 days after the second lambing season ($p < 0.05$). It was observed that the groups differed significantly in terms on the average of magnesium (considering a significance level of 0.05) 30 days before the first lambing and at all times measured from the 30 days after the second lambing ($p < 0.005$). In this study, we showed that Blackbelly sheep raised under free grazing conditions in the Ecuadorian Amazon Region had very low serum calcium values, and supplementation was unable to improve them. Meanwhile, phosphorus and magnesium levels were below the required values, but after supplementation, they exceeded the minimum threshold. Mineral supplementation in the rearing of sheep in grazing systems is necessary during the entire production cycle, but it must be done taking into account the soil–plant–animal relationship specifically for the Amazonian Region systems.

Keywords: minerals; sheep; pre- and postpartum; grazing system

1. Introduction

The Blackbelly breed is of utmost relevance for meat production in the Ecuadorian Amazon Region, as it can adapt to extreme conditions. This fact is reflected in its low mortality, precocity and high reproductive fertility (multiple births), medium size, and productive longevity [1].

Many authors have shown that one of the problems regarding grazing ruminants is that pasture does not cover their protein, energy, or mineral needs, affecting the normal development of their metabolic processes [2], because their requirements depend exclusively on the forage's composition [3]. The soils of the Amazonian Region, with a pH lower than 5.5 during most of the year, generate restrictions on plants. The soil acidity modifies physical and chemical properties. These negative effects are reflected by an excess of certain minerals—such as aluminum—and the deficiency of others, for instance, molybdenum. It should also be considered that forage resources have fluctuations throughout the year, both in quantity and in quality [4].

The intake of minerals in ruminants reared in grazing systems depends on the composition and the total consumption of forage, the consumption and the minerals content of water, and the composition of the soil [5]. Adequate amounts of essential minerals are critical to maximize the productivity and the health of livestock [6]. Furthermore, appropriate mineral levels improve the interaction between production and reproduction [7,8]. A widely used indicator to assess nutritional status, health, and well-being of sheep is the Body Condition Index, as indicated in the Animal Welfare Indicators (AWIN) protocols [9].

A suitable quantity and proportion of essential minerals prevents diseases by activating the immune system, improving growth, production, viability, and fertility [10]. Previously, some authors described the need for supplementation [11,12] and the importance of using concentrated mixtures for sheep according to age groups [13–15]. However, for many years, the supplementation under grazing conditions was not considered a cost-effective alternative. Preliminary studies showed that supplementation is profitable if kilograms of weaned lamb per female per year were analyzed. This variable involves the factors that most influence the profitability of the herd, since it considers prolificity of the female, milk production, maternal instinct, mortality, and weight gain of lambs [16].

The objective of this study was to determine the effect of mineral supplementation on the serum concentration of calcium, phosphorus, and magnesium in pre- and postpartum Blackbelly sheep throughout three successive lambing periods under free grazing conditions in the Ecuadorian Amazon Region.

2. Materials and Methods

The field work was carried out in the Centre for Research, Postgraduate Studies and Conservation of Amazon Biodiversity (CIPCA). It is located in the Arosemena Tola canton, province of Napo, Ecuador (coordinates: 01°14.325′ S; 077°53.134′ W). The altitude varies between 580 and 990 m above sea level. The environment is tropical with an average annual rainfall of 4000 mm, an average relative humidity of 80%, and temperatures ranging between 15 and 25 °C.

The study was performed between January 2015 and February 2018. This interval included three lambing periods by the sheep under study. The same 20 Blackbelly females aged 24 to 32 months and with an average weight and standard deviation of 34 ± 4 kg were used.

The sheep were free to graze from 07:00 to 18:00, then housed overnight with water consumption ad libitum. They were randomly divided into two groups. Group 1 (G1) was fed with forage plus a supplement at a dose of 0.005 kg per animal per day every day from when the firstborn lamb was weaned until the end of the trial (Pecutrin® Mineral supplement plus vitamins A, D3, and E. Bayer HealthCare; parts of the formula components: calcium min. 17% max. 20%; phosphorus min. 18%; magnesium min. 3.0%; relationship calcium–phosphorus 1.3:1) The mineral supplement dose was the minimum recommended by the manufacturer; each sheep had its own feeder space. Group 2 (G2) was fed only with forage without mineral supplementation. Table 1 shows the chemical composition of the predominant species in the pasturage used. The presence of aluminum and hydrogen ions determined by intense washing phenomena as a consequence of high rainfall generates acid soils with a pH of 5.2 [17]. This reduction in the pH affects chemical and biological characteristics of the soil, reducing the growth of plants. It also decreases the availability of nutrients such as calcium, magnesium, phosphorus, and potassium. In turn, acidic soils tend toward the accumulation of toxic elements for plants, such as aluminum and manganese [18].

Table 1. Chemical composition of the pasture.

Pasturage	DM (kg/ha/yr)	Protein (%)	Calcium (%)	Phosphorus (%)	Magnesium (%)	IVD (%)
Brachiaria decumbens	17.585	10.6	0.20	0.18	0.15	44.4
Arachis Pintoi	6.212	19.4	1.7	0.21	0.63	59.2

DM: dry matter; IVD: in vitro digestibility. References: [19,20].

The sanitary management routinely used for the sheep flock of CIPCA was applied. This includes deworming, baths to combat ticks and flies, vaccines for foot and mouth disease, alongside antifungal and antibacterial medicines.

Through the fodder mixture, the animals consumed 0.612 kg/day of dry matter (1.8% of their average weight). This percentage was estimated based on pastures advanced in their physiological maturity [21]. The contribution of Ca^{2+}, P^{3-}, and Mg^{2+} was 2.5 g/day, 0.5 g/day, and 1.8 g/day, respectively. Except for Mg^{2+}, the mineral levels were below the daily minimum allowance recommended, given that the animals should consume 3.1 g/day, 2.9 g/day, and 1.0 g/day of Ca^{2+}, P^{3-}, and Mg^{2+}, respectively [21]. Pecutrin® (Bayer, Leverkusen, Alemania)., at the dose offered, contributed with a total of 0.6 g/day of Ca^{2+}, 0.5 g/day of P^{3-} and 0.9 g/day of Mg^{2+}.

Blood samples were only taken from the female adult sheep. Three blood samples from the coccygeal vein were taken 30 days before lambing as well as 30 days and 60 days post lambing. Ten milliliters of blood was obtained from each animal in a vacuum tube without anticoagulant (BD Vacutainer® red cap, Franklin Lakes, NJ, USA). The blood was centrifuged (3000 rpm × 15–30 min), and the separated serum was stored at −20 °C until analyzed.

Serum minerals were determined, including phosphorus (P^{3-}) and magnesium (Mg^{2+}). These two tests were performed by means of molecular spectrophotometry with the GENESYS 10 UV Series Thermo Scientific Spectrophotometer (Waltham, MA, USA). The reagents used were HUMAN (Germany), and the equipment was used according to its technical specifications for each of the samples. For calcium (Ca^{2+}), we employed the AUDICOM, AC 9801 Electrolytic Analyzer (Maharashtra, India) with specific AUDICOM reagents according to the equipment's technical specifications.

A descriptive statistical analysis was performed by calculating means. A linear regression model for correlated data was adjusted for inferential statistical analysis. This model took into account the correlation between the observations measured at the same unit over time by incorporating an intra-unit correlation structure. First, the correlation structure that best fit the data was chosen using the

Akaike Penalized Likelihood Criterion. Then, we proceeded to interpret the results of the hypothesis tests for the model's terms. If interaction between group and time was significant, contrasts were performed to compare if there were differences on average between the groups for each time. Statistical program R 3.6.3 was used.

3. Results

3.1. Calcium

There was not significant interaction between group and time for the calcium variable ($p = 0.1469$). There were significant differences between the average of calcium between times, and there were also significant differences between groups regarding the average of calcium ($p < 0.0001$ in both cases) (Table 2). Even though the model detected differences in the average calcium between groups, and the means were 2.44 mg/dL and 2.25 mg/dL, they were statistically different but not biologically.

Table 2. Serum calcium levels throughout the three successive lambing seasons.

Ca^{2+} (mg/dL)	Lambing Season								
	First Lambing			Second Lambing			Third Lambing		
Days before or after	−30	+30	+60	−30	+30	+60	−30	+30	+60
Supplemented group	2.17	2.44	2.24	2.48	2.37	2.40	2.66	2.61	2.56
Not supplemented group	2.20	2.20	2.20	2.22	2.28	2.29	2.28	2.36	2.41

Results are expressed as mean values of calcium per days before or after lambing. No significant differences were observed between groups.

3.2. Phosphorus

There was an interaction between group and time for the variable phosphorus ($p = 0.0013$). When performing contrasts to compare the means of the groups at each moment, significant differences were found between groups for all times from 30 days after the second lambing season ($p < 0.05$) (Table 3).

Table 3. Serum phosphorus levels throughout the three successive lambing seasons.

P^{3-} (mg/dL)	Lambing Season								
	First Lambing			Second Lambing			Third Lambing		
Days before or after	−30	+30	+60	−30	+30	+60	−30	+30	+60
Supplemented group	4.50	3.31	4.29	4.67	5.71	5.70	6.80	6.39	6.96
Not supplemented group	3.90	4.24	4.30	4.77	4.16	4.31	5.61	4.37	4.54
	ns	ns	ns	ns	*	ns	ns	*	*

Results are expressed as mean values of phosphorus per days before or after lambing. (*): significant differences ($p < 0.001$); (ns): no significant differences were observed between groups.

3.3. Magnesium

When adjusting the model, it was observed that there was interaction between group and time ($p < 0.0001$). The mean trajectory in time was not the same for both groups with respect to the variable Mg^{2+} (Table 4). When performing contrasts, it was observed that the groups differed significantly in terms of the average of magnesium (considering a significance level of 0.05) 30 days before the first lambing and at all times measured from the 30 days after the second lambing ($p < 0.005$).

Table 4. Serum magnesium levels throughout the three successive lambing seasons.

Mg^{2+} (mg/dL)	Lambing Season								
	First Lambing			Second Lambing			Third Lambing		
Days before or after	−30	+30	+60	−30	+30	+60	−30	+30	+60
Supplemented group	2.18	2.42	2.25	2.70	3.45	4.50	3.49	3.95	3.86
Not supplemented group	2.72	2.90	2.54	2.17	2.39	2.29	2.36	2.50	2.73
	ns	ns	ns	ns	ns	*	*	*	*

Results are expressed as mean values magnesium per days before or after lambing. (*): significant differences ($p < 0.001$); (ns): no significant differences were observed between groups.

4. Discussion

Few studies exist concerning feeding sheep raised in an Amazonian environment. Most regions have adapted their production systems according to knowledge obtained from production systems in temperate climates [22]. In the conditions of the Amazon Region, sheep grow consuming poor quality forage (that has excessive fiber, low protein, and low energy). The low quality of plants in acidic soils is due to the combination of toxicities of aluminum and manganese as well as deficiencies of sodium, phosphorus, potassium, calcium, magnesium, and some micronutrients, such as iron and zinc. Plants are subject to varying degrees of abiotic stress (soil acidity, mineral deficiencies and/or toxicities, droughts/floods, light/shadow quality, and extreme temperatures) and biotic stress (pests, diseases, and weeds). These tensions have an effect on growth and development, causing less absorption, a reduced utilization of absorbed nutrients and, consequently, a reduction in the efficient use of nutrients [23,24].

These characteristics could affect the normal development of metabolic and mineral processes in sheep. Blood is the most important bio-substrate for the estimation of an animal's mineral state [25,26].

Minerals are essential nutrients for small ruminants and their concentrations in blood must oscillate within bounded intervals in order to maintain their health and well-being. Adequate concentrations of macronutrients allow structural, physiological, catalytic, and regulatory functions of the organism to be rightly developed [27,28].

Normal calcemic values in sheep are 11.5–12.8 mg/dL [29]. In this study, calcium levels in the ewes ranged between 2.6 and 2.8 mg/dL, well below the normal range. Quintero-Moreno [30] found values in sheep without supplementation of 8.73 mg/dL, while in supplemented sheep, they obtained values of 15.2 mg/dL. Underwood and Suttle [31] found that normal blood serum calcium values ranged from 7 to 8 mg/dL in young sheep, while Norton [32] described that calcium rarely is a limitation in fodder diets. The lower Ca^{2+} content in Amazonian pastures could be due to the natural dilution of the process by which the production of dry matter exceeds mineral uptake [33]. In this case, supplementation did not help to raise the calcium level; the calcium level in G1 increased (2.44 mg/dL) in comparison to G2 (2.25 mg/dL), thus they were statistically different but not biologically. This result could be explained in part because a minimum dose of 5 g/animal/day was used. However, the interactions with toxic minerals, which could interfere and decrease absorption, should also be considered. No placental retention problems were observed, nor ewes that fell after lambing, spontaneous fractures, or other related pathologies, although the values suggest acute subclinical hypocalcaemia. Herdt [34] suggested that animals would adapt to this condition and that such adaptation would be mainly based on glucose, non-esterified fatty acids (NEFA), and ketone body availability and supply. The improvement in Ca^{2+} levels could be reflected in the general state and surely in higher and better milk production, which should be seen in the weight of lamb weaned. Abarghani [35] found that Ca^{2+} and P^{3-} were deficient during different seasons in grazing animals in all regions and for all sheep categories. These authors concluded that, for sheep in continuous grazing, it was not possible to prevent deficiencies in the concentration of Ca^{2+} and P^{3-} in serum or plasma. They concluded that supplementing these small ruminants with a bioavailable mineral mixture could increase the blood level of these minerals. However, the authors stressed the importance of conducting more studies to determine requirements and economic benefits of mineral supplements.

Normal values of phosphatemia in sheep are 5–7.3 mg/dL [29]. In this study, the ewes' phosphatemia values ranged between 4.03 and 6.71 mg/dL. In the first lambing, the values were lower than normal, but after incorporating the supplementation, the G1 values were within the normal values, as was observed for the G2 in the third lambing. Stojkovic [36] obtained similar results; he found that the phosphorus level in the blood serum of the sheep studied was between 3.84–4.61 mg/dL, which was below the normal limits necessary to meet the sheep's needs for this macroelement. The low phosphorus content could be due to the low presence of P^{3-} in the acidic Amazonian soil and also to the low amount of this mineral in *Arachis pintoi* [23,24,37,38]. In ruminants, the amount of mineral excreted varies according to the type of diet. The amount and the source of phosphorus used in the diet also affect the kinetics of the mineral in the body [39,40]. The age of the animal can influence the use of dietary phosphorus, since it has been observed that young animals have higher absorption efficiency [41]. In this way, there are several aspects that can influence the phosphorus homeostasis in the organism and, therefore, the mineral's kinetics. In recent years, the study of this mineral in the context of animal production has mainly focused on redefining the requirements of the mineral, seeking to minimize excretion without affecting animal performance [42]. In addition, the absorptions of calcium and phosphorus are independent of each other, which allow the sheep to adjust to different physiological demands [43]. The results showed an inverse Ca^{2+}: P^{3-} ratio. For optimal performance in ruminants, it must be between 1.5:1 and 2.5:1 [26]. In our study, it was 1:1.7 at the beginning and 1:2.57 at the end, showing a clear mineral imbalance. However, in ruminants, the ratio of Ca^{2+} to P^{3-} only affects the absorption of these minerals if the diet is inadequate in said minerals [44]. The metabolism of Ca^{2+} is closely related to that of P^{3-}. The excess or the deficit of one mineral can affect the use of the other [44].

In this study, the serum levels of Mg^{2+} found in the sheep studied ranged between 2.5 and 2.7 mg/dL in the first lambing, even when the sheep of both groups were not supplemented. Supplementation allowed Mg^{2+} in blood to reach normal values (2.5–3.5 mg/dL) and to exceed it in the third lactation. The normal values of magnesemia in sheep are 5–7.3 mg/dL [29]. Our results coincided with those reported by Abarghani [35], with values that ranged between 2.9 and 3.7 mg/dL. The organic deficit of an important element such as Mg^{2+} results in less growth and development in the animal. McDowell [2] stated that magnesium is an enzymatic activator involved in the metabolism of carbohydrates and lipids, since it is a catalyst for a wide variety of enzymes. It is also part of protein synthesis through its action in ribosomal aggregation.

5. Conclusions

In this study, we showed that Blackbelly sheep raised in free grazing conditions in the Ecuadorian Amazon Region had very low calcium values in serum, and supplementation failed to improve them. Phosphorus and magnesium values were below the required values, but after supplementation, they exceeded the minimum threshold. Mineral supplementation in the rearing of sheep in grazing systems is necessary throughout the production cycle, but adjustments are essential upon taking into account the soil–plant–animal relationship for systems in the Amazon region.

Author Contributions: Conceptualization, M.L.F. and P.R.M..; methodology, J.C.M.T. and S.A.L.; software, P.R.M.; validation, J.C.M.T., S.A.L., P.R.M. and M.L.F.; formal analysis, P.R.M. and M.L.F.; investigation, J.C.M.T., P.R.M., S.A.L.; resources, S.A.L. and J.C.M.T.; data curation, P.R.M.; writing—original draft preparation, P.R.M. and M.L.F.; writing—review and editing, P.R.M. and M.L.F.; visualization, S.A.L. and J.C.M.T.; supervision, M.L.F. and P.R.M.; project administration, S.A.L. and J.C.M.T.; funding acquisition, J.C.M.T. All authors have read and agreed to the published version of the manuscript.

References

1. Mendives, J.A. Importancia de los ovinos tropicales introducidos al país: Características productivas y reproductivas. *Latinoam. Prod. Anim.* **2007**, *15*, 310–315. Available online: http://www.bioline.org.br/pdf?la07068 (accessed on 13 May 2020).
2. McDowell, L.R. *Minerals in Animal and Human Nutrition*, 2nd ed.; Elsevier: Amsterdam, The Netherlands, 2003; p. 644.
3. Morales, A.E.; Domínguez, V.I.; González Ronquillo, M.; Jaramillo, E.G.; Castelán, O.O.; Pescador, S.N.; Huerta, B.M. Diagnóstico mineral en forraje y suero sanguíneo de bovinos lecheros en dos épocas en el valle central de México. *Técnica Pecu. México* **2007**, *45*, 329–344.
4. Organización de las Naciones Unidas Para la Alimentación y la Agricultura (FAO). Suelos Ácidos. 2019. Available online: http://www.fao.org/soils-portal/soil-management/manejo-de-suelos-problematicos/suelos-acidos/es/ (accessed on 13 May 2020).
5. Rodríguez, A.; Banchero, G. Deficiencia de minerals en ruminates. *Revista INIA* **2007**, *13*, 11–15.
6. Wang, H.; Liu, Z.; Huang, M.; Wang, S.; Cui, D.; Dong, S.; Li, S.; Qi, Z.; Lin, Y. Effects of Long-Term Mineral Block Supplementation on Antioxidants, Immunity, and Health of Tibetan Sheep. *Biol. Trace Elem. Res.* **2016**, *172*, 326–335. [CrossRef] [PubMed]
7. Almeida, A.M.; Schwalbach, L.M.J.; Cardoso, L.A.; Greyling, J.P.C. Scrotal, testicular and semen characteristics of young Boer bucks fed winter veld hay: The effect of nutritional supplementation. *Small Rumin. Res.* **2007**, *73*, 216–220. [CrossRef]
8. Griffiths, L.M.; Loeffler, S.H.; Socha, M.T.; Tomlinson, D.J.; Johnson, A.B. Effects of supplementing complexed zinc, manganese, copper and cobalt on lactation and reproductive performance of intensively grazed lactating dairy cattle on the South Island of New Zealand. *Anim. Feed Sci. Technol.* **2007**, *137*, 69–83. [CrossRef]
9. Animal Welfare Indicators (AWIN). AWIN Welfare Assessment Protocol for Sheep. 2015. Available online: https://www.researchgate.net/publication/275887069_AWIN_Welfare_Assessment_Protocol_for_Sheep (accessed on 13 May 2020). [CrossRef]
10. Underwood, E.J. *The Mineral Nutrition of Livestock*, 2nd ed.; Commonwealth Agriculture Bureau: Slough, UK, 1981.
11. Yadav, K.K.; Mandokhot, V.M. Effect of sulfur supplementation on the performance of stall-fed Nali lambs, growth responses, nutrient and minerals. *Indian J. Anim. Sci.* **1988**, *58*, 843–848.
12. Samanta, A.; Samanta, G. Mineral profile of different feed and fodders and their effect on plasma profile in ruminants of West Bengal. *Indian J. Anim. Nutr.* **2002**, *19*, 278–281.
13. Karim, S.A.; Verma, D.L. Growth performance and carcass characteristics of finisher lambs maintained on intensive feeding or grazing with supplementation. *Indian J. Anim. Sci.* **2001**, *71*, 959–961.
14. Karim, S.A.; Santra, A.; Sharma, V.K. Milk yield, its composition and performance of Malpura ewews under protocol of grazing and supplementation. *Indian J. Anim. Sci.* **2001**, *71*, 258–260.
15. Santra, A.; Karim, S.A.; Chaturvedi, O.H. Effect of concentrate supplementation on nutrient intake and performance of lambs of two genotypes grazing a semiarid rangeland. *Small Rumin. Res.* **2002**, *44*, 37–45. [CrossRef]
16. Moyano, J.C.; Marini, P.R.; Fischman, M.L. Biological Efficiency in Hair Sheep Reared in a Sustainable Farming System in the Ecuadorian Amazon Region. *Dairy Vet. Sci. J.* **2019**, *11*. [CrossRef]
17. Calva, C.; Espinosa, J. Effect of four liming materials on acidity control of a soil from Loreto, Orellana. *Siembra* **2017**, *4*, 110–120. [CrossRef]
18. Sumner, M.; Pavan, M. Alleviating Soil Acidity through Organic Matter Management. 2005. Available online: http://brasil.ipni.net/ipniweb/region/brasil.nsf/e0f085ed5f091b1b852579000057902e/5dfb8bcc4691864983257b0c004a898f/$FILE/Palestra%20Malcolm%20Sumner.pdf (accessed on 13 May 2020).
19. Instituto Nacional Autónomo de Investigaciones Agropecuarias (INIAP). *Manual de Pastos Tropicales Para la Amazonia Ecuatoriana*; Manual N33, pp. 66; Napo-Ecuador; 1997; Available online: http://repositorio.iniap.gob.ec/bitstream/41000/1622/1/Manual%20n%C2%BA%2011%20de%20pastos%20tropicales%20reducido%20ultimo.pdf (accessed on 13 May 2020).
20. Leonard, I. Recursos forrajeros autóctonos y promisorios para la ganadería en la provincia de Pastaza. In *Retos y Posibilidades para una Ganadería Sostenible en la Provincia de Pastaza de la Amazonia Ecuatoriana*; Universidad Estatal Amazónica: Puyo, Ecuador, 2015; Chapter IV; pp. 46–69.

21. Romero Yáñez, O.; Bravo Marchan, S. Fundamentos de la producción ovina en la Región de La Araucanía. Ministerio de Agricultura Instituto de Investigaciones Agropecuarias Centro Regional de Investigación Carillanca. *BOLETÍN INIA* **2012**, *245*, 204.
22. Pereira, E.S.; Fontenele, R.M.; Medeiros, A.N.; Lopes, R.O.; Campos, A.C.N.; Heinzen, E.L.; Bezerra, L.R. Requirements of protein for maintenance and growth in ram hair lambs. *Trop. Anim. Health Prod.* **2016**, *48*, 1323–1330. [CrossRef] [PubMed]
23. Fageria, N.K.; Baligar, V.C.; Li, Y.C. The role of nutrient efficient plants in improving crop yields in twenty first century. *J. Plant Nutr.* **2008**, *31*, 1121–1157. [CrossRef]
24. Lyda, S.D. Alleviating pathogen stress. In *Modifying the Root Environment to Reduce Crop Stress*; ASAE Monograph No. 4; Arkin, G.F., Taylor, H.M., Eds.; American Society of Agricultural Engineers: St. Joseph, MI, USA, 1981; pp. 195–214.
25. Khan, Z.I.; Ashraf, M.; Ahmad, K.; Valeem, E.E.; McDowell, L.R. Mineral Status of Forage and Its Relationship with that of Plasma of Farm Animals in Southern Punjab; Pakistan. *Pak. J. Bot.* **2009**, *36*, 851–856.
26. Khan, Z.I.; Hussain, A.M.; Ashraf, M.; McDowell, L.R. Mineral Status of Soils and Forages in Southwestern Punjab; Pakistan: Micro Minerals. *Asian-Aust. J. Anim. Sci.* **2006**, *19*, 1139–1147. [CrossRef]
27. *NRC: Nutrient Requirements of Small Ruminants: Sheep; Goats; Cervids and New World Camelids*, 1st ed.; National Research Council, National Academy Press: Washington, DC, USA, 2007.
28. Suttle, N.F. *Mineral Nutrition of Livestock*, 4th ed.; CABI Publishing: Wallingford, UK, 2010; p. 595.
29. Kaneko, J.J.; Harvey, J.W.; Bruss, M.L. Appendix. In *Clinical Biochemistry of Domestic Animals*, 5th ed.; Academics Press: Cambridge, MA, USA, 1997; pp. 885–905.
30. Quintero-Moreno, A.; Miranda, S.; López, R.; Dean, D.; Rojas, N.; González, A.; Palomares, R.; Boscan, J. Crecimiento, niveles de calcio; fósforo y magnesio y perfiles séricos de progesterona en corderas prepúberes mestizas west african suplementadas con tres fuentes de minerales. *Rev. Científica FCV-LUZ* **2000**, *3*, 205–211.
31. Underwood, E.J.; Suttle, N.F. *Los Minerales en la Nutrición del Ganado*, 3rd ed.; Editorial Acribia, S.A.: Zaragoza, Spain, 2003; p. 637.
32. Norton, B.W. The nutritive value of tree legumes. In *Forage Tree Legumes in Tropical Agricultura*; Gutteridge, R.C., Shelton, H.M., Eds.; CAB International: Oxon, UK, 1994; pp. 177–191.
33. Fleming, G.A. Mineral composition of herbage. In *Chemistry and Biochemistry of Herbage*; Butler, G.W., Bailey, W., Eds.; Academic Press: London, UK, 1973; pp. 529–566.
34. Herdt, T.H. Ruminant adaptation to negative energy balance. Influences on the etiology of ketosis and fatty liver. *Vet. Clin. N. Am. Food Anim. Pract.* **2000**, *16*, 215–230. [CrossRef]
35. Abarghani, A.; Mostafaei, M.; Alamisaed, K.; Ghanbari, A.; Sahraee, M.; Ebrahimi, R. Investigation of Calcium, Phosphorous and Magnesium Status of Grazing Sheep in Sabalan Region. *Iran. J. Agric. Sci. Tech.* **2013**, *15*, 65–76.
36. Stojković, J. The contents of some mineral elements in the blood serum of sheep in different physiological cycles and with the different level of minerals in a meal. *Biotechnol. Anim. Husb.* **2009**, *25*, 979–984.
37. Masters, D.G.; Purser, D.B.; Yu, S.X.; Wang, Z.S.; Yang, R.Z.; Liu, N.; Lu, D.X.; Wu, L.H.; Ren, J.K.; Li, G.H. Mineral nutrition of grazing sheep in Northern China. 1 Macro-minerals on pasture feed supplement and sheep. *Asian-Aust. J. Anim. Sci.* **1993**, *6*, 99–105. [CrossRef]
38. Alvaro Rincón, C. Maní forrajero (*Arachis pintoi*), la leguminosa para sistemas sostenibles en producción agropecuaria. *Corpoica Inf. Técnica Año* **1999**, *3*, 1–8.
39. Vitti, D.M.S.S.; Kebreab, E.; Lopes, J.B.; Abdalla, A.L.; De Carvalhos, F.F.R.; De Resende, K.T.; Crompton, L.A.; France, J. A kinetic model of phosphorus metabolism in growing goats. *J. Anim. Sci.* **2000**, *78*, 2706–2712. [CrossRef]
40. Dias, R.S.; Lopes, S.; Silva, T.; Pardo, R.M.P.; Silva Filho, J.C.; Vitti, D.M.S.S.; Kebreb, E.; France, J. Rumen phosphorus metabolism in sheep. *J. Agric. Sci.* **2009**, *147*, 391–398. [CrossRef]
41. Braithwaite, G.D. Studies on the absorption and retention of calcium and phosphorus by young and mature Ca-deficient sheep. *Br. J. Nutr.* **1975**, *34*, 311–324. [CrossRef] [PubMed]
42. Pinares-Patiño, C.S.; McEwan, J.C.; Dodds, K.G.; Cárdenas, E.A.; Hegarty, R.S.; Koolaard, J.P.; Clark, H. Repeatability of methane emissions from sheep. *Anim. Feed Sci. Technol.* **2011**, *166*, 210–218.
43. Pfeffer, E.; Beede, D.K.; Valk, H. Phosphorus metablolism in ruminants and requirements of cattle. In *Nitrogen and Phosphorus Nutrition of Cattle and the Environment*; Pfeffer, E., Hristov, A., Eds.; CAB International: Wallingford, UK, 2005; pp. 195–231.

8

The Effects of Fungal Feed Additives in Animals

Wen Yang Chuang [1], Yun Chen Hsieh [1] and Tzu-Tai Lee [1,2,*]

[1] Department of Animal Science, National Chung Hsing University, Taichung 402, Taiwan; xssaazxssaaz@yahoo.com.tw (W.Y.C.); richard840909@gmail.com (Y.C.H.)

[2] The iEGG and Animal Biotechnology Center, National Chung Hsing University, Taichung 402, Taiwan

* Correspondence: ttlee@dragon.nchu.edu.tw.

Simple Summary: Fungal probiotics and ferments have potential as feed additives, but their use has long been ignored. The main goal of this review article is to report on the potential benefits and hazards of fungal feed additives. Previous research indicates that fungal feed additives enhance antioxidant capacity and decrease the inflammatory response in animals through polysaccharides, triterpenes, polyphenols, ergosterol, and adenosine. Accordingly, fungal feed additives could further enhance growth performance and animal health and could be of functional use.

Abstract: As probiotics, fungi enhance animal health and are suitable animal feed additives. In addition to brewing fungi, there are also edible and medicinal fungi. Common fungi utilized in feeding programs include *Saccharomyces cerevisiae, Aspergillus oryzae, Pleurotus* spp., *Antrodia cinnamomea*, and *Cordyceps militaris*. These fungi are rich in glucans, polysaccharides, polyphenols, triterpenes, ergosterol, adenosine, and laccases. These functional components play important roles in antioxidant, anti-inflammatory, anti-obesity, and immune system regulation. As such, fungal feed additives could be of potential use when breeding livestock. In previous studies, fungal feed additives enhanced body weight and egg production in poultry and improved the feed conversion rate. Several mycotoxins can be produced by hazardous fungi but fortunately, the cell walls constituents and enzymes of fungal probiotics can also act to decrease the toxicity of mycotoxins. Overall, fungal feed additives are of value, but their safety and usage must be studied further, including cost-benefit economic analyses.

Keywords: fungi; probiotic; feed additive; mushroom waste compost

---

1. Introduction

Fungi, as ancient eukaryotes, have been present on Earth for at least 2.4 billion years [1], and include 120,000 different species [2]. Fungi have traditionally been classified according to their morphology but can now be identified by DNA. The existence of fungi is inseparable from human history and development. Fungi can be classified as either edible or fermenting, based on their usage. The former includes common fungi such as *Pleurotus eryngii, Flammulina velutipes, P. ostreatus*, and mushrooms; the latter includes probiotics such as *Aspergillus oryzae*. Currently, although studies have shown that *Pennisetum* can be used as a culture medium for edible fungi, most are still planted in space packages filled with wood chips [3,4]. Mature fungi are rich in minerals, vitamins, and probiotics and, due to their powerful antioxidative functions, are increasingly attractive to consumers [5]. However, due to the widespread cultivation of mushrooms, increasing attention has been paid to the treatment of mushroom waste compost, the residual culture medium [4,6].

Saccharomyces cerevisiae and *A. oryzae* have been integrated into human life since ancient times. The former is related mainly to alcohol production, while the latter is involved in the fermentation process of soy sauce. Chuang et al. [7] and Khempaka et al. [8] both pointed out that, following fermentation by *A. oryzae* or *Saccharomyces cerevisiae*, neutral detergent fibers decrease and hemicellulose and crude protein content increase in the fermented product. In addition, our research shows that *S. cerevisiae* and *A. oryzae* can be used as feed additives and assist in reducing the inflammatory response in animals [7].

In addition to probiotics and edible mushrooms, the earliest discovered antibiotic—penicillin—is also produced by fungi (*Penicillium*). With the discovery and use of antibiotics, animals can grow rapidly and be free of many diseases. However, the abuse of antibiotics in recent years has caused many microorganisms to develop resistance, therefore reducing antibiotics in the diet has become increasingly important. Since 2006, the European Union has banned antibiotic use in animal feed (Commission Implementing Regulation (EU) 1831/2003 and 1463/2004, European Parliament) and other countries are also heading in this direction. In order to reduce the use of antibiotics, beneficial fungi whose ferments do not produce antibiotics or mycotoxins have been investigated, due to their increased antioxidant capacity [9].

Mycotoxins are fungal toxins often found in improperly preserved feed whose negative effects on animals include decreased growth performance and intestinal and liver damage [10,11]. However, other fungal probiotics could reduce damage by encasing mycotoxins in their cell walls or even degrading them [7,12,13].

Although most of the discussion on probiotics has focused on bacteria, fungi have considerable potential as probiotics and have been ignored for too long. Over the past five years, our research team has investigated the effects of fungal feed supplements on poultry health [4,7,9,14–21]. Although there have been many studies on the use of fungi and their ferments as feed supplements and their effect on animal health, most review articles have focused on the addition of probiotics or the effects of mycotoxins. This review aims instead to explore the use of fungal feed additives and their potential risks.

2. Edible Fungi and Their Potential Uses

There are more than 2000 known species of edible fungi worldwide, several of which have been commercially cultivated, including medicinal fungi. Common medicinal fungi include *Cordyceps* spp. and *Antrodia* spp., while edible fungi include *Pleurotus* spp., *Lentinula* spp., *Agaricus* spp., and *Flammulina* spp. *Cordyceps militaris*, *Pleurotus eryngii*, *Pleurotus sajor-caju*, and *Flammulina velutipes* are 4 fungi renowned for their nutritional benefits to animals.

Pleurotus eryngii, *Pleurotus sajor-caju* and *Flammulina velutipes* are the most commonly cultured mushrooms in Taiwan (Taiwan Agricultural Research Institute, Council of Agriculture, Executive Yuan). According to the estimates of the Taiwan Agricultural Research Institute, the annual production volume of fresh mushrooms is 140,000 tons, and the output value exceeds 10 billion new Taiwan dollars. It is therefore important to exploit any added value of mushroom usage. Although wild *Cordyceps sinensis* has many functional components and strong biological activity, it takes a long time to grow under strict conditions in order to form fruiting bodies and its yield is insufficient for human needs. Furthermore, *Cordyceps sinensis* cannot be cultivated artificially and can only be harvested once. In contrast, *Cordyceps militaris* only takes one to three months to mature and therefore has greater potential for development and use. In addition, *Cordyceps militaris* can be cultivated in media and has functional components similar to *Cordyceps sinensis*, garnering it greater attention [22]. *Cordyceps militaris* is rich in a variety of bioactive compounds, such as cordycepin, polysaccharides, ergosterol, and mannitol [23]. Some studies have also extracted polysaccharides with antioxidative or immunomodulatory activities from *Cordyceps militaris* [24–26].

Antrodia cinnamomea, an endemic species in Taiwan, has been used for centuries due to its high antioxidative and anti-inflammatory capacities [27,28]. The major functional components in *A. cinnamomea* are polysaccharides, triterpenes, sterols, benzenoids, benzoquinone derivatives, and maleic acid. [29]. These components could enhance antioxidative capacities and reduce the damage from tumors [30]. According to Lee et al. [20], the addition of 0.1%, 0.2%, or 0.4% *A. cinnamomea* powder not only decreased the coliform count and increased lactobacilli in broiler intestines, but it also enhanced the antioxidative capacity in serum. As a result, *A. cinnamomea* was also considered as a potential feed additive in chicken.

Pleurotus eryngii contains many functional ingredients, such as polysaccharides and peptides, is high in fiber, and has high anti-inflammatory and antioxidative capacities [31]. Widely used by humans as an important edible mushroom [32], the whole plant can be processed and is used widely as a health food [32]. *Pleurotus sajor-caju*, a mushroom similar to *Pleurotus eryngii*, is low in calories with very low lipid and starch levels but is rich in protein, fiber, minerals, and vitamins [5]. *Pleurotus sajor-caju* has a variety of biologically active compounds, including polysaccharides, phenols, terpenes, and sterols. [5]. It also has antiviral, antibacterial, antifungal, antioxidative, and anti-inflammatory activities, and is therefore widely used in traditional medicine and nutritional research [33].

In addition to its delightful aroma and taste, *Flammulina velutipes* has a variety of pharmacological properties [34]. It is also well known for its curative properties in liver disease and gastroenteric ulcers [35]. Yang et al. [36] indicated that the extractable polysaccharides in *Flammulina velutipes* have high antioxidative capacities, while the mushroom itself has anti-cancer, antibacterial, and immunomodulatory properties [36]. Based on these characteristics, *Flammulina velutipes* is very promising for further research. However, no matter the species, one must keep in mind that when the mature mushroom is removed, the waste compost remains, potentially causing new environmental problems [4]. Previous research shows that mushroom byproducts have a similar nutritional composition as edible parts [37,38]. As such, based on its low cost, rich nutritional composition, large supply, and numerous biologically active functions, it has potential as a feed additive.

3. Hazardous Fungi Species

Mycotoxins are secondary metabolites produced by some pathogenic fungi. Mycotoxins can affect animal health and are one of the main detriments to growth [10]. Unfortunately, corn and soybeans, the main ingredients in animal feed, are prone to mold when improperly stored and may accumulate mycotoxins [11]. Common mycotoxins include aflatoxins, citrinin, ochratoxin, T-2 toxin, and vomitoxin [39]. Mycotoxins reduce growth performance in animals and are also nephrotoxic, hepatotoxic, and neurotoxic [10,40]. At present, the harm from mycotoxins can be reduced by mold adsorbents, mycotoxin-degrading enzymes, and probiotic degradation [12,40]. Interestingly, although produced by pathogenic fungi, mycotoxins can be treated with yeast cell walls or enzymes produced by edible fungi [12,40]. The main components in yeast cell walls, glucan, and mannan, are mycotoxin-adsorbing substances, while laccase and oxidase produced by *Pleurotus eryngii* reduce the toxicity of mycotoxins [12,13]. The main concern regarding the use of mushroom waste compost is whether its culture medium is conducive to mold growth and could produce mycotoxins. Fortunately, according to Loi et al. [12] and Haque et al. [13], the mushroom mycelium and enzymes contained in waste compost are not conducive to mycotoxin production. In addition, under proper storage conditions (including rapid drying and crushing), the risk of mycotoxins in mushroom waste compost can be further minimized.

4. The Functional Components of Fungi

During fungus cultivation, several enzymes assist the fungi in degrading and utilizing medium nutrients [7]. During this process, many fungal metabolites, secondary metabolites, and synthetics are produced [5,41] (Figure 1). These substances include fungal cell walls, polysaccharides, ergosterol, adenosine, and triterpenes. [5]. In addition to anti-inflammatory, antimicrobial, anti-fatigue, and anti-malarial effects, these functional substances can also protect the lungs and liver and enhance immunity and antioxidant capacity [41].

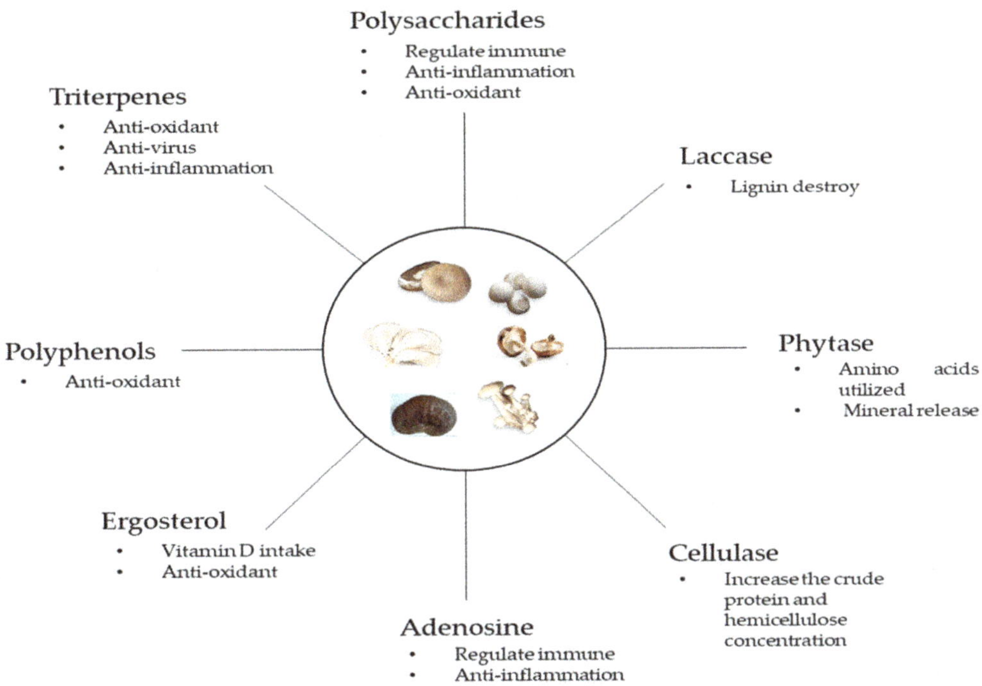

Figure 1. Fungal compounds and their potential functions. Mushroom images were adapted from Du et al. [42], and the photograph of *Ganoderma* was taken by the author.

4.1. *Triterpenes*

Triterpenes are functional sterol metabolites commonly found in mushrooms, that are mainly present in lanostane [5,43]. However, although Chuang et al. [4] reported that triterpenes can be found in mushroom waste compost, most research has focused on triterpenes in medicinal fungi. Mushroom triterpenes can inhibit α-glucosidase activity [44] and thereby reduce blood sugar concentration in animals [45]. Triterpenoids present in *Ganoderma lucidum* can reduce LPS/ D-Galactosamine-induced liver damage by reducing tumor necrosis factor-alpha (TNF-α) and interleukin-6 (IL-6) expression [46]. These triterpenes can also inhibit TLR4-MyD88 (Toll-like receptor 4-myeloid differentiation primary response 88) activity and therefore decrease NF-κB expression [46,47]. In addition, Choi et al. [48] indicated that the anostane triterpene content in *G. lucidum* enhanced heme oxygenase-1 expression and decreased the inflammatory response in RAW264.7 cells. Novak [49] reported on the antibacterial activity of diterpenes in *P. mutilis*, while Ma et al. [43] described the functionality of other terpenoids in mushrooms, including antioxidant and anti-virus capacities.

4.2. *Polyphenols and Flavonoids*

Polyphenols are functional substances commonly found in nature, especially plants and fungi [50,51], with flavonoids being among the most effective [50]. Gil-Ramirez et al. [52] indicated there were no flavonoids in mushrooms, due to the absence of flavonoid absorption and synthesis enzymes. This claim was challenged four years later by Mohanta [53], who pointed to evidence that a large number of fungi contain genes and enzymes related to flavonoid synthesis.

According to Gil-Ramirez et al. [52], it comes down to four points: (1) the traditional method of measuring flavonoid content via absorbance is wrong; (2) flavonoids in the medium cannot be absorbed by the mushrooms; (3) mushrooms do not have genes or enzymes related to flavonoid synthesis; and (4) something went wrong in about 9% of the reports that used HPLC analysis to determine flavonoid content in mushrooms. However, although Gil-Ramirez et al. [52] provided sufficient evidence to prove the first and second points, the third point was challenged by Mohanta [53]. In addition, there is no evidence for the fourth point, only the authors' speculations. For these reasons, we agree that we should avoid "considering fungi as plants" but the components of fungi should not be ignored. Therefore, without further scientific proof to the contrary, we assume fungi contain flavonoids.

The feed additives used with animals are usually agriculture byproducts or ferments. It was confirmed that fermented products, including waste compost and fungi-fermented products, contain high amounts of flavonoids, which come from the plant-based medium [4]. The phenolic components in fungi play important roles in the antioxidant capacities of animals [51,54–56]. It is well known that phenols can chelate free radicals and decrease oxidative stress in animal cells [57,58]. In vivo, phenols enhance antioxidant systems, such as Nrf-2 and glutamate-cysteine ligase catalytic (GCLC) expression, and therefore improve lipid metabolism and meat quality [4,20,59,60].

4.3. Ergosterol

First discovered in yeast, ergosterol is commonly found in fungi and has anti-inflammatory effects [5]. Kuo et al. [61] found that ergosterol inhibits the expression of RAW264.7 NF–κB induced by lipopolysaccharides (LPS) and reduces the inflammatory response of TNF-α and other substances (IC50 = 24.5 μg/mL). Ergosterol inhibits the performance of cyclooxygenase-2 (COX-2) but has no effect on nitric oxide production [61]. Not only can ergosterol reduce the inflammatory response in animals, but it is also converted into vitamin D following exposure to ultraviolet light [62]. Additional vitamin D intake increases serum levels in animals [63] and may therefore enhance antioxidant and fat metabolism capabilities.

4.4. Adenosine

Adenosine, an end product from the breakdown of ATP, is common in fungi and animal cells [64,65]. It regulates energy utilization in animal cells and reduces the inflammatory response. Adnosine binds to G proteins (A1, A2A, A2B, and A3) to regulate the immune capacity and inhibit the production of harmful enzymes and proteins [64,66]. When combined with different G protein receptors, adenosine improves the growth of macrophages (A1), suppresses the inflammatory response by inhibiting Th1 activities and promoting Th2 activities (A2A), upregulates Th2 activities (A2B), and suppresses immunity (A3) [66]. In *Cordyceps militaris*, there is another special adenosine, 3′-deoxyadenosine, also known as cordycepin [65]. Cordycepin upregulates pro-apoptosis genes such as P53, BCL2-associated X protein, caspase-3, and caspase-9, while downregulating antioxidant gene expression. Apoptosis is triggered by the destruction of mitochondria in the tumor cell, thereby inhibiting the growth of cancer cells in the brain [67,68].

4.5. Fungal Cell Wall and Polysaccharides

The fungal cell wall consists mainly of β-glucan, mannoprotein, and chitin [69]. Of these, the first two are very effective at coating LPS and mycotoxins and can reduce their damage [69–71]. In addition, β-glucan, mannoprotein, and chitin can regulate immunity, increase antioxidant capacity, and reduce the inflammatory response [70,72]. A common fungal polysaccharide is yeast cell walls.

Yeast cell walls are dominated by mannan-oligosaccharides and β-glucans, with less chitin [70]. Mannan-oligosaccharides and β-glucans reduce the harm of ochratoxin A to broilers and regulate their immunity [70]. β-glucans also reduce the inflammatory response and the harm from LPS [73,74]. The concentration of polysaccharides in *Cordyceps*, as with other functional metabolites, is significantly affected by medium composition (moisture, geographical environment, and climate) [37,75]. In general, *Cordyceps militaris* contains about 3–8% of dry matter and is stored in fruiting bodies and mycelium in solid and liquid media [37,75]. *Cordyceps* polysaccharides enhance immune cell phagocytosis, improve humoral and cell immunity, enhance the activity of macrophages, monocytes, and lymphocytes, and reduce oxidative stress by enhancing the activity of SOD and Gpx. In addition, crude extracts of *Cordyceps militaris* can reduce the amount of lipid oxidation [75].

Polysaccharides derived from *Flammulina velutipes* decrease the pH value in mice intestines and increase both the amount of short-chain fatty acids in mice intestines and serum immunoglobin [34]. In addition to acting as a kind of prebiotic, polysaccharides in edible mushrooms are also well known for their antioxidant and anti-inflammatory capacities [76–81]. In mice, the polysaccharides in *L. edodes* and *Phallus atrovolvatus* decrease TNF-α, IL-6, IL-1β, blood urea nitrogen, and uric acid levels in blood serum while increasing antioxidant enzymes such as SOD, CAT, and Gpx in the kidney [77,79]. Accordingly, due to the antioxidant and anti-inflammatory capacities of their polysaccharides, mushrooms have potential as a feed additive which improves animal health.

4.6. Enzymes

Fungi produce multiple enzymes during cultivation, which can be divided into three categories: peroxidases, carbohydrases, and phytases. Peroxidases assist fungi in facing the challenge of toxic phenols and increase their environmental competitiveness [80]. Laccase, one of the most well-known peroxidases in mushrooms, degrades lignin in the culture medium [81]. By degrading lignin, fungi effectively use nutrients that are difficult for other microorganisms to consume [82]. Based on this, the fermentation of plant-based feed supplements can reduce lignin content and increase the concentration of hemicellulose [7,8]. Fang et al. [83] further indicated that fungi could transform fiber into volatile fatty acids as carbohydrate enzymes such as laccase efficiently destroyed the structure of lignin.

Once laccase destroys the structure of lignin, cellulase, and hemicellulase further degrade the fiber in plant-based ingredients. As such, fungi reduce the concentration of difficult-to-digest fiber and produce more hemicellulose-based prebiotics in the fermented medium [7,8]. In previous research, cellulose enzyme groups produced by fungi included cellulase, β-glucanase, xylanase, and β-glucosidase [7,84]. In addition to degrading fiber in the culture medium and providing a carbon source for fungal growth, these enzymes increase crude protein and hemicellulose concentrations in plant-based ingredients following fungal fermentation thanks to the destruction of cellulose and condensed nutrients [7,8]. These additional nutrients can be further utilized by animals, suggesting that fungi-fermented plant-based ingredients have potential as feed additives [4,7].

Another common fungal enzyme is phytase. Phytase degrades phytate in plant materials and increases the animal's use of amino acids and minerals, especially phosphorus [85]. In general, phytase is found in *Aspergillus niger* and *Escherichia coli* [86]. However, Wanzenbock et al. [87] indicated that the phytate content in wheat bran decreased after *P. eryngii* fermentation, which reveals potential phytase activities in *P. eryngii*. Phytase significantly increases the use of feed nutrients by animals and reduces the excretion of inorganic phosphorus [88]. Interestingly, co-fermentation of phytase with fungal microorganisms enhances the efficiency of phytase by increasing the release of phytate from plant-based ingredients [7]. This indirectly reduces the harm to the environment when raising livestock [7,88].

5. The Potential Use of Fungal Feed Additives

T.T. Lee's research team has discussed in detail the effects of fungal feed additives on poultry health. Fungal species investigated include *S. cerevisiae, A. oryzae, A. cinnamomea, Trichoderma pseudokoningii, Cordyceps militaris, Pleurotus ostreatus, Pleurotus eryngii,* and *Aureobasidium pullulans*. Additives were added directly to the diet or a portion of an ingredient was replaced with either pure probiotic powder, the fermented product, or mushroom waste compost. Fungal feed additives may enhance animal production performance; the results are listed in Table 1.

Table 1. The effects of fungal feed additives on animal growth [1].

Animal Type	Treatment	Rearing Period	Body (Egg) Weight	Feed Conversion Rate	References
Broilers	0.5% *A. pullulans* fermentation [2]	22–35 d 1–35 d	+12% * +8% *	−4% −3%	[14]
Hendrix laying hens	2% *C. militaris* wastes residue	5–8 weeks 9–12 weeks 0–12 weeks	+4% * +3% * +7% *	−10% * −5% * −6% *	[15]
White Roman geese	5% mushroom waste compost	5–8 weeks 9–12 weeks	+2% −1%	0% +31%	[16]
Male broilers	0.1% *T. pseudokoningii* powder	1–21 d 22–35 d	+10% * +6%	−4% −3%	[18]
Broilers	10% *P. eryngii* mushroom compost fermented wheat bran	1–21 d 22–35 d 1–35 d	+2% +3% +2%	−8% 0% −3%	[6]
Male broilers	0.1% *A. cinnamomea*	1–21 d 22–35 d 1–35 d	+14% * +10% * +11% *	−1% −5% −7%	[20]
Male broilers	10% *A. cinnamomea* fermented wheat bran	1–21 d 22–35 d 1–35 d	+7% * −1% −2%	−5% * +1% +2%	[21]
Male broilers	0.5% mushroom waste compost	1–21 d 22–35 d 1–35 d	+3% +5% +4%	−6% * +5% −7%	[4]

[1] All percentage changes calculated by the difference between the treatment group and control group. [2] 75% *Aureobasidium pullulans* fermented soybean hulls in combination with 25% *Pleurotus eryngii* stalk residue * Significant difference between control and treatment group ($p < 0.05$).

In addition to improving production performance, fungal feed additives could further enhance animal health (Table 2). Both *S. cerevisiae* and *A. oryzae* are ancient probiotics used in the brewing industry. However, they could also enhance broiler health as feed additives. Chuang et al. [7] indicated that the product from co-fermenting 0.1% *S. cerevisiae* or *A. oryzae* with phytase could reduce the inflammatory response in broilers by decreasing the number of *Clostridium perfringens* in the ileum and suppressing inflammation-related mRNA expression, including NF-κB and iNOS. Furthermore, the *A. oryzae* and phytase co-fermented product enhanced villus height in the jejunum. According to the research of Teng et al. [19], villus height and lactic acid concentration in the ileum increased significantly after supplementation with 10% *S. cerevisiae* wheat bran. Previous research revealed that *A. oryzae* contains multiple enzymes, especially cellulase and hemicellulase, which improves the fiber type in feed and enhances nutrient absorption [8,89,90]. A specific kind of yeast, *Aureobasidium pullulans* increases the hemicellulose concentration of soybean hulls and *Pleurotus eryngii* stalk residue by about 1.5 times following fermentation [60]. *Aureobasidium pullulans* fermented soybean hulls and *Pleurotus eryngii* stalk residue could also decrease the ammonia nitrogen concentration in the cecum [14]. Furthermore, Lai et al. [14] indicated that the *Aureobasidium pullulans* ferment contained high amounts of xylanase and mannanase, and these enzymes proportionally increased relative to the days of fermentation.

Table 2. The effects of fungal feed additives on animal health.

Animal Type	Treatment	Functional Components	Functions	References
Broilers	0.5% *A. pullulans* ferment [1]	- [2]	Increased SOD and CAT activities, ileum villus height and lactic acid bacteria number in cecum	[14]
Hendrix laying hens	2% *C. militaris* waste residue	Cordycepin, cordycepic acid, crude polysaccharides, flavonoid, adenosine, and crude triterpenoid	Increased egg mass and eggshell strength; improved feed conversion rate throughout the entire experimental period; decreased cholesterol content in egg yolk	[15]
White Roman geese	5% mushroom waste compost	-	Increased SOD activities and decreased MDA content in serum; improved flavor, color, and acceptability on sensory evaluation	[16]
Male broilers	10% *T. pseudokoningii* fermented wheat bran	Cellulase, xylanase, and reducing sugar	Decreased coliform count and increased villus height in ileum	[17]
Male broilers	0.1% *T. pseudokoningii* powder	Phenols, xylanase, and cellulase	Increased SOD and CAT activities, jejunum villus height, and lactic acid bacteria levels in cecum	[18]
Broilers	10% *S. cerevisiae* fermented wheat bran	-	Increased villus height and lactic acid content in ileum	[19]
Male broilers	0.4% *A. cinnamomea* addition	Crude triterpenoids, crude polysaccharides, and phenols	Enhanced Nrf-2, GCLC, SOD, CAT and HO-1 mRNA expression and decreased NF-κB and IL-1β mRNA expression	[20]
Male broilers	10% *A. cinnamomea* fermented wheat bran	Crude triterpenoids, crude polysaccharides, and phenols	Enhanced SOD and CAT activities and decreased total cholesterol and low-density lipoprotein content in serum	[21]
Male broilers	0.1% *S. cerevisiae* or *A. oryzae powder*	Xylanase, protease, cellulase, and ß-glucanase	Decreased the number of *Clostridium perfringens* in ileum and suppressed inflammation-related mRNA expression	[7]
Male broilers	0.5% mushroom waste compost	Crude triterpenes, phenols, flavonoids, gallocatechin, and epigallocatechin	Increased antioxidant capacity, adipolysis, and gut barrier expression	[4]
Female broilers	1% *C. militaris* waste residue	Polysaccharides, triterpenes, phenols, and flavonoids	Enhanced antioxidant-related mRNA expression	[9]

[1] 75% *Aureobasidium pullulans* fermented soybean hulls in combination with 25% *Pleurotus eryngii* stalk residue;
[2] The symbol "-" indicates the author did not mention the functional components.

Among fungal microbes, there is one special species: *Trichoderma pseudokoningii*. Although *T. pseudokoningii* is not a legal feed additive, its high potential in the poultry industry was reported by both Lin et al. [18] and Chu et al. [17]. Supplementation with 10% *T. pseudokoningii* fermented wheat bran enhanced the villus height, thereby increasing the villus height:crypt depth ratio, and the coliform count in the ileum decreased [17]. The addition of 0.1%, 0.2%, or 0.4% *T. pseudokoningii* powder improved animal growth performance in the first 21 days and improved the feed conversion ratio [18]. Furthermore, supplementation with *T. pseudokoningii* powder increased lactate concentration and lactic acid bacteria levels in the cecum [18].

C. militaris and *A. cinnamomea* are both well-known traditional Chinese medicines, due to their antioxidative, anti-inflammatory, and antitumor capacities [9,20]. Lee et al. [20,21] reported on the effects of *A. cinnamomea* powder and fermentation (fermented wheat bran) supplements on antioxidant, anti-inflammation, and fat metabolism in broilers. *A. cinnamomea* powder decreased the coliform count and increased lactobacilli in an in vitro test [20]. Supplementation with 0.1%, 0.2%, or 0.4% *A. cinnamomea* powder enhanced SOD and CAT activities in mRNA as well as protein levels in

21- and 35-day-old animals. The expression of inflammation-related mRNA, such as NF-κB and IL-1β, also decreased [20]. Furthermore, the addition of *A. cinnamomea*-fermented wheat bran to the broiler diet enhanced SOD and CAT activities and reduced total cholesterol and low-density lipoprotein levels in broiler serum [21]. Due to its ease of cultivation, *C. militaris* is more popular than *C. sinensis*. However, as with other types of mushrooms, its waste residue is rich in mycelium and other functional compounds, suggesting it has potential as a feed additive. Wang et al. [15] indicated that supplementation with 0.5%, 1%, or 2% *C. militaris* waste residue increased egg mass and eggshell strength and improved the feed conversion rate throughout the entire experimental period. The cholesterol content in egg yolk also decreased after supplementation with 1% and 2 % *C. militaris* waste residue [15]. Furthermore, Hsieh et al. [9] revealed that *C. militaris* waste residue was about 9 times higher in polysaccharide content than *Pleurotus eryngii, Pleurotus sajor-caju*, or *Fammulina velutipes* waste residues. As such, supplementation with *C. militaris* waste residue could significantly increase antioxidant capacity via activated Keap-1 and Nrf-2 mRNA expression [9].

Another common fungal feed additive is mushroom waste compost. As the residue left after cultivation, mushroom waste compost is high in mycelium and wood fiber, which enhances animal antioxidants and adipolysis [4]. Although the real effect of mushroom waste compost on adipose metabolism is unclear, it appears to increase KCTD15 and adiponectin expression, thereby enhancing adipolysis-related mRNA expression [4]. Interestingly, in recent years the anti-obesity capacity of vitamin D, which is converted from ergosterol, has attracted attention [62]. The addition of vitamin D could enhance antioxidant capacity, decrease the total cholesterol and triglyceride levels in serum, and further enhance the peroxisome proliferator-activated receptor γ (PPARγ) and perilipin-2 expression to enhance adipolysis [91–93]. Bindhu and Arunave [94] also indicated that *Pleurotus ostreatus* and its bioactive anthraquinone could enhance the PPARγ and CCAAT enhancer-binding protein α (CEBPα), thereby decreasing adipose accumulation. This suggests that the anti-obesity capacity of mushrooms may be due to ergosterol and anthraquinone intake.

Several papers point out that mushroom waste compost could enhance antioxidant capacity [4,6,16]. By enhancing Nrf-2 and GCLC mRNA expression, mushroom waste compost could upregulate the antioxidant capacity of poultry and further decrease MDA concentration [4]. Based on its high fiber content, the gut barrier would also be improved by supplementation with mushroom waste compost [4]. Interestingly, Wang et al. [6] indicated that a 10% supplementation of wheat bran fermented by the remaining mycelium in *P. eryngii* mushroom compost could also enhance CAT activities. Overall, whether they're used in traditional Chinese medicine or are common edible species, fungi play an important role in upregulating animal health and could protect against stress. Based on the effects of the different fungal feed additives on growth performance and animal health, supplementation with 0.1% *A. cinnamomea*, 0.5% mushroom waste compost, or 2% *C. militaris* waste residue are recommended [4,15,20]. Furthermore, in addition to avoiding illegal or harmful fungi, it is not recommended to add more than 5% of any fungal feed additive since better results were achieved with a small amount (<2%) (Table 1).

6. Future Perspectives

In general, probiotic discussions have focused on bacterial microorganisms and the potential use of fungi has long been ignored. Through this review, we have tried our best to fully explain the potential uses of fungi. Fungi can enhance antioxidant capacity and fat metabolism in animals and maintain intestinal health. However, pathogenic fungi should be avoided, even if probiotic fungi can offset their negative effects. The most suitable amount depends on the species and type of fungi. Overall, the value of fungi as probiotics or functional feed additives has been highlighted in this review.

Author Contributions: Conceptualization, W.Y.C. and T.-T.L.; supervision, T.-T.L.; investigation, W.Y.C. and T.-T.L.; writing—original draft preparation, W.Y.C., Y.C.H. and T.-T.L.; writing—review and editing, W.Y.C. and T.-T.L. All authors have read and agreed to the published version of the manuscript.

Acknowledgments: The authors thank the Ministry of Science and Technology (MOST 107-2313-B-005-037-MY2) and the iEGG and Animal Biotechnology Center from The Feature Areas Research Center Program within the framework of the Higher Education Sprout Project by the Ministry of Education (MOE) in Taiwan for supporting this study.

Abbreviations

TNF-α	Tumor necrosis factor alpha
Gpx	Glutathione peroxidase
SOD	Superoxidase dismutase
MDA	Malondialdehyde
TLR4	Toll-like receptor 4
MyD88	Myeloid differentiation primary response 88
COX-2	Cyclooxygenase-2
CAT	Catalase
NFκB	Nuclear factor kappa B p 65
iNOS	Inducible nitric oxide synthases
IFN-γ	Interferon-γ
IL-1ß	Interleukin-1ß
Nrf-2	Nuclear factor erythroid 2–related factor 2
GCLC	Glutamate-cysteine ligase catalytic
KCTD-15	Potassium channel tetramerization domain-containing 15
CEBPα	CCAAT-enhancer-binding proteins-alpha
CPT-1	Carnitine palmitoyltransferase I
PPAR-γ	Peroxisome proliferator-activated receptor gamma

References

1. Bengtson, S.; Rasmussen, B.; Ivarsson, M.; Muhling, J.; Broman, C.; Marone, F.; Stampanoni, M.; Bekker, A. Fungus-like mycelial fossils in 2.4-billion-year-old vesicular basalt. *Nat. Ecol. Evol.* **2017**, *1*, 141. [CrossRef] [PubMed]
2. Mueller, G.M.; Schmit, J.P. Fungal biodiversity: What do we know? What can we predict? *Biodivers. Conserv.* **2006**, *16*, 1–5. [CrossRef]
3. Maleko, D.; Mwilawa, A.; Msalya, G.; Pasape, L.; Mtei, K. Forage growth, yield and nutritional characteristics of four varieties of napier grass (*Pennisetum purpureum Schumach*) in the west Usambara highlands. *Afr. Crop Sci. J.* **2019**, *6*, e00214. [CrossRef]
4. Chuang, W.Y.; Liu, C.L.; Tsai, C.F.; Lin, W.C.; Chang, S.C.; Shih, H.D.; Shy, Y.M.; Lee, T.T. Evaluation of waste mushroom compost as a feed supplement and its effects on the fat metabolism and anti-oxidant capacity of broilers. *Animals* **2020**, *10*, 445. [CrossRef] [PubMed]
5. Finimundy, T.C.; Barros, L.; Calhelha, R.C.; Alves, M.J.; Prieto, M.A.; Abreu, R.M.V.; Dilon, A.J.P.; Henriques, J.A.P.; Roesch, M.; Ferreira, I.C.F.R. Multifunctions of *pleurotus sajor-caju* (fr.) singer: A highly nutritious food and a source for bioactive compounds. *Food Chem.* **2018**, *245*, 150–158. [CrossRef] [PubMed]
6. Wang, C.C.; Lin, L.J.; Chao, Y.P.; Chiang, C.J.; Lee, M.T.; Chang, S.C.; Yu, B.; Lee, T.T. Anti-oxidant molecular targets of wheat bran fermented by white rot fungi and its potential modulation of antioxidative status in broiler chickens. *Br. Poult. Sci.* **2017**, *58*, 262–271. [CrossRef] [PubMed]
7. Chuang, W.Y.; Lin, W.C.; Hsieh, Y.C.; Huang, C.M.; Chang, S.C.; Lee, T.T. Evaluation of the combined use of *Saccharomyces Cerevisiae* and *Aspergillus Oryzae* with phytase fermentation products on growth, inflammatory, and intestinal morphology in broilers. *Animals* **2019**, *9*, E1051. [CrossRef]
8. Khempaka, S.; Thongkratok, R.; Okrathok, S.; Molee, W. An evaluation of cassava pulp feedstuff fermented with *A. oryzae*, on growth performance, nutrient digestibility and carcass quality of broilers. *J. Poult. Sci.* **2013**, *51*, 71–79. [CrossRef]
9. Hsieh, Y.C.; Lin, W.C.; Chuang, W.Y.; Chen, M.H.; Chang, S.C.; Lee, T.T. Effects of mushroom waster medium and stalk residues on the growth performance and oxidative status in broilers. *Asian-Australas. J. Anim. Sci.* **2020**. [CrossRef]

10. Maciorowski, K.G.; Herrera, P.; Jones, F.T.; Pillai, S.D.; Ricke, S. C Effects on poultry and livestock of feed contamination with bacteria and fungi. *Anim. Feed Sci. Technol.* **2007**, *133*, 109–136. [CrossRef]
11. Munkvold, G.P.; Arias, S.; Taschl, I.; Gruber-Dorninger, C. Chapter 9—Mycotoxins in corn: Occurrence, impacts, and management. In *Corn*; Elsevier Inc.: Amsterdam, The Netherlands, 2019; pp. 235–287. [CrossRef]
12. Loi, M.; Fanelli, F.; Cimmarusti, M.T.; Mirabelli, V.; Haidukowski, M.; Logrieco, A.F.; Caliandro, R.; Mule, G. In vitro single and combined mycotoxins degradation by Ery4 laccase from *Pleurotus eryngii* and redox mediators. *Food Control* **2018**, *90*, 401–406. [CrossRef]
13. Haque, M.A.; Wang, Y.; Shen, Z.; Li, X.; Saleemi, M.K.; He, C. Mycotoxin contamination and control strategy in human, domestic animal and poultry: A review. *Microb. Pathog.* **2020**, *142*, 104095. [CrossRef] [PubMed]
14. Lai, L.P.; Lee, M.T.; Chen, C.S.; Yu, B.; Lee, T.T. Effects of co-fermented *Pleurotus eryngii* stalk residues and soybean hulls by *Aureobasidium pullulans* on performance and intestinal morphology in broiler chickens. *Poult. Sci.* **2015**, *94*, 2959–2969. [CrossRef] [PubMed]
15. Wang, C.L.; Chiang, C.J.; Chao, Y.P.; Yu, B.; Lee, T.T. Effect of *Cordyceps Militaris* waster medium on production performance, egg traits and egg yolk cholesterol of laying hens. *J. Poult. Sci.* **2015**, *52*, 188–196. [CrossRef]
16. Chang, S.C.; Lin, M.J.; Chao, Y.P.; Chiang, C.J.; Jea, Y.S.; Lee, T.T. Effects of spent mushroom compost meal on growth performance and meat characteristics of grower geese. *R. Bras. Zootec.* **2016**, *45*, 281–287. [CrossRef]
17. Chu, Y.T.; Lo, C.T.; Chang, S.C.; Lee, T.T. Effects of *Trichoderma* fermented wheat bran on growth performance, intestinal morphology and histological findings in broiler chickens. *Ital. J. Anim. Sci.* **2017**, *16*, 82–92. [CrossRef]
18. Lin, W.C.; Lee, M.T.; Lo, C.T.; Chang, S.C.; Lee, T.T. Effects of dietary supplementation of *Trichoderma pseudokoningii* fermented enzyme powder on growth performance, intestinal morphology, microflora and serum anti-oxidantive status in broiler chickens. *Ital. J. Anim. Sci.* **2017**, *17*, 153–164.
19. Teng, P.Y.; Chang, C.L.; Huang, C.M.; Chang, S.C.; Lee, T.T. Effects of solid-state fermented wheat bran by *Bacillus amyloliquefaciens* and *Saccharomyces cerevisiae* on growth performance and intestinal microbiota in broiler chickens. *Ital. J. Poult. Sci.* **2017**, *54*, 134–141. [CrossRef]
20. Lee, M.T.; Lin, W.C.; Wang, S.Y.; Lin, L.J.; Yu, B.; Lee, T.T. Evaluation of potential anti-oxidant and anti-inflammatory effects of *Antrodia cinnamomea* powder and the underlying molecular mechanisms via Nrf2- and NF-κB-dominated pathways in broiler chickens. *Poult. Sci.* **2018**, *97*, 2419–2434. [CrossRef]
21. Lee, M.T.; Lin, W.C.; Lin, L.J.; Wang, S.Y.; Chang, S.C.; Lee, T.T. Effects of dietary *Antrodia cinnamomea* fermented product supplementation on antioxidation, anti-inflammation, and lipid metabolism in broiler chickens. *Asian-Australas. J. Anim. Sci.* **2019**. In press. [CrossRef]
22. Wu, L.; Sun, H.; Hao, Y.; Zheng, X.; Song, Q.; Dai, S.; Zhu, Z. Chemical structure and inhibition on α-glucosidase of the polysaccharides from *Cordyceps militaris* with different developmental stages. *Int. J. Biol. Macromol.* **2020**, *148*, 722–736. [CrossRef] [PubMed]
23. Ng, T.B.; Wang, H.X. Pharmacological actions of cordyceps, a prized folk medicine. *J. Pharm. Pharmacol.* **2005**, *57*, 1509–1519. [CrossRef] [PubMed]
24. Chen, X.; Wu, G.; Huang, Z. Structural analysis and anti-oxidant activities of polysaccharides from cultured *Cordyceps militaris*. *Int. J. Biol. Macromol.* **2013**, *58*, 18–22. [CrossRef] [PubMed]
25. Wang, M.; Meng, X.Y.; Yang, R.L.; Qin, T.; Wang, X.Y.; Zhang, K.Y.; Fei, C.Z.; Li, Y.; Hu, Y.L.; Xue, F.Q. *Cordyceps militaris* polysaccharides can enhance the immunity and antioxidation activity in immunosuppressed mice. *Carbohydr. Polym.* **2012**, *89*, 461–466. [CrossRef]
26. Won, S.Y.; Park, E.H. Anti-inflammatory and related pharmacological activities of cultured mycelia and fruiting bodies of cordyceps militaris. *J. Ethnopharmacol.* **2005**, *96*, 555–561. [CrossRef]
27. Wang, H.C.; Chu, F.H.; Chien, S.C.; Liao, J.W.; Hsieh, H.W.; Li, W.H.; Lin, C.C.; Shaw, J.F.; Kuo, Y.H.; Wang, S.Y. Establishment of the metabolite profile for an *Antrodia cinnamomea* health food product and investigation of its chemoprevention activity. *J. Agric. Food Chem.* **2013**, *61*, 8556–8564. [CrossRef]
28. Yang, F.C.; Yang, Y.H.; Lu, H.C. Enhanced anti-oxidant and antitumor activities of *Antrodia cinnamomea* cultured with cereal substrates in solid state fermentation. *Biochem. Eng. J.* **2013**, *78*, 108–113. [CrossRef]
29. Lee, I.H.; Huang, R.L.; Chen, C.T.; Chen, H.C.; Hsu, W.C.; Lu, M.K. Antrodia camphorate polysaccharides exhibit anti-hepatitis B virus effects. *FEMS Microbiol. Lett.* **2002**, *209*, 63–67. [CrossRef]
30. Wu, H.; Pan, C.L.; Yao, Y.C.; Chang, S.S.; Li, S.L.; Wu, T.F. Proteomic analysis of the effect of Antrodia camphorate extract on human lung cancer A549 cell. *Proteomics* **2006**, *6*, 826–835. [CrossRef]

31. Li, S.; Shah, N.P. Anti-inflammatory and anti-proliferative activities of natural and sulphonated polysaccharides from *Pleurotus eryngii*. *J. Funct. Foods* **2016**, *23*, 80–86. [CrossRef]
32. Zhang, C.; Li, S.; Zhang, J.; Hu, C.; Che, G.; Zhou, M.; Jia, L. anti-oxidant and hepatoprotective activities of intracellular polysaccharide from *Pleurotus eryngii* si-04. *Int. J. Biol. Macromol.* **2016**, *91*, 568–577. [CrossRef] [PubMed]
33. Duru, M.E.; Cayan, G.T. Biologically active terpenoids from mushroom origin: A review. *Rec. Nat. Prod.* **2015**, *9*, 456–483.
34. Zhao, R.; Hu, Q.; Ma, G.; Su, A.; Xie, M.; Li, X.; Chen, G.; Zhao, L. Effects of *Flammulina velutipes* polysaccharide on immune response and intestinal microbiota in mice. *J. Funct. Foods* **2019**, *56*, 255–264. [CrossRef]
35. Tang, C.; Hoo, P.C.; Tan, L.T.; Pusparajah, P.; Khan, T.M.; Lee, L.H.; Goh, B.H.; Chan, K.G. Golden Needle Mushroom: A culinary medicine with evidenced-based biological activities and health promoting properties. *Front. Pharmacol.* **2016**, *7*, 474–501. [CrossRef]
36. Yang, W.; Yu, J.; Zhao, L.; Ma, N.; Fang, Y.; Pei, F.; Mariga, M.; Hu, Q. Polysaccharides from *Flammulina velutipes* improve scopolamine-induced impairment of learning and memory of rats. *J. Funct. Foods* **2015**, *18*, 411–422. [CrossRef]
37. Liu, L.; Chen, Y.; Luo, Q.; Xu, N.; Zhou, M.; Gao, B.; Wang, C.; Shi, Y. Fermenting liquid vinegar with higher taste, flavor and healthy value by using discarded *Cordyceps militaris* solid culture medium. *LWT-Food Sci. Technol.* **2018**, *98*, 654–660. [CrossRef]
38. Mahfuz, S.; Song, H.; Liu, Z.; Liu, X.; Diao, Z.; Ren, G.; Guo, Z.; Cui, Y. Effect of golden needle mushroom (*Flammulina velutipes*) stem waste on laying performance, calcium utilization, immune response and serum immunity at early phase of production. *Asian-Australas. J. Anim. Sci.* **2018**, *31*, 705–711. [CrossRef]
39. Whitlow, L.W.; Hagler, W.M.J. Mycotoxins in Dairy Cattle: Occurrence, Toxicity, Prevention and Treatment. *Proc. Southwest Nutr. Conf.* **2005**, 124–138.
40. Peng, W.X.; Marchal, J.L.M.; vanderPoel, A.F.B. Strategies to prevent and reduce mycotoxins for compound feed manufacturing. *Anim. Feed Sci. Technol.* **2018**, *237*, 129–153. [CrossRef]
41. Das, S.K.; Masuda, M.; Sakurai, A.; Sakakibara, M. Medicinal uses of the mushroom *Cordyceps militaris*: Current state and prospects. *Fitote* **2010**, 961–968. [CrossRef]
42. Du, B.; Zhu, F.; Xu, B. An insight into the anti-inflammatory properties of edible and medicinal mushrooms. *J. Funct. Foods* **2018**, *47*, 334–342. [CrossRef]
43. Ma, G.; Yang, W.; Zhao, L.; Pei, F.; Fang, D.; Hu, Q. A critical review on the health promoting effects of mushrooms nutraceuticals. *Food Sci. Hum. Wellness* **2018**, *7*, 125–133. [CrossRef]
44. Chen, X.Q.; Zhao, J.; Chen, L.X.; Wang, S.F.; Wang, Y.; Li, S.P. Lanostane triterpenes from the mushroom Ganoderma resinaceum and their inhibitory activities against a-glucosidase. *Phytochemistry* **2018**, *149*, 103–115. [CrossRef]
45. Ali, K.M.; Chatterjee, K.; De, D.; Jana, K.; Bera, T.K.; Ghosh, D. Inhibitory effect of hydro-methanolic extract of seed of *Holarrhena antidysenterica* on alpha-glucosidase activity and postprandial blood glucose level in normoglycemic rat. *J. Ethnopharmacol.* **2011**, *135*, 194–196. [CrossRef] [PubMed]
46. Hu, Z.; Du, R.; Xiu, L.; Bian, Z.; Ma, C.; Sato, N.; Hattori, M.; Zhang, H.; Liang, Y.; Yu, S.; et al. Protective effect of triterpenes of *Ganoderma lucidum* on lipopolysaccharide induced inflammatory responses and acute liver injury. *Cytokine* **2020**, *127*, 154917. [CrossRef] [PubMed]
47. Dudhgaonkar, S.; Thyagarajan, A.; Sliva, D. Suppression of the inflammatory response by triterpenes isolated from the mushroom *Ganoderma lucidum*. *Int. Immunopharmacol.* **2009**, *9*, 1272–1280. [CrossRef]
48. Choi, S.; Nguyen, V.T.; Tae, N.; Lee, S.; Ryoo, S.; Min, B.S.; Lee, J.H. Anti-inflammatory and heme oxygenase-1 inducing activities of lanostane triterpenes isolated from mushroom *Ganoderma lucidum* in RAW264.7 cells. *Toxicol. Appl. Pharmacol.* **2014**, *280*, 434–442. [CrossRef]
49. Novak, R. Are pleuromutilin antibiotics finally fit for human use? *Ann. N.Y. Acad. Sci.* **2011**, *1241*, 71–81. [CrossRef]
50. Chiang, Y.M.; Chuang, D.Y.; Wang, S.Y.; Guo, Y.H.; Tsai, P.W.; Shyur, L.F. Metabolite profiling and chemopreventive bioactivity of plant extracts from *Bidens Pilosa*. *J. Ethnopharmacol.* **2004**, *95*, 409–419. [CrossRef]
51. Palacios, I.; Lozano, M.; Moro, C.; Arrigo, M.D.; Rostagno, M.A.; Martinez, J.A.; GarciaLafuente, A.; Guillamon, E.; Villares, A. Anti-oxidant properties of phenolic compounds occurring in edible mushrooms. *Food Chem.* **2011**, *128*, 674–678. [CrossRef]

52. Gil-Ramírez, A.; Pavo-Caballero, C.; Baeza, E.; Baenas, N.; Garcia-Viguera, C.; Marín, F.R.; Soler-Rivas, C. Mushrooms do not contain flavonoids. *J. Funct. Food* **2016**, *25*, 1–13. [CrossRef]
53. Mohanta, T.K. Fungi contain genes associated with flavonoid biosynthesis pathway. *J. Funct. Foods* **2020**, *68*, 103910. [CrossRef]
54. Vaz, J.A.; Barros, L.; Martins, A.; Morais, J.S.; Vasconcelos, M.H.; Ferreira, I.C.F.R. Phenolic profile of seventeen Portuguese wild mushrooms. *Food Sci. Technol.* **2011**, *44*, 343–346. [CrossRef]
55. 56Gasecka, M.; Mleczek, M.; Siwulski, M.; Niedzielski, P. Phenolic composition and anti-oxidant properties of *Pleurotus ostreatus* and *Pleurotus eryngii* enriched with selenium and zinc. *Eur. Food Res. Technol.* **2016**, *242*, 723–732. [CrossRef]
56. Sun, Y.; Zhang, M.; Fang, Z. Efficient physical extraction of active constituents from edible fungi and their potential bioactivities: A review. *Trends Food Sci. Technol.* **2019**. In press. [CrossRef]
57. Yan, J.; Zhao, Y.; Suo, S.; Liu, Y.; Zhao, B. Green tea catechins ameliorate adipose insulin resistance by improving oxidative stress. *Free Radic. Biol. Med.* **2012**, *52*, 1648–1657. [CrossRef]
58. Guo, H.; Saravanakumar, K.; Wang, M.H. Total phenolic, flavonoid contents and free radical scavenging capacity of extracts from tubers of *Stachys affinis*. *Biocatal. Agric. Biotechnol.* **2018**, *15*, 235–239. [CrossRef]
59. Starčević, K.; Krstulović, L.; Brozić, D.; Maurić, M.; Stojević, Z.; Mikulec, Ž.; Bajić, M.; Mašek, T. Production performance, meat composition and oxidative susceptibility in broiler chicken fed with different phenolic compounds. *J. Sci. Food Agric.* **2015**, *95*, 1172–1178. [CrossRef]
60. Lee, M.T.; Lai, L.P.; Lin, W.C.; Ciou, J.Y.; Chang, S.C.; Yu, B.; Lee, T.T. Improving nutrition utilization and meat Quality of broiler chickens through solid-state fermentation of agricultural by-products by *Aureobasidium Pullulans*. *Rev. Bras. Cienc. Avic.* **2017**, *19*, 645–654. [CrossRef]
61. Kuo, C.F.; Hsieh, C.H.; Lin, W.Y. Proteomic response of LAP-activated RAW 264.7 macrophages to the anti-inflammatory property of fungal ergosterol. *Food Chem.* **2011**, *126*, 207–212. [CrossRef]
62. Papoutsis, K.; Grasso, S.; Menon, A.; Brunton, N.P.; Lyng, J.G.; Jacquier, J.C.; Bhuyan, D.J. Recovery of ergosterol and vitamin D2 from mushroom waste - Potential valorization by food and pharmaceutical industries. *Trends Food Sci. Technol.* **2020**, *99*, 351–366. [CrossRef]
63. Baur, A.C.; Kühn, J.; Brandsch, C.; Hirche, F.; Stangl, G.I. Intake of ergosterol increases the vitamin D concentrations in serum and liver of mice. *J. Steroid Biochem.* **2019**, *194*, 105435. [CrossRef] [PubMed]
64. Caiazzo, E.; Maione, F.; Morello, S.; Lapucci, A.; Paccosi, S.; Steckel, B.; Lavecchia, A.; Parenti, A.; Iuvone, T.; Schrader, J.; et al. Adenosine signaling mediates the anti-inflammatory effects of the COX-2 inhibitor nimesulide. *Biochem. Pharmacol.* **2016**, *112*, 72–81. [CrossRef]
65. Chiu, C.P.; Hwang, T.L.; Chan, Y.; El-Shazly, M.; Wug, T.Y.; Lo, I.W.; Hsu, Y.M.; Lai, K.H.; Houh, M.F.; Yuani, S.S.; et al. Research and development of *Cordyceps* in Taiwan. *Food Sci. Hum. Wellness* **2016**, *5*, 177–185. [CrossRef]
66. Haskó, G.; Antonioli, L.; Cronstein, B.N. Adenosine metabolism, immunity and joint health. *Biochem. Pharmacol.* **2018**, *151*, 307–313. [CrossRef]
67. Nakamura, K.; Shinozuka, K.; Yoshikawa, N. Anticancer and antimetastatic effects of cordycepin, an active component of *Cordyceps sinensis*. *J. Pharma. Sci.* **2014**, *127*, 53–56. [CrossRef]
68. Chaicharoenaudomrung, N.; Jaroonwitchawan, T.; Noisa, P. *Cordycepin* induces apoptotic cell death of human brain cancer through the modulation of autophagy. *Toxicol. In Vitro* **2018**, *46*, 113–121. [CrossRef]
69. Rodríguez, A.; Cuesta, A.; Ortuño, J.; Esteban, M.A.; Meseguer, J. Immunostimulant properties of a cell wall-modified whole *Saccharomyces cerevisiae* strain administered by diet to seabream (*Sparus aurata* L.). *Vet. Immunol. Immunopathol.* **2003**, *96*, 183–192. [CrossRef]
70. Awaad, M.; Atta, A.M.; El-Ghany, W.A.; Elmenawey, M.; Ahmed, A.; Hassan, A.A.; Nada, A.; Abdelaleem, G.A.; Kawkab, A.A. Effect of a specific combination of mannan-oligosaccharides and β-glucans extracted from yeast cell wall on the health status and growth performance of ochratoxicated broiler chickens. *J. Anim. Sci.* **2011**, *7*, 82–96.
71. Al-Khalaifah, H.S. Benefits of probiotics and/or prebiotics for antibiotic-reduced poultry. *Poult. Sci.* **2018**, *11*, 3807–3815. [CrossRef]
72. Carrasco-González, J.A.; Serna-Saldívar, S.O.; Gutierrez-Uribe, J.A. Mycochemical changes induced by selenium enrichment in *P. ostreatus* fruiting bodies. *J. Agric. Food Chem.* **2017**, *65*, 4074–4082. [CrossRef] [PubMed]

73. Penney, J.; Lu, Y.; Pan, B.; Feng, Y.; Walk, C.; Li, J. Pure yeast beta-glucan and two types of yeast cell wall extracts enhance cell migration in porcine intestine model. *J. Funct. Foods* **2019**, *59*, 129–137. [CrossRef]
74. Peng, Q.H.; Cheng, L.; Kang, K.; Tian, G.; Mohammad, A.M.; Xue, B.; Wang, L.Z.; Zou, H.W.; Mathew, G.G.; Wang, Z.S. Effects of yeast and yeast cell wall polysaccharides supplementation on beef cattle growth performance, rumen microbial populations and lipopolysaccharides production. *J. Integr. Agric.* **2020**, *19*, 810–819. [CrossRef]
75. Nie, S.; Cui, S.; Xie, M. Bioactive polysaccharides from *Cordyceps sinensis*: Isolation, structure features and bioactivities. *Bioact. Carbohydr. Diet. Fibre.* **2017**, *1*, 38–52. [CrossRef]
76. Zhang, L.; Hu, Y.; Duan, X.; Tang, T.; Shen, Y.; Hu, B.; Liu, A.; Chen, H.; Li, C.; Liu, Y. Characterization and anti-oxidant activities of polysaccharides from thirteen boletus mushrooms. *Int. J. Biol. Macromol.* **2018**, *113*, 1–7. [CrossRef] [PubMed]
77. Song, X.; Ren, Z.; Wang, X.; Jia, L.; Zhang, C. Anti-oxidant, anti-inflammatory and renoprotective effects of acidichydrolytic polysaccharides by spent mushroom compost (*Lentinula edodes*) on LPS-induced kidney injury. *Int. J. Biol. Macromol.* **2019**. [CrossRef]
78. Yan, J.; Zhu, L.; Qu, Y.; Qu, X.; Mu, M.; Zhang, M.; Muneer, G.; Zhou, Y.; Sun, L. Analyses of active anti-oxidant polysaccharides from four edible mushrooms. *Int. J. Biol. Macromol.* **2019**, *123*, 945–956. [CrossRef]
79. Chaiyama, V.; Keawsompong, S.; LeBlanc, J.G.; deLeBlanc, A.M.; Chatel, J.M.; Chanput, W. Action modes of the immune modulating activities of crude mushroom polysaccharide from *Phallus atrovolvatus*. *Bioact. Carbohydr. Diet. Fibre.* **2020**, *23*, 100216. [CrossRef]
80. Bettin, F.; Cousseau, F.; Martins, K.; Boff, N.A.; Zaccaria, S.; daSilveira, M.M.; Dillon, A.J.P. Phenol removal by laccases and other phenol oxidases of *Pleurotus sajor-caju* PS-2001 in submerged cultivations and aqueous mixtures. *J. Environ. Manag.* **2019**, *236*, 581–590. [CrossRef]
81. Junior, J.A.; Vieira, Y.A.; Cruz, I.A.; Vilar, D.S.; Aguiar, M.M.; Torres, N.H.; Bharagava, R.N.; Lima, Á.S.; deSouza, R.L.; Ferreira, L.F.R. Sequential degradation of raw vinasse by a laccase enzyme producing fungus *Pleurotus sajor-caju* and its ATPS purification. *Biotechnol. Rep.* **2020**, *25*, e00411. [CrossRef]
82. Abdel-Hamid, A.M.; Solbiati, J.O.; Cann, I.K.O. Chapter One—Insights into lignin degradation and its potential industrial applications. *Adv. Appl. Microbiol.* **2013**, *82*, 1–28. [PubMed]
83. Fang, W.; Zhang, P.; Zhang, X.; Zhu, X.; van Lier, J.B.; Spanjers, H. White rot fungi pretreatment to advance volatile fatty acid production from solid-state fermentation of solid digestate: Efficiency and mechanisms. *Energy* **2018**, *162*, 534–541. [CrossRef]
84. Hu, T.; Wang, X.; Zhen, L.; Gu, J.; Zhang, K.; Wang, Q.; Ma, J.; Peng, H.; Lei, L.; Zhao, W. Effects of inoculating with lignocellulose-degrading consortium on cellulose degrading genes and fungal community during co-composting of spent mushroom substrate with swine manure. *Bioresour. Technol.* **2019**, *291*, 121876. [CrossRef] [PubMed]
85. Cowieson, A.J.; Ruckebusch, J.P.; Sorbara, J.O.B.; Wilson, J.W.; Guggenbuhl, P.; Roos, F.F. A systematic view on the effect of phytase on ileal amino acid digestibility in broilers. *Anim. Feed Sci. Technol.* **2017**, *225*, 182–194. [CrossRef]
86. Menezes-Blackburn, D.; Jorquera, M.; Gianfreda, L.; Rao, M.; Greiner, R.; Garrido, E.; Mora, M.L. Activity stabilization of *Aspergillus niger* and *Escherichia coli* phytases immobilized on allophanic synthetic compounds and montmorillonite nanoclays. *Bioresour. Technol.* **2011**, *102*, 9360–9367. [CrossRef]
87. Wanzenbock, E.; Apprich, S.; Tirpanalan, O.; Zitz, U.; Kracher, D.; Schedle, K.; Kneifel, W. Wheat bran biodegradation by edible *Pleurotus* fungi e A sustainable perspective for food and feed. *Food Sci. Technol.* **2017**, *86*, 123–131. [CrossRef]
88. Hamdi, M.; Perez, J.F.; L´etourneau-Montminy, M.P.; Franco-Rossell'o, R.; Aligue, R.; Sol'a-Oriol, D. The effects of microbisal phytase and dietary calcium and phosphorus levels on the productive performance and bone mineralization of broilers. *Anim. Feed Sci. Technol.* **2018**, *243*, 41–51. [CrossRef]
89. Ndrepepa, G. Uric acid and cardiovascular disease. *Eur. Caediol.* **2018**, *484*, 150–163. [CrossRef]
90. Sandhya, C.; Alagarsamy, S.; Szakacs, G.; Pandey, A. Comparative evaluation of neutral protease production by *Aspergillus oryzae* in submerged and solid-state fermentation. *Process. Biochem.* **2005**, *40*, 2689–2694. [CrossRef]
91. Li, J.; Mihalcioiu, M.; Li, L.; Zakikhani, M.; Camirand, A.; Kremer, R. Vitamin D prevents lipid accumulation in murine muscle through regulation of PPARγ and perilipin-2 expression. *J. Steroid Biochem.* **2018**, *177*, 116–124. [CrossRef]

92. Barzegari, M.; Sarbakhsh, P.; Mobasseri, M.; Noshad, H.; Esfandiari, A.; Khodadadi, B.; Gargari, B.P. The effects of vitamin D supplementation on lipid profiles and oxidative indices among diabetic nephropathy patients with marginal vitamin D status. *Diabetes Metab. Syndr.* **2019**, *13*, 542–547. [CrossRef] [PubMed]
93. Asbaghi, O.; Kashkooli, S.; Choghakhori, R.; Hasanvand, A.; Abbasnezhad, A. Effect of calcium and vitamin D co-supplementation on lipid profile of overweight/obese subjects: A systematic review and meta-analysis of the randomized clinical trials. *Obes. Med.* **2019**, *15*, 100124. [CrossRef]
94. Bindhu, J.; Arunava, D. An edible fungi *Pleurotus ostreatus* inhibits adipogenesis via suppressing expression of PPAR γ and C/EBP α in 3T3-L1 cells: In vitro validation of gene knock out of RNAs in PPAR γ using CRISPR spcas9. *Biomed. Pharmacother.* **2019**, *116*, 109030.

9

Honey Bee Pollen in Meagre (*Argyrosomus regius*) Juvenile Diets: Effects on Growth, Diet Digestibility, Intestinal Traits, and Biochemical Markers Related to Health and Stress

Valentina Panettieri [1], Stavros Chatzifotis [2], Concetta Maria Messina [3], Ike Olivotto [4], Simona Manuguerra [3], Basilio Randazzo [4], Andrea Ariano [1], Fulvia Bovera [1,*], Andrea Santulli [3,5], Lorella Severino [1] and Giovanni Piccolo [1]

1 Department of Veterinary Medicine and Animal Production, University of Napoli Federico II, Via F. Delpino 1, 80137 Napoli, Italy; valentina.panettieri@unina.it (V.P.); andrea.ariano@unina.it (A.A.); lorella.severino@unina.it (L.S.); giovanni.piccolo@unina.it (G.P.)

2 Institute of Marine Biology, Biotechnology and Aquaculture, Hellenic Centre for Marine Research, Gournes Pediados P.O. Box 2214,71003 Heraklion, Crete, Greece; stavros@hcmr.gr

3 DiSTeM, Marine Biochemistry and Ecotoxicology Laboratory, University of Palermo, Via G. Barlotta 4, 91100 Trapani, Italy; concetta.messina@unipa.it (C.M.M.); simona.manuguerra@unipa.it (S.M.); andrea.santulli@unipa.it (A.S.)

4 Department of sea science, University Polytechnic of Marche, via Brecce Bianche, 60100 Ancona, Italy; i.olivotto@staff.univpm.it (I.O.); b.randazzo@staff.univpm.it (B.R.)

5 Consorzio Universitario della Provincia di Trapani, Institute of Marine Biology, Via G. Barlotta 4, 91100 Trapani, Italy

* Correspondence: bovera@unina.it.

Simple Summary: Recently, several studies have focused on the use of nutraceuticals and honey bee products to improve the welfare and sustainability of animal husbandry. Honey bee pollen is rich in bioactive substances, presenting a strong antioxidant activity with possible positive effects on growth performance and non-specific immune responses in reared fish. Despite its favorable characteristics, the addition of honey bee pollen to a meagre (*Argyrosomus regius*) diet in our trial resulted in a reduction of growth performances and diet digestibility, histological alterations of intestinal morphology, and high levels of biomolecular stress markers, probably due to its complex ultrastructure, which is indigestible for monogastric animals. These negative effects could be overcome by using bioactive component extraction methods and thus eliminating the indigestible fractions. Our results confirmed the general assumption that it should always be considered necessary to test nutraceutical additives of natural origin in a given species in order to verify the effective positive action and exclude any negative repercussions on animal health.

Abstract: This research aimed to evaluate the effects of the inclusion of honey bee pollen (HBP) in meagre (*Argyrosoumus regius*) juveniles' diets on growth performance, diet digestibility, intestinal morphology, and immunohistochemistry. Furthermore, stress-related molecular markers and biochemical blood profile of fish were evaluated, together with mineral trace and toxic element concentration in pollen and diets. Specimens of meagre (360) of 3.34 ± 0.14 g initial body weight, were randomly allocated to twelve 500 L circular tanks (30 fish per tank). Four diets were formulated: a control diet and three experimental diets with 1%, 2.5%, and 4% of HBP inclusion. All the growth parameters and crude protein and ether extract digestibility coefficients were negatively linearly affected by increased HBP inclusion ($p < 0.0001$). Histology of medium intestine showed slight signs of alterations in group HPB1 and HPB2.5 compared to control. Fish from HBP4 group showed severe alterations at the intestinal mucosa level. Immunohistochemical detection of TNF-α in the medium intestine showed the presence of TNF-α+ cells in the lamina propria, which resulted in accordance with the increased level of the TNF-α protein detected by immunoblotting in the liver. This stress situation was confirmed by the increased hepatic level of HSP70 ($p < 0.05$) in fish fed the HBP4 diet and by the linear decrease of total serum protein levels in HBP-containing diets ($p < 0.0001$). These negative effects can be related to the ultrastructure of the bee pollen grain walls, which make the bioactive substances unavailable and can irritate the intestine of a carnivorous fish such as meagre.

Keywords: meagre; honey bee pollen; growth trial; digestibility trial; TNF-α; HSP70; intestinal immunohistochemistry; toxic elements; trace elements; total serum protein

---

1. Introduction

In the past 20 years, researchers have focused their attention on several natural molecules to be used as possible antioxidant therapeutic and preventive agents [1].

A growing amount of scientific evidence is demonstrating that supplementing diets with natural compounds that act as protective factors or immunostimulants can improve growth performance, product quality, fish health, and physiological response to stress situations and diseases [2–6].

Bee-derived products have been widely investigated for their positive characteristics, such as their strong antioxidant activity, their positive effects on non-specific immune responses, and their capacity to increase growth performance in various animal species [7–9]. Research has been carried out to test the inclusion of these nutraceutical products in aquafeeds because of their antioxidant potential to prevent or treat aquatic animal diseases and to increase their performance [8,10].

Honey bee pollen (HBP) consists of the male generative cells gathered by honeybees from flower stamens and the anthers of flowers, which are collected by foraging bees and carried to the hives, where pollen agglutinates with bee secretions and the addition of nectar [11,12]. The consumption of HBP has increased in the recent years because it has been considered a healthy and therapeutic product because of its nutritional properties, source of proteins, lipids, vitamins, minerals, amino acids, carotenoids, flavonoids, phenolic compounds, antioxidants, carotenes, and xanthophylls [13,14]. Phenolic acid, flavonoids, and tannins act as potent antioxidants and protective agents and their free radical scavenging activity encourages their application in the biomedical field [15,16].

The introduction of HBP in freshwater fish diets has led to improved growth and immune status [17], increase in the number of phagocytic cells, (i.e., neutrophils and monocytes), and reduction of the mortality caused by *Aeromonas hydrophila* in Nile tilapia (*Oreochromis niloticus*) [8].

Despite these favourable characteristics, no research has been carried out to evaluate the effects of the inclusion of HBP in diets for marine fish species. The present research intended to fill this gap, evaluating growth performance, diet digestibility, intestinal morphology and immunohistochemistry, stress-related molecular markers, and biochemical blood profile in meagre (*Argyrosoumus regius*) juveniles fed diets containing increasing levels of commercial HBP. In addition, the research aimed to contribute to the knowledge of the mineral trace element concentration in honey bee pollen and in fish diets.

2. Materials and Methods

A chestnut honey bee pollen (HBP) purchased from a local organic farm located in the city of Naples (Napoli, Italy) was chosen for the research. Both growth and digestibility trials were conducted at the Institute of Marine Biology, Biotechnology and Aquaculture (IMBBC) of the Hellenic Centre for Marine Research (Crete, Greece). The experimental protocol was designed according to the guidelines of European Directive (2010/63/EU) on the protection of animals used for scientific purposes. The digestibility trial was approved by the experts of ASSEMBLE Plus project (Contract number: SR020220189047). The growth trial procedures and the blood sampling were approved by ethics experts of the Trans National Access (TNA) selection panel of the AQUAEXCEL 2020 (project number: AE040069).

2.1. *Growth Trial*

2.1.1. Fish and Experimental Conditions

Meagre juveniles were obtained from the IMBBC hatchery. After a 2 week period of acclimation to the experimental conditions, 360 fish of 3.34 ± 0.14 g initial body weight were individually weighed under moderate anaesthesia conditions (anaesthetic: MS222-Tricain Methanesulphonate at 50 mg/L dosage) and randomly allocated to twelve 500 L circular tanks (30 fish per tank) supplied by flow-through borehole aerated seawater (renewal 200% per hour).

The trial was conducted from July to October 2018 under constant temperature (19.0 ± 0.5 °C), salinity (36‰), and dissolved oxygen (DO 8 ± 0.5 ppm) and a dark:light cycle of 12:12 h.

2.1.2. Fish Diets

To determine the effects of the inclusion of HBP, four isonitrogenous (53% crude protein (CP) on as fed basis), isolipidic (15% ether extract (EE) on as fed basis), and isocaloric diets were formulated to meet the meagre nutrient and energy requirements according to Chatzifotis et al. [18]: a control diet with no addition of HBP (HBP0) and three experimental diets in which HBP was included at 1% (HBP1), 2.5% (HBP2.5), and 4% (HBP4) on as fed basis.

The experimental feeds were prepared at the IMBBC laboratory. All dietary ingredients were finely ground and thoroughly mixed; water was then blended into the mixture to obtain an appropriate consistency for pelleting using a 2.5 mm die meat grinder. After pelleting, the diets were dried in a ventilated oven at 40 °C for 24 h and stored in plastic bags until use.

Chemical composition of the experimental diets was determined as follows: dry matter (DM), ash, CP, EE, and crude fiber (CF) (procedure numbers 934.01, 942.05, 954.01, 920.39, and 962.09, respectively) according to AOAC [19]. Before starting the analysis, the pellets were finely ground using a cutting mill. The ingredients and approximate composition of the diets are reported in Table 1 and the approximate composition of the bee pollen is reported in Table 2.

Table 1. Ingredients and chemical composition of experimental diets for meagre.

Ingredients (g/kg)	HBP0	HBP1	HBP2.5	HBP4
Fish meal	440	438	435	430
Soybean meal	240	238	235	230
Corn gluten meal	100	98	96	95
Wheat flour	100	98	93	90
Fish oil	100	98	96	95
Honey bee pollen	0	10	25	40
Mineral mix	10	10	10	10
Vitamin mix	10	10	10	10
Chemical Composition				
Dry matter (%)	90.0	92.7	90.6	93.0
Ash (%)	9.4	9.8	9.0	9.4
Crude protein (%)	53.0	53.6	52.5	52.5
Ether extract (%)	15.0	14.9	15.0	14.8
Crude fiber (%)	7.7	7.7	7.7	7.5
Na (%)	0.086	0.075	0.058	0.075
Mg (%)	0.130	0.117	0.131	0.125
K (ppb)	381.65	351.92	381.93	377.90
Ca (ppb)	870.39	696.49	456.83	562.36

Table 2. Chemical composition of honey bee pollen.

	DM (%)	Ash (%)	CP (%)	EE (%)	CF (%)	Na (%)	Mg (%)	K (%)	Ca (%)
HBP	85.02	3.79	23.18	7.82	19.23	0.067	0.062	0.019	0.061

Abbreviations: HBP: honey bee pollen; DM: dry matter; CP: crude protein; EE: ether extract; CF: crude fiber.

The experimental diets were randomly assigned to triplicate groups of 30 fish.

Each diet was administered three times per day (09.00 am; 12.00 pm, and 4.00 pm) to apparent satiety (until the first pellet was refused), 7 days per week. Any not-ingested pellet was recovered, dried, and weighed. The exact quantity of feed distributed in each tank was recorded daily. The trial lasted 88 days.

2.1.3. Trace Elements

The trace elements contained in the honey bee pollen and in the diets were also determined. Before analysis, the samples (0.5 ± 0.02 g) were placed in a Teflon vessel with 5.0 mL of 65% HNO_3 and 2.0 mL of 30% H_2O_2 (Romil UpA). The vessel was sealed and placed in a microwave digestion system (Milestone, Bergamo, Italy). Microwave-assisted digestion was performed with a mineralization program for 15 min at 200 °C. The vessel was then cooled at 30 °C, the digestion mixture was transferred into a 50.0 mL flask, and the final volume was obtained by adding Milli-Q water [20]. Trace element concentrations were determined by inductively coupled plasma optical emission spectroscopy (ICP-OES) technique using an Optima 2100 DV instrument (PerkinElmer Inc., Wellesley, MA, USA) coupled with a CETAC U5000AT (CETAC Technologies, Omaha, NE, USA). The calibration curve and two blanks were run during each set of analyses, to check the purity of the chemicals. A reference material (CRM DORM-4, National Research Council of Canada (NRC-CNRC), Ottawa (Ontario), Canada) was also included for quality control. All the values of the reference materials were within the certified limits.

The instrumental detection limits are expressed as wet weight (w.w.) and were determined following the protocol described by Perkin Elmer ICP, application study number 57 [21].

2.1.4. Growth Performance

At the end of the trial, fish were starved for 1 day, lightly anesthetized (MS222-Tricain methanesulphonate at 50 mg/L dosage), and group weighed.

The following growth performance indexes and protein efficiency ratios (PER) were calculated according to the following formulas [22]:

- Weight gain (WG%) = 100 × ((FBW, final body weight (g)—IBW, initial body weight (g))/initial live weight (g))
- Daily intake rate (DIR, %/day) = 100 × ((feed intake (g)/mean weight (g))/days)
- Specific growth rate (SGR, %/day) = ((lnFBW − lnIBW)/number of feeding days) × 100
- Feed conversion ratio (FCR) = (total feed supplied (g)/weight gain (g))
- Protein efficiency ratio (PER) = (weight gain (g)/total protein fed (g))

2.2. *Somatic Traits*

At the end of the feeding trial, 45 fish per treatment (15 fish per replicate) were randomly chosen, weighed, and sacrificed by over-anaesthesia (MS222-Tricain methanesulphonate at 250 mg/L dosage) and dissected. Liver and gut were weighed to determine hepatosomatic (HSI) and viscerosomatic (VSI) indexes as described in Piccolo et al. [23]. Liver, muscle, and intestine of four fish for each tank were sampled.

2.3. Digestibility Trial

Dry matter, protein, and lipid digestibility of the tested diets was evaluated in a separate trial. One hundred and eighty meagre of 23.53 ± 2.16 g initial weight obtained from the IMBBC hatchery were distributed in 12 circular fiberglass tanks (3 tanks per treatment; 15 fish per tank) of 270 L equipped with a settling column. The water and environmental conditions were the same as described for the growth trial.

Diets were prepared at the IMBBC laboratory following the same procedure described above and using the same inclusion levels of HBP (1%, 2.5%, 4%). The apparent digestibility coefficients were measured using the indirect acid-insoluble ash (AIA) method; for this reason, 1% celite® (Fluka, St. Gallen, Switzerland) was added to the diets used in the growth trial as an inert marker.

Fish were fed the experimental diets ad libitum three times a day for 4 weeks. Feces accumulated in the settling column was collected daily before the morning meal, centrifuged at 7000 rpm for 10 min, and stored at −20 °C until analysis.

Thirty minutes after the last feeding in the afternoon, tanks were cleaned to remove excess of feces, and then the settling column was placed.

Feeds and feces were analyzed according to AOAC [19]. The AIA contents of feeds and feces were determined according to Vogtmann et al. [24].

The apparent digestibility coefficients of dry matter (ADC DM), crude protein (ADC CP), and ether extract (ADC EE) were calculated following Palmegiano et al. [25].

2.4. Histology

For the histological analysis, segments of medium intestine were isolated from growth trial fish and immediately fixed in modified buffered Karnovsk fixative for at least 24 h. After fixation, samples were dehydrated in graded alcohol solutions, cleared in xylene, and embedded in solid paraffin. Cross-sections of 5 μm, cut with a LEICA microtome (Leica Microsystems Srl, Buccinasco (MI), Italy), were stained with Mayer's haematoxylin and eosin Y and examined under a Zeiss Axio Imager.A2 microscope (Carl Zeiss Microscopy GmbH, Jena, Germany); images were acquired by means of a combined colour digital camera Axiocam 503 (Carl Zeiss Microscopy GmbH, Jena, Germany). For the morphometric evaluation of intestinal fold height, and mucous cells abundance, nine fish for each treatment (three for each tank) were considered. Three sections for each fish were observed at intervals of about 200 μm in order to avoid repetitions in morphometric measurements of folds and quantification of mucous cells. Intestinal folds were measured from the apex to the base, excluding the underlying connective layer and values have been expressed by mean and standard deviation (SD).

2.5. Immunohistochemistry

For the immunodetection of TNF-α in the medium intestine, nine fish for each treatment (three for each tank) were used and sections (5 μm) were placed on gelatinized slides (0.5% fish gelatin, 0.05% chromopotassium sulfate) in order to prevent section detachment during the reaction. Sections were deparaffinized and rehydrated through serial graded ethanol solutions. Endogenous peroxidase activity was blocked by treating sections with 0.3% hydrogen peroxide for 10 min at room temperature. To prevent aspecific antibody binding, slides were rinsed in 0.01 M phosphate-buffered saline (PBS), pH 7.4, for 15 min and blocked using 5% BSA in Tris for 20 min. After rinsing in PBS for 15 min, sections were incubated overnight at 4 °C with anti-TNF-α rabbit polyclonal primary antibody specific for zebrafish (ANASPEC). After rinsing the slides with PBS, the sections were incubated with the secondary fluorescent antibody (Goat Anti-Rabbit IgG H&L, Alexa Fluor 488 No. ab150077; Abcam, dilution 1:400) at room temperature for 1 h and 30 min. Following incubation, the slides were mounted with Fluoreshield Mounding Medium with DAPI (ab104139) for nucleus staining. The sections were observed under a Zeiss Axio Imager M2 microscope and images were captured with a high-resolution camera Zeiss Axiocam 105 color (Carl Zeiss Microscopy GmbH, Jena, Germany). Negative controls were obtained by incubation without the primary antibody. For the relative quantification of TNF-α+ cells, three sections from three different fish for each treatment were analyzed. Observations were performed counting TNF-α+-labeled cells in three undamaged folds for each section. An arbitrary unit was adopted on the mean of the count: 0–20 = +, 20–50 = ++, 50–>100 = +++.

2.6. Extraction of Total Proteins and Detection of Hepatic Biomolecular Markers by Immunoblotting

For total protein extraction, aliquots of lyophilized livers were homogenized on ice (1:3 w/v) with 100 mM HEPES pH 7.4, 10 mM EDTA, 10 mM EGTA, and 4 M NaCl 23.37 g / 100 mL, in the presence of 10 μL protease inhibitor mix (NaF 42 mg/mL, Aprotinine, Na_3PO_4 183.9 mg/mL, Leupeptine 4×). The homogenate was centrifuged at 6000 rpm at 4 °C for 15 min and the supernatant was stored on ice for protein quantification according to Lowry et al. [26]. Biomolecular markers in liver were analyzed by immunoblotting.

Equivalent amounts of proteins (30 μg) diluted with Laemmli buffer (1970) (Sigma-Aldrich, St. Louis, MO, USA) were loaded, after protein denaturation for 5 min at 90 °C, on pre-cast gel for SDS–polyacrylamide electrophoresis (SDS-PAGE) (Bio-Rad, Hercules, CA, USA) and blotted using a Trans Blot Turbo Transfer System (Bio-Rad, Hercules, CA, USA). The correct amount of protein loaded was confirmed by red Ponceau staining. Filters were used for protein detection by primary antibodies (AbI) specifics for TNF-α (Ab polyclonal from rabbit) and HSP70 (Ab monoclonal from rabbit). (Sigma-Aldrich, Dorset, UK; Santa Cruz, CA, USA). The primary antibodies were diluted in buffer at the concentrations suggested by the manufactures for each AbI. In relation to the origin of the AbI, the appropriate secondary antibodies were used (anti-mouse or anti-rabbit secondary); as means of detection, the secondary antibodies were conjugated with horseradish peroxidase (GAR/M-HRP Bio-Rad, Hercules, CA, USA) or alkaline phosphatase. The signals originated by immunoreaction were detected using enhanced chemo-luminescent (ECL) reagents (Clarity Western ECL Blotting Substrate, Bio-Rad, Hercules, CA, USA) and a BCIP/NBT substrate system (Bio-Rad). Images were obtained, photographed, and digitalized with Chemi Doc XRS (Bio-Rad, Hercules, CA, USA), and further analyzed with Image Lab software (Bio-Rad, Hercules, CA, USA). The results were expressed as fold increase of each treatment in relation to the respective control; the images shown are representative of three separate experiments, for which the mean quantification is reported in each figure, together with the significance of the differences ($p < 0.05$).

2.7. Assessment of Biochemical Parameters in Blood Samples

Biochemical measurements were performed on blood samples from 2-month-old fish of the species *Argyrosomus regius*, within a weight 20–25 g. The purpose of the test was to evaluate the effect of bee pollen enrichment in accordance with the requirements of the AquaExcel2020 program.

On the day of the sampling, fish were euthanized, and their weight and length was measured. Immediately, blood was collected from the caudal vessel via heparinized syringes. Hematocrit was measured, using capillary tubes and a specialized centrifuge to determine the percentage of red blood cells by volume. Additionally, hemoglobin was determined using a colorimetric assay kit (Spinreact, Girona, Spain).

Following these procedures, blood samples were centrifuged at 2000× *g* for 10 min and the resulting plasma was stored at −20 °C until further analysis. Plasma samples were used to assess the following biochemical parameters: cholesterol, triglycerides, albumin, total proteins, glucose, creatinine, sodium, potassium, calcium, phosphorus, lactate, glutamic oxaloacetic transaminase (GOT), and glutamic pyruvic transaminase (GTP).

More specifically, plasma cholesterol (CO/PAP, Biosis, Greece), triglycerides (GPO/PAP, Biosis, Athens, Greece), albumin (BCG, Biosis, Greece), total serum proteins (Biuret, Biosis, Greece), glucose (GOD/PAP, Biosis, Greece), calcium (Arsenazo III, Biosis, Greece), phosphorus (UV, Biosis, Greece), lactate (LO-POD, Spinreact, Girona, Spain), and creatinine (Cayman Chemical Company, Ann Arbor, MI, USA) concentrations were assessed using commercial enzymatic colorimetric kits. Moreover,

GOT-AST (IFCC/LIQUID, Biosis, Greece), GPT-ALT (IFCC/LIQUID, Biosis, Greece) concentrations were estimated using enzymatic kinetic assays. Finally, a flame-photometer was used to determine plasma sodium and potassium concentrations.

2.8. Statistical Analysis

The data were tested for normal distribution and equal variances before analysis (SAS, 2000, SAS Institute Inc., Cary, NC, USA). All the data were analyzed by one-way ANOVA, using the GLM procedure of SAS (2000), according to the model

$$Yij = m + Di + eij, \quad (1)$$

For growth and digestibility trials, the experimental unit was the tank and each value was obtained as the average of the 30 fish in the tank. For somatic indexes, the experimental unit was the individual fish (15 fish for each tank). Finally, for the blood parameters, the unit was derived from the average of four fish per tank.

In addition, the mean comparison was performed using orthogonal contrast analysis. The examined components were linear and quadratic (SAS, 2000, SAS Institute Inc., Cary, NC, USA). $P < 0.05$ was considered the threshold for statistical significance.

3. Results

3.1. Trace Elements

Table 3 shows the content of toxic and essential trace elements in experimental feeds and honey bee pollen.

Table 3. Toxic and essential trace elements in experimental feeds and honey bee pollen.

Trace Elements (ppb)	HBP0	HBP1	HBP2.5	HBP4	Honey Bee Pollen
Fe	499.27	490.53	511.96	512.95	134.63
Cu	40.17	38.13	44.73	38.86	6.16
Zn	186.06	178.50	193.95	191.28	29.77
Co	0.15	0.15	0.15	0.15	0.02
Al	62.98	46.01	56.82	66.21	58.13
Se	0.55	0.53	0.54	0.57	1.85
Mn	259.04	262.06	257.79	268.78	31,73
Pb	0.22	0.23	0.23	0.24	0.16
Cd	0.12	0.10	0.11	0.11	0.03
Cr	0.81	0.79	0.83	0.82	0.30
Ni	1.92	1.81	1.90	1.88	0.34
As	2.45	2.23	2.41	2.35	0.59

3.2. Growth Performance

In Table 4, the growth parameters measured during the trial are reported.

Table 4. Growth performance of meagre fed experimental diets.

	HBP0	HBP1	HBP2.5	HBP4	p-Values			
					RMSE	Linear [1]	Quadratic [1]	Cubic [1]
Initial BW (g)	3.35	3.34	3.34	3.34				
Final BW (g)	26.78	23.29	22.69	21.36	0.66	<0.0001	<0.0001	0.0659
FCR	0.96	1.04	1.18	1.23	0.06	0.0007	0.0003	0.3186
SGR	2.37	2.21	2.18	2.11	0.08	0.0044	0.0062	0.4158
Intake‰ ABW/d	16.48	17.23	19.08	19.43	0.98	0.0061	0.0025	0.336
WG%	705.3	606.9	579.7	539.7	52.9	0.0050	0.0059	0.5564
PER	1.96	1.80	1.61	1.55	0.10	0.0008	0.0004	0.5619

Abbreviations: Initial BW: initial body weight; Final BW: final body weight; FCR: feed conversion ratio; SGR: specific growth rate; Intake‰ ABW/d: intake per kg adjusted body weight; WG%: weight gain percentage per initial body weight; PER: protein efficiency ratio. [1] Contrast analysis.

All the growth parameters were significantly influenced by the diet. Daily feed intake increased as bee pollen inclusion in the diet increased ($p < 0.01$), while final weight ($p < 0.0001$), FCR ($p < 0.001$), SGR ($p < 0.01$), PER ($p < 0.001$), and WG %ABW ($p < 0.01$) were negatively linearly affected by bee pollen inclusion in the diet. The growth parameters reached the worst values in the HBP4 diet.

3.3. Somatic Indexes, Slaughter Traits

In Table 5, the viscerosomatic indexes and the slaughter traits are shown.

Table 5. Somatic indexes of meagre fed experimental diets.

	HBP0	HBP1	HBP2.5	HBP4	*p*-Values			
					RMSE	Linear [1]	Quadratic [1]	Cubic [1]
Final Weight	37.4	33.9	36.8	33.9	4.41	0.0584	0.316	0.037
VSI (%)	4.37	4.27	4.20	4.11	0.70	0.3299	0.333	0.9185
HIS (%)	1.62	1.69	1.67	1.72	0.36	0.4245	0.517	0.6901
K	1.17	1.05	1.06	1.06	0.13	0.0397	0.078	0.3916
Gut length %BW	37.50	45.64	40.09	44.05	7.46	0.0413	0.282	0.0206

Final weight: final body weight; HIS: hepatosomatic index; VSI: viscerosomatic index; K: condiction factor; [1] Contrast analysis.

No significant differences in somatic indexes emerged among the fish fed the experimental diets, except for the gut length, which showed a linear increase with increasing honey bee pollen inclusion in the diet.

3.4. Digestibility Trial

The estimated ADC of DM, CP, and EE of the four diets are reported in Table 6.

Table 6. Digestibility trial results.

	HBP0	HBP1	HBP2.5	HBP4	*p*-Values			
					RMSE	Linear [1]	Quadratic [1]	Cubic [1]
ADC DM	64.24	74.28	36.03	45.63	0.82	<0.0001	<0.0001	0.0001
ADC CP	90.26	91.36	78.71	78.56	0.76	<0.0001	<0.0001	0,0003
ADC EE	93.45	92.15	85.725	85.91	0.91	<0.0001	<0.0001	<0.0001

ADC DM: dry matter apparent digestibility coefficient; ADC CP: crude protein apparent digestibility coefficient; ADCEE: ether extract apparent digestibility coefficient. [1] Contrast analysis.

ADC DM linearly decreased as the HBP content in the diets increased. Furthermore, linear and quadratic components indicated a gradual reduction of crude protein and ether extract digestibility coefficients going from the control to HBP4 diet. Nevertheless, the ADCs of HBP1 diet resulted higher than those shown by HBP2.5 and HBP4 groups, as indicated by the quadratic component of the variance ($p < 0.001$).

3.5. Histology

Histology of the medium intestine showed slight signs of alterations in group HPB1 and HPB2.5 when compared with the HBP0 group. No significant differences were shown by the morphometric evaluation of intestinal folds among experimental groups (Table 7).

Table 7. Morphometric evaluation of mucosal folds.

	MF Height (μm)
HBP0	462.3 ± 18.7
HBP1	487.8 ± 23.5
HBP2.5	453.1 ± 24.8
HBP4	476.6 ± 46.4

MF height: mucosal fold height. Data are expressed as mean and SD.

A significant linear increase in mucous cells in response to pollen inclusion was observed (Figure 1).

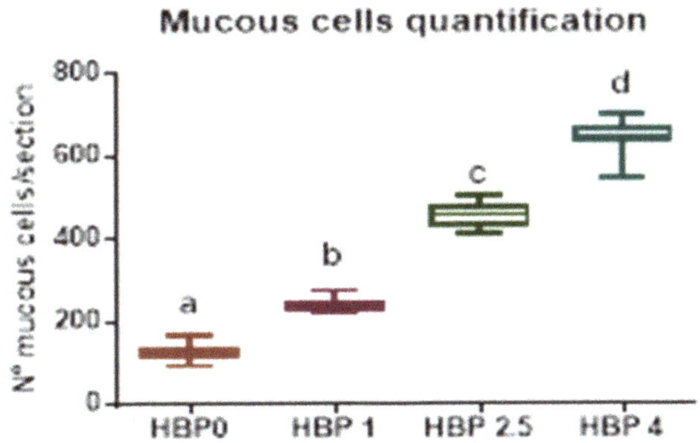

Figure 1. Abundance of mucous cell in medium intestine. The number of mucous cells is reported as mean of the observations performed on three transversal intestinal sections from individual fish (n = 9) from the different experimental groups. Different letters indicate significant differences among groups ($p < 0.05$).

The most frequently encountered alterations consisted of leucocyte infiltration at the level of the lamina propria and thickening of the submucosa (Figure 2g). Intestine samples from HBP4 group showed severe alteration at the level of the mucosa, even preserving epithelial integrity. In this group, a substantial increase in the number of mucous cells was observed (Figures 1 and 2). Moreover, melano macrophage-like, melanin-containing cells were often observed at the base of the epithelial layer and infiltrating lamina propria and submucosa (Figure 2h).

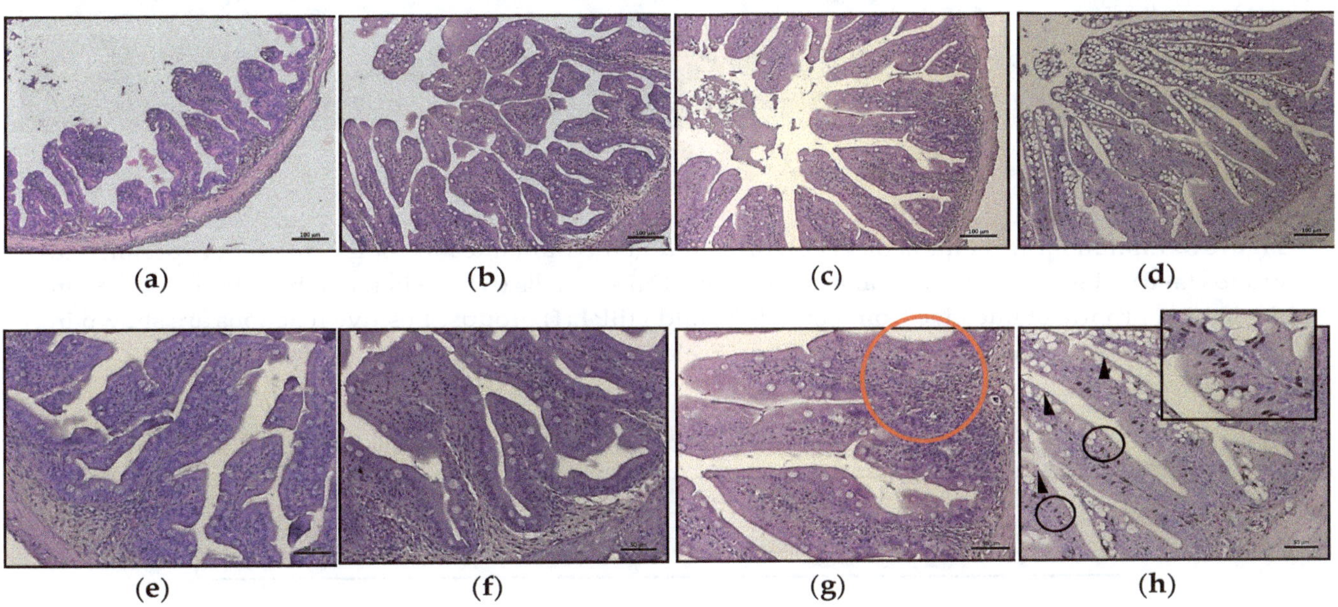

Figure 2. Intestinal histology of meagre fed experimental diets after 88 days of trial. Histology of *A. regius* medium intestine from the different experimental groups: (**a**,**e**) normal histology from HBP0 group; (**b**,**f**) HBP1 group did not show signs of pathological alteration; (**c**,**g**) in the HBP2.5 group, it was possible to detect inflammatory influx in some portions of the intestinal submucosa (red circle); (**d**,**h**) the HBP4 group showed a severe degree of mucipar hyperplasia of the mucosa (arrowhead) with an abundant presence of melano macrophages at the base of the enterocytes (black circle and box). Scale bar: a, b, c, d = 200 µm; e, f, g, h = 50 µm.

3.6. Immunohistochemistry

Immunohistochemical detection of TNF-α in medium intestine showed the presence of TNF-α+ cells in the lamina propria and submucosa of bee-pollen-treated fish. While a weak reaction was recorded in the epithelial layer of intestinal folds in all treated groups, a moderate influx of TNF-α+ cells was recorded in particular in groups HBP2.5 and HBP4 (Figure 3). In particular, round macrophage/neutrophil-like TNF-α+ cells were observed in the submucosa and lamina propria of HBP1, HBP2.5, and HBP4 groups. The relative abundance evaluation of TNF-α+ cells infiltrating submucosa and lamina propria showed low values in HBP1, while HBP2.5 and HBP4 showed high values with no significant differences between the two groups (Table 8) at the base of the epithelial layer and infiltrating lamina propria and submucosa.

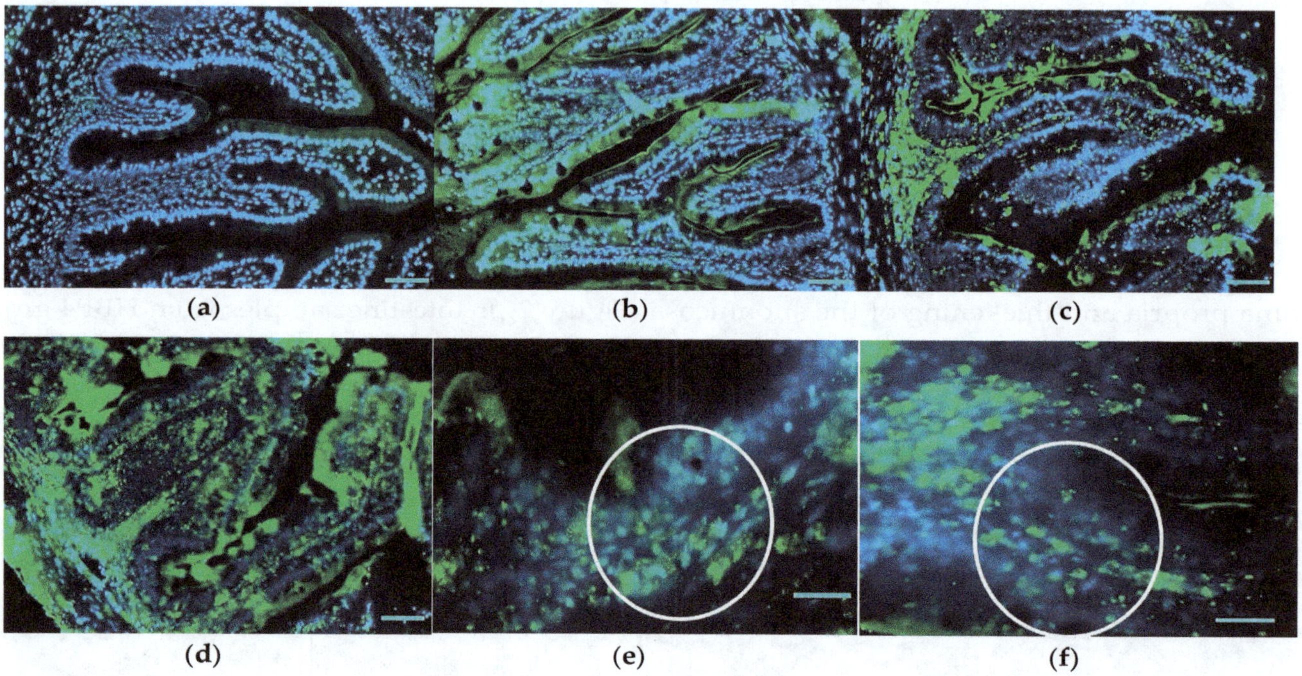

Figure 3. Immunohistochemical detection of TNF-α in medium intestine of the different experimental groups (**a–d**). High magnification images showing TNF-α+ cells (white circle) in the submucosa and in the lamina propria of intestine from HBP2.5 (**e**) and HBP4 (**f**) groups. Positive reactions are shown in green. Nuclei were stained with DAPI (blue). Scale bar: a,b,c,d = 50 μm; e,f = 20 μm.

Table 8. Relative density of TNF-α+ cells in medium intestine.

CTRL	HBP1	HBP2.5	HBP4
+	+	+++	+++

3.7. Hepatic Biomolecular Markers

The evaluation of the molecular markers related to immunostimulation and stress resistance showed that the levels of the protein TNF-α presented a significant increase only in fish fed the diet with the highest level of HBP inclusion (HBP4) ($p < 0.05$), while HSP70 resulted increased both in HBP2.5 and HBP4 groups compared to the control group (Figure 4).

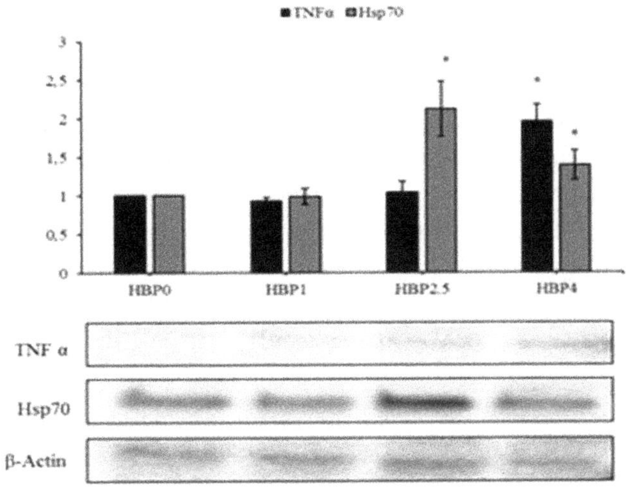

Figure 4. Immunoblot of TNFα, HSP70, evaluated on samples of the liver of *A. regius* fed on three different diets containing increasing levels of commercial HBP (HBP1, HHBP2.5, HBP4) and in standard condition (HBP0). β-Actin was used as the internal control. The images are representative of at least three separate experiments. The relative protein quantification is represented in the graphic (* $p < 0.05$ vs. HBP0).

3.8. *Assessment of Blood Samples*

The results obtained from the blood analysis are shown in Table 9.

Table 9. Biochemical assessment of blood samples.

	HBP0	HBP1	HBP2.5	HBP4	*p*-Value			
					RMSE	Linear [1]	Quadratic [1]	Cubic [1]
HCT (%)	28.67	26.00	27.00	26.67	3.19	0.46	0.62	0.56
HGB (g/dL)	3.83	3.80	3.76	3.56	0.40	0.43	0.45	0.88
Glucose (mmol/L)	4.71	5.21	4.92	5.96	1.20	0.24	0.34	0.51
TSP (g/dL)	10.20	8.15	6.53	3.73	0.98	<0.0001	<0.0001	0.58
Albumin (mg/dL)	1.61	1.91	1.38	1.64	0.60	0.94	0.60	0.33
Globulin (mg/dL)	8.56	5.72	5.14	2.07	1.27	0.0004	0.0007	0.2331
A/G	0.19	0.43	0.36	0.83	0.24	0.01	0.03	0.27
Creatinine (μmol/L)	80.00	132.50	94.67	104.33	16.51	0.16	0.76	0.03
GOT (UI/L)	360.00	345.00	355.67	243.33	67.7	0.11	0.16	0.48
GPT (U/L)	335.50	298.50	273.33	386.67	49.45	0.30	0.48	0.41
Triglyceride (mmol/L)	13.83	16.43	15.10	17.10	2.45	0.14	0.28	0.28
Cholesterol (mg/dL)	6.63	7.63	9.55	6.90	0.91	0.73	0.19	0.07
P (mg/dL)	7.57	9.50	9.77	11.90	4.63	0.29	0.32	0.78
Ca (mg/dL)	27.07	30.87	22.53	33.47	7.31	0.32	0.75	0.14
Na (mmol/L)	210.33	179.67	206.67	187.00	24.35	0.27	0.67	0.14
K (mmol/L)	2.70	2.17	2.07	2.53	0.73	0.79	0.75	0.95
Lactate (mmol/L)	3.17	4.15	2.75	3.3	0.73	0.83	0.45	0.11

HCT: hematocrit; HGB: hemoglobin; TSP: total serum protein; A/G: albumin to globulin ratio; GOT: glutamic oxaloacetic transaminase; GTP: glutamic pyruvic transaminase. statistically different. [1] Contrast analysis.

This table highlights the differences among the groups. Total serum protein, globulin, and albumin to globulin ratio linearly and quadratically decreased as the HBP content in the diets increased.

4. Discussion

The trial results showed that honey bee pollen induced negative effects on meagre growth performance and diet digestibility. These data were not in line with the current, although limited bibliography, in which bee pollen had notable beneficial effects in other fish species, as described by El-Asely for Nile tilapia and Choobkar for rainbow trout [8,10], even if it must be remarked that these latter authors used a bee pollen alcoholic extract.

The bad results obtained with the in vivo results can be explained by looking at the other performed investigations. The intestinal histology evidenced pathological alteration of the mucosa morphology, showing a severe mucipar hyperplasia and an abundant presence of melano macrophage-like cells at the base of the enterocytes. These alterations became more evident as the inclusion of bee pollen in the diets increased, where, besides the inflammatory influx, the migration and infiltration of melanin-containing cells indicated a further, extreme index of an immune response. The presence of TNF-α cells, resembling in shape neutrophil cells or macrophages, in fish fed different HBP inclusion corroborated the histological results. Tumor necrosis factor-alpha is an important pro-inflammatory cytokine playing an important role in cell survival through the activation, proliferation, and differentiation of macrophages. It is released by several immune cells during infection and tissue damage and its inhibition increases susceptibility to disease and reduces capacity to resolve infection [27–30]. TNF-α isoforms have been identified in various fish species [27,28,31], showing high levels of conservative regions with mammals and a constitutive expression in healthy fish tissues [28]. Inflammation is a highly energy-demanding process and it has been widely shown that intestinal inflammation correlated to anti-nutritional factors in fish diet is able to affect growth in fish [32,33].

Regarding the variation of the liver biomolecular markers related to general stress response and immunostimulation, in this case the increase of TNF-α could also be ascribed to a significant production of mediators of inflammation, acting as a strategy of protection [34]. The positive modulation of TNF-α has been observed in specimens of *Cyprinus carpio* treated, during vaccination, with *Aloe vera*, and also on immunity response in poultry, mice, and humans [35]. In that case, the effects were attributed to some poorly defined compounds of the plant [35].

Members of the HSP70 family are widely used as biomarkers of environmental stress in ecological and toxicological studies in fish [36–38]. The functions of the different HSP70 family members depend on their cellular localization, acting both as chaperons and immunomodulators [38], with the general aim of maintaining protein homeostasis [39] in normal and stressful conditions. HSP70 can be induced in response to thermal stress, hypoxia, oxidative stress, ultraviolet radiation, nutrient deprivation, osmotic pressure, heavy metals, chemical agents, microbial infections, and inflammation [37–41].

The levels of HSP70 protein increased in fish fed with the highest doses of bee pollen inclusion (HBP2.5 and HBP4) (Figure 4), indicating a stress condition that could be related to metabolic or oxidative stress. In fact, this protein is normally present at a low level in organisms with a good balance of antioxidants and over-expressed in a situation of oxidative stress [42–44]. Our previous study on the inclusion of dehydrated lemon peel in the diet for seabream showed a reduction of the HSP70 levels in fish fed on increased levels of bioactive compounds, due to the antioxidant composition of this resources, which is able to guarantee protection against oxidative stress [2]. Accordingly, Di Giancamillo et al. [45] described a decrease of HSP70 in pigs fed a diet supplemented with a natural verbascoside extract, compared with pig fed a high-fat diet. These authors suggested a role of HSP70 in modulating hepatic oxidative stress [45].

The main evidence that emerged from the blood analyses was the linear reduction of the total serum protein levels, concomitant with the increase of HBP inclusion in diets. TSP is a non-destructive parameter that is robust, easy to measure everywhere, and cheap, representing a suitable way of monitoring the overall welfare of fish by its regular increase [46]. It has been reported that this parameter tends to decrease in fish in response to various stress conditions such as hyperoxia, hypercapnia, stocking density, transfer to another tank, or nodavirus injection [46]. Many explanations have been suggested to explain TSP decrease, such as changes in blood volume and plasma hydration, alterations in permeability of environmental barriers, tissue destruction, malabsorption, and tissue injury by parasites or other pathogens [47]. These results confirm what emerged in the intestinal histology and immunohistochemistry and in hepatic-stress-related biomolecular markers.

Given that the fish of the different groups were subjected to the same environmental conditions, and received isoproteic and isoenergetic diets that varied only according to the level of pollen inclusion, our results can be explained through a negative action of the pollen in terms of intestinal inflammation.

Concerning the mineral content, the macro elements and the trace element contents in our HBP were in line with the values reported by Taha [48]. Compared to the maximum levels (MLs) of heavy metals set by the EU Commission (Directive (EC) No 2002/32 [49]), Cd, Pb, and as levels in honey bee pollen were lower than the maximum values established for feedstuff and feed materials, indicating that the risk of exposure to heavy metals deriving from consumption of the honey bee pollen used in this trial was relatively low and in compliance with EU regulations. However, in all the diets used in the trial, the levels of potentially toxic elements resulted lower than the MLs stated by EU regulation establishing the following MLs of heavy metals content in mg/kg (ppm) relative to a feed with a moisture content of 12%: the Cd MLs in the feed materials and complete feed are 2.0 and 0.5 mg/kg, respectively; the Pb MLs in the feed materials and complete feed are 10.0 and 5.0 mg/kg, respectively; the As MLs in the feed materials and complete feed are both 2 mg/kg. These elements, therefore, may not have been important in determining the observed deterioration of the growth performance and health status of the intestine in such fish species.

To deeply comprehend the reason for such negative results, it is necessary to dwell on the microscopic structure of bee pollen. Honey bee pollen is composed of a multitude of microscopic particles (6–200 μm in diameter) with variable shape, usually spherical or oval. The pollen cell walls consist of a series of stratified concentric layers [50]. As Roulstone described, the outermost layer of the pollen wall is the pollenkitt [51], a semi-solid coating comprised primarily of neutral lipids, hydrocarbons, terpenoids, and carotenoid pigments [52]. Inside the pollenkitt is the exine, an often intricately ridged matrix of the complex carbohydrate sporopollenin. Sporopollenin is a compound that provides chemical resistance to bee pollen and preserves the compounds within it [53]. The chemical analysis of sporopollenin has been limited due to the difficulty of obtaining large quantities of exine without affecting or modifying the structure of this molecule. Atkin et al. [53] recently found that this molecule has an empirical formula $C_{90}H_{144}O_{27}$. Nowadays, it is suggested that the sporopollenin consists of a series of biopolymers, homologous to the compounds cutin, suberin, and lignin.

The exine greatly resists decay and digestion, but is commonly perforated by one-to-several pores or slits (germination pores) that lead to the inner wall layer, known as the intine. The intine, composed primarily of cellulose and pectin, is also very resistant, and forms the final barrier to the nutrient-rich cytoplasm. From reports based on literature, it is known that the exine is rich in antioxidant compounds, while the intine has the rest of the nutritional and bioactive compounds [54]. Thus, any animal consuming pollen consumes pollenkitt nutrients through external probing of pollen grains or ingestion of pollen grains, but must penetrate or dismantle two resistant pollen wall layers in order to access cytoplasmic nutrients [51]. As the pollen has such complex ultrastructure, it is not certain that the beneficial substances contained in it are available for monogastric animals. Previous studies have shown that the exine and intine are indigestible by the human digestive tract [55]. Furthermore, administering pollen suspended in water to mice by means of gastric intubation resulted in grains being found in the feces [56]. The artificial digestion at various pH values and after various incubation times appeared unaffected after acid treatment (corresponding to gastric digestion), with the exception of the digestion of substances located outside the walls. Partial digestion of grains occurred only during alkaline treatment in the presence of pancreatic enzymes [57]; grains may be partially emptied if the enzymes are able to penetrate the exine pores, and it depends also on the thickness of the intin layer.

Taking into account the difficulties of monogastric animals in digesting bee pollen grains and the similarity of pollen walls with indigestible fibrous substances such as suberin and lignin, it can be assumed that bee pollen had a pro-inflammatory action on the intestinal mucosa of meagre, a fish with carnivorous feeding habits [19,58]. In contrast, the bee pollen structure may not represent a problem to herbivorous fish species such as the Nile tilapia, with an enzymatic profile more suitable for the digestion of fibrous substances and in which positive effects of honey bee pollen addition to the diet have been reported [8].

5. Conclusions

In conclusion, the results of the present trial showed that the addition of honey bee pollen to juvenile meagre diets not only did not have the positive effects described in the literature for other fish species, but it even had negative effects on both growth performance and diet digestibility. Moreover, the inflammatory state of the intestine progressively worsened as the level of HBP inclusion increased. These effects could be ascribed to the ultrastructure of the bee pollen grain walls (exine and intine), which made the bioactive substances unavailable for the fish, while the intestinal inflammation could have been due to the bee pollen grains' chemical composition rich in lignin- and suberin-like substances, which can irritate the intestine of monogastric animals such as carnivorous fish. These negative HBP effects could be overcome by using an extraction method able to concentrate the bioactive substances and eliminate the indigestible fractions. As a general conclusion, it is worth highlighting that it should always be considered necessary to test nutraceutical additives of natural origin in each species in order to verify the effective positive action and exclude any negative repercussions on animal health.

Author Contributions: V.P., G.P., F.B., contributed to the conception of the experimental design; V.P. and S.C., conducted the digestibility and the growth trials and performed the samples collection; S.C. performed the blood analyses (at "Biology—Physiology of Marine Organisms", University of Crete, Prof. M. Pavlidis); C.M.M., S.M., and A.S., performed the biomolecular stress related markers analyses; A.A. and L.S., performed the trace and toxic elements analyses; I.O. and B.R., performed the intestinal histology and immunohistochemistry analyses; F.B., C.M.M., and I.O., performed the statistical analysis; V.P., C.M.M., I.O., S.C., S.M., B.R., F.B., A.A., G.P., supported the acquisition and interpretation of data and contributed in writing the article. All authors have read and agreed to the published version of the manuscript.

References

1. Ferreira, A.; Proença, C.; Serralheiro, M.L.M.; Araújo, M.E.M. The in vitro screening for acetylcholinesterase inhibition and antioxidant activity of medicinal plants from Portugal. *J. Ethnopharmacol.* **2006**, *108*, 31–37. [CrossRef]
2. García Beltrán, J.M.; Espinosa, C.; Guardiola, F.A.; Manuguerra, S.; Santulli, A.; Messina, C.M.; Esteban, M.Á. Effects of dietary dehydrated lemon peel on some biochemical markers related to general metabolism, welfare and stress in gilthead seabream (*Sparus aurata* L.). *Aquac. Res.* **2019**, *50*, 3181–3191. [CrossRef]
3. Guardiola, F.A.; Bahi, A.; Messina, C.M.; Mahdhi, A.; Santulli, A.; Arena, R.C.; Bakhrouf, A.; Esteban, M.A. Quality and antioxidant response of gilthead seabream (*Sparus aurata* L.) to dietary supplements of fenugreek (*Trigonella foenum graecum*) alone or combined with probiotic strains. *Fish Shellfish Immun.* **2017**, *63*, 277–284. [CrossRef] [PubMed]
4. Bahi, A.; Guardiola, F.A.; Messina, C.M.; Mahdhi, A.; Cerezuela, R.; Santulli, A.; Bakhrouf, A.; Esteban, M.A. Effects of dietary administration of fenugreek seeds, alone or in combination with probiotics, on growth performance parameters, humoral immune response and gene expression of gilthead seabream (*Sparus aurata* L.). *Fish Shellfish Immun.* **2017**, *60*, 50–58. [CrossRef] [PubMed]
5. Bulfon, C.; Volpatti, D.; Galeotti, M. Current research on the use of plant-derived products in farmed fish. *Aquac. Res.* **2015**, *46*, 513–551. [CrossRef]
6. Messina, M.; Piccolo, G.; Tulli, F.; Messina, C.M.; Cardinaletti, G.; Tibaldi, E. Lipid composition and metabolism of European sea bass (*Dicentrarchus labrax* L.) fed diets containing wheat gluten and legumes meals as substitutes for fish meal. *Aquaculture* **2013**, *376*, 6–14. [CrossRef]
7. Babaei, S.; Shaban, R.; Torshizi, K.M.A.; Gholam, T.H.; Miran, K.S.N. Effect of Honey, royal jelly and bee pollen on performance, immune system and blood parameters in Japanese quail. *J. Anim. Prod.* **2016**, *17*, 311–320.
8. El-Asely, A.M.; Abbass, A.A.; Austin, B. Honey bee pollen improves growth, immunity and protection of Nile tilapia (*Oreochromis niloticus*) against infection with *Aeromonas hydrophila*. *Fish Shellfish Immun.* **2014**, *40*, 500–506. [CrossRef]
9. Attia, Y.A.; Al-Hanoun, A.; Tag El-Din, A.E.; Bovera, F.; Shewika, Y.E. Effect of bee pollen levels on productive, reproductive and blood traits of NZW rabbits. *J. Anim. Physiol. Anim. Nutr.* **2010**, *95*, 294–303. [CrossRef]
10. Choobkar, N.; Kakoolaki, S.; Mohammadi, F.; Rezaeimanesh, M. The effect of dietary propolis and pollen extracts on growth performance and haematological responses of rainbow trout (*Onchorhynchus mykiss*). *J. Aquat. Anim. Health* **2017**, *3*, 16–25. [CrossRef]

11. LeBlanc, B.W.; Davis, O.K.; Boue, S.; DeLucca, A.; Deeby, T. Antioxidant activity of Sonoran Desert bee pollen. *Food Chem.* **2009**, *115*, 1299–1305. [CrossRef]
12. Thorp, R.W. The collection of pollen by bees. In *Pollen and Pollination*; Dafni, A., Hesse, M., Pacini, E., Eds.; Springer: Vienna, Austria, 2000; pp. 211–223. [CrossRef]
13. Campos, M.G.R.; Bogdanov, S.; de Almeida-Muradian, L.B.; Szczesna, T.; Mancebo, Y.; Frigerio, C.; Ferreira, F. Pollen composition and standardisation of analytical methods. *J. Apic. Res.* **2008**, *47*, 154–161. [CrossRef]
14. Jean-Prost, P.; Le Conte, Y. *Apicultura: Conocimiento de la Abeja. Manejo de la Colmena*, 4th ed.; Ediciones Paraninfo: Madrid, Spain, 2007.
15. Attia, Y.A.; Al-Hamid, A.A.; Ibrahim, M.S.; Al-Harthi, M.A.; Bovera, F.; Elnaggar, A.S. Productive performance, biochemical and hematological traits of broiler chickens supplemented with propolis, bee pollen, and mannan oligosaccharides continuously or intermittently. *Livest. Sci.* **2014**, *164*, 87–95. [CrossRef]
16. Mărghitaş, L.A.; Stanciu, O.G.; Dezmirean, D.S.; Bobiş, O.; Popescu, O.; Bogdanov, S.; Campos, M. In vitro antioxidant capacity of honeybee-collected pollen of selected floral origin harvested from Romania. *Food Chem.* **2009**, *115*, 878–883. [CrossRef]
17. Abbass, A.A.; El-Asely, A.M.; Kandiel, M.M. Effects of Dietary Propolis and Pollen on Growth Performance, Fecundity and Some Hematological Parameters of Oreochromis niloticus. *Turk. J. Fish. Aquat. Sci.* **2012**, *12*, 851–859. [CrossRef]
18. Chatzifotis, S.; Panagiotidou, M.; Papaioannou, N.; Pavlidis, M.; Nengas, I.; Mylonas, C. Effect of dietary lipid levels on growth, feed utilization, body composition and serum metabolites of meagre (*Argyrosomus regius*) juveniles. *Aquaculture* **2010**, *307*, 65–70. [CrossRef]
19. AOAC. *Official Methods of Analysis*, 18th ed.; Association of Official Analytical Chemists: Arlington, VA, USA, 2004; Available online: https://www.aoac.org/ (accessed on 30 January 2020).
20. Ariano, A.; Marrone, R.; Andreini, R.; Smaldone, G.; Velotto, S.; Montagnaro, S.; Anastasio, A.; Severino, L. Metal concentration in muscle and digestive gland of common octopus (*Octopus vulgaris*) from two coastal site in Southern Tyrrhenian Sea (Italy). *Molecules* **2019**, *24*, 2401. [CrossRef]
21. Barnard, T.W.; Crockett, M.I.; Ivaldi, J.C.; Lundberg, P.L. Design and evaluation of an echelle grating optical system for ICP-OES. *Anal. Chem.* **1993**, *65*, 1225–1230. [CrossRef]
22. Piccolo, G.; Iaconisi, V.; Marono, S.; Gasco, L.; Loponte, R.; Nizza, S.; Bovera, F.; Parisi, G. Effect of *Tenebrio molitor* larvae meal on growth performance, in vivo nutrients digestibility, somatic and marketable indexes of gilthead sea bream (*Sparus aurata*). *Anim. Feed Sci. Technol.* **2017**, *226*, 12–20. [CrossRef]
23. Piccolo, G.; Centoducati, G.; Bovera, F.; Marrone, R.; Nizza, A. Effects of Mannan Oligosaccharide and Inulin on Sharpsnout Seabream (*Diplodus Puntazzo*) in the Context of Partial Fish Meal Substitution by Soybean Meal. *Ital. J. Anim. Sci.* **2013**, *12*, 132–138. [CrossRef]
24. Vogtmann, H.; Pfirter, H.P.; Prabucki, A.L. A new method of determining metabolisability of energy and digestibility of fatty acids in broiler diets. *Br. Poult. Sci.* **1975**, *16*, 531–534. [CrossRef]
25. Palmegiano, G.B.; Dapra, F.; Forneris, G.; Gai, F.; Gasco, L.; Guo, K.; Peiretti, P.G.; Sicuro, B.; Zoccarato, I. Rice protein concentrate meal as a potential ingredient in practical diets for rainbow trout (*Oncorhynchus mykiss*). *Aquaculture* **2006**, *258*, 357–367. [CrossRef]
26. Lowry, O.H.; Rosebrough, N.J.; Farr, A.L.; Randall, R.J. Protein measurement with the Folin phenol reagent. *J. Biol. Chem.* **1951**, *193*, 265–275.
27. Hong, S.; Li, R.; Xu, Q.; Secombes, C.J.; Wang, T. Two Types of TNF-α Exist in Teleost Fish: Phylogeny, Expression, and Bioactivity Analysis of Type-II TNF-α3 in Rainbow Trout *Oncorhynchus mykiss*. *J. Immunol.* **2013**, *15*, 5959–5972. [CrossRef]
28. Roca, F.J.; Mulero, I.; López-Muñoz, A.; Sepulcre, M.P.; Renshaw, S.A.; Meseguer, J.; Mulero, V. Evolution of the Inflammatory Response in Vertebrates: Fish TNF-α Is a Powerful Activator of Endothelial Cells but Hardly Activates Phagocytes. *J. Immunol.* **2008**, *181*, 5071–5081. [CrossRef]
29. Johnston, B.L.; Conly, J.M. Tumour necrosis factor inhibitors and infection: What is there to know for infectious diseases physicians? *Can. J. Infect. Dis. Med. Microbiol.* **2006**, *17*, 209–212. [CrossRef]
30. Piccolo, G.; Bovera, F.; Lombardi, P.; Mastellone, V.; Nizza, S.; Di Meo, C.; Marono, S.; Nizza, A. Effect of *Lactobacillus plantarum* on growth performance and hematological traits of European sea bass (*Dicentrarchus labrax*). *Aquac. Int.* **2015**, *23*, 1025–1032. [CrossRef]

31. García-Castillo, J.; Pelegrín, P.; Mulero, V.; Meseguer, J. Molecular cloning and expression analysis of tumor necrosis factor α from a marine fish reveal its constitutive expression and ubiquitous nature. *Immunogenetics* **2002**, *54*, 200–207. [CrossRef]
32. Wang, Y.; Wang, L.; Zhang, C.; Song, K. Effects of substituting fishmeal with soybean meal on growth performance and intestinal morphology in orange-spotted grouper (*Epinephelus coioides*). *Aquac. Rep.* **2017**, *5*, 52–57. [CrossRef]
33. Zhong, J.; Feng, L.; Jiang, W.; Wu, P.; Liu, Y.; Jiang, J.; Kuang, S.; Tang, L.; Zhou, X. Phytic acid disrupted intestinal immune status and suppressed growth performance in on-growing grass carp (*Ctenopharyngodon idella*). *Fish Shellfish Immunol.* **2019**, *92*, 536–551. [CrossRef]
34. Chang, H.R.; Arsenijevic, D.; Pechère, J.C.; Piguet, P.F.; Mensi, N.; Girardier, L.; Dulloo, A.G. Dietary supplementation with fish oil enhances in vivo synthesis of tumor necrosis factor. *Immunol. Lett.* **1992**, *34*, 13–17. [CrossRef]
35. Abdy, E.; Alishahi, M.; Tabandeh, M.R.; Ghorbanpoor, M.; Jafari, H. Comparative effect of Freund's adjuvant and *Aloe vera* L. gel on the expression of TNF-α and IL-1β in vaccinated *Cyprinus carpio* L. with *Aeromonas hydrophila* C. bacterin. *Aquac. Res.* **2016**, *47*, 2543–2552. [CrossRef]
36. Girilal, M.; Krishnakumar, V.; Poornima, P.; Fayaz, A.M.; Kalaichelvan, P.T. A comparative study on biologically and chemically synthesized silver nanoparticles induced Heat Shock Proteins on fresh water fish *Oreochromis niloticus*. *Chemosphere* **2015**, *139*, 461–468. [CrossRef]
37. Metzger, D.C.H.; Hemmer-Hansen, J.; Schulte, P.M. Conserved structure and expression of *hsp70* paralogs in teleost fishes. *Comp. Biochem. Physiol Part D Genomics Proteomics* **2016**, *18*, 10–20. [CrossRef]
38. Radons, J. The human HSP70 family of chaperones: Where do we stand? *Cell Stress Chaperones* **2016**, *21*, 379–404. [CrossRef]
39. Zuo, D.; Subjeck, J.; Wang, X.Y. Unfolding the role of large heat shock proteins: New insights and therapeutic implications. *Front. Immunol.* **2016**, *7*, 75–90. [CrossRef]
40. Han, Y.L.; Yang, W.X.; Long, L.L.; Sheng, Z.; Zhou, Y.; Zhao, Y.Q.; Wang, Y.F.; Zhu, J.Q. Molecular cloning, expression pattern, and chemical analysis of heat shock protein 70 (HSP70) in the mudskipper *Boleophthalmus pectinirostris:* Evidence for its role in regulating spermatogenesis. *Gene* **2016**, *575*, 331–338. [CrossRef]
41. Teigen, L.E.; Orczewska, J.I.; McLaughlin, J.; O'Brien, K.M. Cold acclimation increases levels of some heat shock protein and sirtuin isoforms in threespine stickleback. *Comp. Biochem. Physiol. Part A Mol. Integr. Physiol.* **2015**, *188*, 139–147. [CrossRef]
42. Messina, C.M.; Pizzo, F.; Santulli, A.; Bušelić, I.; Boban, M.; Orhanović, S.; Mladineo, I. *Anisakis pegreffii* (Nematoda: Anisakidae) products modulate oxidative stress and apoptosis-related biomarkers in human cell lines. *Parasit. Vectors* **2016**, *9*, 1–10. [CrossRef]
43. Messina, C.M.; Faggio, C.; Laudicella, V.A.; Sanfilippo, M.; Trischitta, F.; Santulli, A. Effect of sodium dodecyl sulfate (SDS) on stress response in the Mediterranean mussel (*Mytilus Galloprovincialis*): Regulatory volume decrease (Rvd) and modulation of biochemical markers related to oxidative stress. *Aquat. Toxicol.* **2014**, *157*, 94–100. [CrossRef]
44. Chopra, A.S.; Kuratnik, A.; Scocchera, E.W.; Wright, D.L.; Giardina, C. Identification of novel compounds that enhance colon cancer cell sensitivity to inflammatory apoptotic ligands. *Cancer Biol. Ther.* **2013**, *14*, 436–449. [CrossRef]
45. Di Giancamillo, A.; Rossi, R.; Pastorelli, G.; Deponti, D.; Carollo, V.; Casamassima, D.; Domeneghini, C.; Corino, C. The effects of dietary verbascoside on blood and liver oxidative stress status induced by a high *n*-6 polyunsaturated fatty acids diet in piglets. *J. Anim. Sci.* **2015**, *93*, 2849–2859. [CrossRef] [PubMed]
46. Coeurdacier, J.L.; Dutto, G.; Gasset, E.; Blancheton, J.P. Is total serum protein a good indicator for welfare in reared sea bass (*Dicentrarchus labrax*)? *Aquat. Living Resour.* **2011**, *24*, 121–127. [CrossRef]
47. Steinhagen, D.; Oesterreich, B.; Korting, W. Carp coccidiosis: Clinical and hematological observations of carp infected with *Goussia carpelli*. *Dis. Aquat. Org.* **1997**, *30*, 137–143. [CrossRef]
48. Taha, E.A. Chemical composition and amounts of mineral elements in honey bee-collected pollen in relation to botanic origin. *J. Apic. Sci.* **2015**, *59*, 75–81. [CrossRef]
49. Directive (EC). No 2002/32 of the European Parliament and of the Council of 7 May 2002 on Undesirable Substances in Animal Feed (OJ L 140, 30.5.2002, p. 10). Available online: https://eur-lex.europa.eu/legal-content/EN/TXT/?uri=CELEX%3A32002L0032 (accessed on 30 January 2020).

50. Domínguez, C.M.; Bermudez, J.C.; de Cuenca, M.C. Valorization alternatives of Colombian bee pollen for its use as food resource. *Vitae* **2014**, *21*, 237–247.
51. T'ai, H.R.; Cane, J.H. Pollen nutritional content and digestibility for animals. In *Plant Systematics and Evolution*; Springer: Vienna, Austria, 1999; pp. 187–209.
52. Dobson, H.E.M. Survey of pollen and pollenkitt lipids-chemical cues to flower visitors? *Am. J. Bot.* **1988**, *75*, 170–182. [CrossRef]
53. Atkin, S.L.; Barrier, S.; Cui, Z.; Fletcher, P.D.I.; Mackenzie, G.; Panel, V.; Sol, V.; Zhang, X. UV and visible light screening by individual sporopollenin exines derived from *Lycopodium clavatum* (club moss) and *Ambrosia trifida* (giant ragweed). *J. Photochem. Photobiol. B Biol.* **2011**, *102*, 209–217. [CrossRef] [PubMed]
54. Mackenzie, G.; Beckett, S.; Atkin, S.; Taboada, A.D. Pollen and Spore Shells. In *Microencapsulation in the Food Industry*, 1st ed.; Gaonkar, A.G., Vasisht, N., Khare, A.R., Sobel, R., Eds.; Academic Press: Cambridge, MA, USA, 2014; pp. 283–297. [CrossRef]
55. Franchi, G.; Corti, P.; Pompella, A. Microspectrophotometric evaluation of digestibility of pollen grains. *Plant Foods Hum. Nutr.* **1997**, *50*, 115–126. [CrossRef]
56. Franchi, G.G. Research on pollen digestibility. *Atti. Soc. Tosc. Sci. Nat.* **1987**, *B 94*, 43–52.
57. Franchi, G.G.; Franchi, G.; Corti, P. *Indagini Preliminari Sulla Cessione del Contenuto del Polline Nell'apparato Digerente*; Summaries of papers presented at 3° Congresso Nazionale della Società Italiana di Farrnacognosia; Università degli Studi: Sassari, Italy, 1985.
58. Chatzifotis, S.; Panagiotidou, M.; Divanach, P. Effect of protein and lipid dietary levels on growth of juveniles meagre (*Argyrosomus regius*). *Aquac. Int.* **2012**, *307*, 65–70. [CrossRef]

10

Effect of Two Nutritional Strategies to Balance Energy and Protein Supply in Fattening Heifers on Performance, Ruminal Metabolism and Carcass Characteristics

Rodrigo A. Arias [1,2,*], Gonzalo Guajardo [3], Stefan Kunick [3], Christian Alvarado-Gilis [1] and Juan Pablo Keim [1]

1. Instituto de Producción Animal, Universidad Austral de Chile, Valdivia-Chile, Valdivia 5090000, Chile; calvarado@uach.cl (C.A.-G.); juan.keim@uach.cl (J.P.K.)
2. Centro de Investigación de Suelos Volcánicos, Universidad Austral de Chile, Valdivia 5090000, Chile
3. Escuela de Graduados, Facultad de Ciencias Agrarias y Alimentarias, Universidad Austral de Chile, Valdivia 5090000, Chile; gonzalo.guajardo.c@gmail.com (G.G.); stefan93_@live.cl (S.K.)

* Correspondence: rodrigo.arias@uach.cl.

Simple Summary: Beef production has been under strong scrutiny during the last half-century. First, because of its supposed negative impact on human health, and more recently, due to the negative impact on the environment, mainly from nitrogen and greenhouse gases. We conducted an experiment to assess the effects of a diet formulated based on the metabolizable protein system and synchronicity between energy and protein, on the nitrogen losses to the environment, performance, and carcass characteristics of the fattening heifers. Our results show that a diet combining a high synchrony index with the metabolizable protein system increases the nitrogen use efficiency without negatively affecting animal performance or carcass characteristics compared to heifers fed a diet without a balanced protein but with a high synchrony index.

Abstract: Latin America is an important contributor to the worldwide beef business, but in general, there are limited studies considering strategies to reduce nitrogen contamination in their production systems. The study's goal was to assess the effect of two nutritional strategies to balance energy and protein supply in fattening heifers on performance, ruminal metabolism, and carcass characteristics. A total of 24 crossbred heifers (initial body weight 'BW 'of 372 ± 36 kg) were used to create two blocks (based on live weight) of two pens each, that were equipped with individual feeders. Within each block, half of the animals were assigned to a diet based on tabular Crude Protein (CP) requirements denominated Crude Protein Diet 'CPD' but without a ruminal degradable protein balance. The other half received a diet denominated Metabolizable Protein Diet 'MPD', formulated with the metabolizable protein system, balanced for the ruminal degradable protein. Both diets had the same ingredients and as well as similar synchrony indexes (0.80 and 0.83, respectively). For nitrogen concentration in feces and urine as well as microbial crude protein synthesis, a total of 12 heifers (three per pen) were randomly selected to collect samples. The dataset was analyzed as a randomized complete block design with a 5% significance. No diet × time interaction was observed for Average Daily Gain 'ADG' ($p = 0.89$), but there was an effect of the time on ADG ($p \leq 0.001$). No differences were observed neither for final weight, dry matter intake 'DMI', and feed conversion rate ($p > 0.05$). Heifers fed with CPD showed greater cold carcass weight ($p = 0.041$), but without differences in ribeye area, backfat thickness, pH, dressing %, and marbling ($p > 0.05$). Differences between diets were observed for the in vitro parameters as well as for the Total Volatile Fatty Acids 'VFA' and NH_3 ($p < 0.05$). Total N concentrations (urine + feces) of heifers fed with MDP was lower than in those fed with the CPD ($p < 0.01$), but no differences were observed in microbial protein, purine derivatives, and creatinine ($p > 0.05$). We conclude that the combination of synchrony and the metabolizable protein system achieve greater efficiency in the use of nitrogen, without negatively affecting animals' performance or the quality of the carcass.

Keywords: nitrogen efficiency; synchrony index; metabolizable protein system

---

1. Introduction

In many countries, traditional beef production systems have a low level of intensification. Though, the implementation of new technologies in herd management, such as the type and form of feeding, would have a positive response in productive efficiency, which directly impacts the final utility of the producer [1]. Nevertheless, due to the growing world beef demand in conjunction with more environmental restrictions, it is expected that the increase in production is mainly due to more intensification [2]. This new scenario should aim not only to achieve more beef and of better quality but also improve animal welfare and, at the same time, cause minimum environmental impact. However, there is a risk that greater intensification is a source of environmental pollution since it has been associated with highly intensive animal production systems [3]. This is mainly due to nitrogen (N) and phosphorus (P), which pollute groundwater, surface water, and air. The same authors detail that the main reasons for contamination are excretions (urine and feces) that are discharged into surface waters, or by volatilization of N in the form of NH_3. It also has been suggested that new feeding strategies should be adopted to improve the use of N and reduce its losses. In this sense, Klemesrud et al. [4] state that satisfying the animal's amino acid requirements, without falling into excess protein in the rations, it allows to reduce N excretions. Thus, adequate management of nutrition should not only allow digesting and utilizing the fiber from forages and promote microbial protein synthesis but also reduce N losses at the ruminal level, which may be excreted to the environment. Farmers and nutritionists should try to maximize microbial protein production in the rumen because it has been established that microbial protein can provide from 50 to 100% of the metabolizable protein (MP) required by the animal, which depends on the amount of ruminal degradable protein and fermentable carbohydrates in the diet [5].

Owen and Sapienza [6] pointed out that to achieve maximum efficiency and production of microbial protein, factors that affect its synthesis should be considered, such as the availability of energy, ammonium or other N sources, minerals, vitamins, and factors related to animal growth. The synchrony between energy and degradable protein in the rumen is a very important factor in the optimization of microbial protein synthesis [7]. Therefore, not only the contribution or availability of energy and nitrogen sources at the ruminal level should be necessary [8,9], but also synchrony [10]. Other authors consider that the amount or availability of energy in the rumen, in a certain nitrogen range, is usually more important than synchrony as such for the synthesis of microbial protein [9,11]. According to Sinclair [12], ruminal synchrony would improve the rumen fermentation, digestibility of nutrients, and the microorganism's population [13], as well as bacterial protein synthesis, nitrogen retention, the animal's productive response, and consequently, the N use efficiency (NUE) [14].

Ruminal synchrony could be affected by different factors such as diet, which is the main determinant of the quantity and quality of nutrients supplied to microorganisms [10]. Mention has also been made to the chemical composition of the feedstuffs [15], the ingredients used in the ration [16], grains processing [17], and the degradation characteristics of organic matter and protein in food [13,18]. Other factors related to the animal, such as health, nutritional requirements, and the interaction with the type of management and the environment, also alter the use of nutrients, causing ruminal asynchrony [10]. As mentioned, the synchrony between energy and protein available in the rumen theoretically should allow more efficient use of nutrients, improving the production of microbial products to increase the supply of nutrients, and thus improve the animal's performance. However, after years of studies and analysis of this concept, it has not been fully demonstrated in practice.

Likewise, asynchronous nutrient diets have generated results as good or higher than synchronous diets in growing beef and dairy cattle in confinement [10]. Duarte et al. [19] evaluated the effects of different levels of energy and non-degradable protein in the rumen on consumption, growth, carcass characteristics, and meat quality in fattening heifers. The authors concluded that there was a higher average daily gain (ADG) with high levels of rumen undegradable protein (RUP). However, levels of RUP had no effect on carcass characteristics and composition, but there was an effect on the energy level with a greater area of the eye of the back (longissimus muscle).

At present, the MP system widely used by beef nutritionists, predicts protein requirements accurately, being effective in providing protein levels even at or near the predicted requirements, ensuring good animal performance [20]. Erickson et al. [21] evaluated the effect of the use of diets balanced with the conventional method (based on CP) vs. others based on the MP system in fattening of steers and calves under highly intensive systems in the USA. They determined that there were no differences in the animal's productive responses. However, the excretion of N was reduced with experimental diets. Nowadays, there is limited information about the implementation of the MP system in the South American beef production systems. Most of these production systems are much less intensive than North American feedlots, such as those studied by Erickson et al. [21]. In this context, we hypothesize that a diet formulated with the MP system will decrease the environmental impact by reducing nitrogen losses to the environment, without negatively affecting in vitro rumen fermentation, the productive performance and the carcass characteristics of the fattening heifers.

2. Materials and Methods

All the procedures, including animal care and handling procedures, followed national legislation (Law No. 20,380 on Protection of Animals; Decree No. 29 about regulation on the protection of animals during their industrial production, their commercialization and in other areas to hold animals), whose application is supervised by the National Service of Agriculture and Livestock (SAG), the competent authority in this matter.

2.1. Animals and Facilities

The study was conducted at the Austral Agricultural Research Station (EEAA) of the Universidad Austral de Chile, located 8 km North of Valdivia, Los Ríos region (39° 46′28′′ S 73° 14′11′′ W), from September 6th, 2016 until December 6th, 2016. A total of 24 crossbred heifers (Hereford × Angus) with an average initial BW of 372 ± 36 kg were used (68% of mature weight). All heifers were dewormed with 1% ivermectin for the control of gastrointestinal parasites. Upon arrival, animals were grouped based on their initial live weight conforming two blocks (light: 345 ± 23.32 kg and heavy: 399 ± 25.10 kg) and were located in four pens that were randomly assigned to each block. Each block consisted of two pens with a space allowance of 26.25 m^2 per head. In addition, each pen was equipped with six individual semi-automatic feeders (American Calan Inc., Northwood, NH, USA), which opened through an electronic key that hangs on the neck of each animal. This allowed the control of individual food intake, considering each animal as an experimental and observational unit. Feeders were under a shed to protect feeds from rain, with a cement surface (platform) of 3.5 m wide. Cattle had access to water through two shared drinking fountains of 600 L, each with a float system to ensure their filling, each located between two adjacent pens.

Heifers were weighed approximately every 20 d with an electronic scale (Iconix FX31, Iconix NZ Ltd., Palmerston North, New Zealand). The weighing was performed during the morning before the feed was supplied to avoid variations due to gastrointestinal content. Average daily weight gain (ADG) was estimated at intervals corresponding to weighing dates. Additionally, feed conversion of the whole trial was determined by dividing the total amount of food consumed by the total kg of body live weight (LW) gained. Finally, the LW of heifers was used to calculate the synthesis of microbial protein.

There was a training period of approximately 20 d that was carried out so that the animals got used to the facilities and the feeders, as was previously described by Arias et al. [22]. During this time, heifers were fed ad libitum with a 100% pasture haylage and water. Animals were fed once a day (09:00 to 11:00 am). The ingredients were weighed and mixed daily before supplying the diet in each feeder, and orts were collected and weighed individually (per feeder) once a week. Simultaneously, the preference of each animal for the feeders was observed in order to assign the respective key. Finally, once assigned and the keys placed on the neck of the heifers, the electrical system was connected to block them and, the experimental phase was started, which took place over a period of 64 d. Animals were fed once a day (09:00 to 11:00 am).

2.2. Dietary Treatments and Nutrient Concentration Analyses

The chemical composition of feed ingredients and their inclusion in the dietary treatments are reported in Tables 1 and 2. The animals within each block were randomly assigned to one of two dietary nutritional strategies. Diets were formulated to be isoenergetic. Each diet was designed according to the ruminal degradation parameters of their respective ingredients in order to obtain similar synchrony between energy and protein in the rumen based on the synchrony index (SI) proposed by Sinclair et al. [12]. The SI of both diets was calculated based on the ruminal degradation parameters of OM and CP of each ingredient of the diets adopting the methodology of Verbič et al. [23]. Both diets were calculated to obtain an ADG of 1.0 kg. The first treatment was formulated based on heifers CP requirements and was defined as the 'Crude Protein diet' (CPD). It was designed to provide the amount of daily metabolizable energy and crude protein needed for heifers with an SI = 0.80. This diet was analyzed using the Beef Cattle Nutrient Requirement Model (BCNRM) software [24], contrasting the values obtained in the balance for predicted ruminal degradable protein (RDP). According to the National Academies of Sciences, Engineering, and Medicine (NASEM) [24], there was an imbalance of the RDP supply (+240 g d^{-1}) since it was calculated based on the CP system. The second dietary treatment 'Metabolizable Protein diet' (MPD) was formulated by using the MP system, providing an adequate supply of ruminal undegradable protein (RUP; −10 g d^{-1}) and with a similar SI than the CPD (0.83).

Table 1. Nutritional characteristics of the ingredients utilized in the study.

Item	Pasture Haylage *	Canola Meal	Sugar Beet Pulp	Triticale Grain
Dry Matter, %	49.67	91.70	88.90	90.35
Crude Protein, %	14.31	33.48	8.69	8.13
Soluble Protein, %	8.10	8.56	0.17	1.79
Neutral Detergent Fiber, %	53.68	32.15	50.22	20.32
Acid Detergent Fiber, %	30.58	21.51	24.65	3.79
DOMD, %	76.20	74.40	90.12	92.52
TDN, %	76.50	74.83	88.97	91.17
Organic matter, %	92.98	94.19	94.75	98.40
NFC, %	21.34	16.97	35.11	68.43
Ether Extract, %	3.65	11.59	0.73	1.52
Ash, %	7.02	5.81	5.25	1.60
ME, Mcal kg^{-1} DM	2.76	2.70	3.21	3.29
NEm, Mcal kg^{-1} DM	1.83	1.78	2.20	2.27
NEg, Mcal kg^{-1} DM	1.20	1.16	1.52	1.57

* Purchased to local farmers, made mainly of *Lolium spp.* but also included a lower proportion of *Dactylis glomerata*, *Bromus spp.*, and *Trifolium spp.* DOMD = digestible organic matter on a dry matter basis; TDN = Total Digestible Nutrients estimated as ((ME/0.82)/4.4) × 100; NFC = Non-fiber carbohydrates estimated as 100—(%NDF + %CP + %Ash + %EE); ME = Metabolizable energy; NEm = Net energy of maintenance, and NEg = Net energy of gain.

Dry matter content was measured by weighing the samples before and after drying with a forced-air oven, initially at 60 °C for 48 h, and then at 105 °C for 12 h. The CP concentration was determined by combustion (Leco Model FP-428, Leco Corporation, St Joseph, MI, USA) based on the DUMAS method (N × 6.25), digestible organic matter on a dry matter basis (DOMD) was measured according to Tilley and Terry [25], neutral detergent fiber (aNDF) was measured by using a heat-stable amylase [26], and ash and ether extract (EE) were analyzed according to AOAC [27] (Methods ID 942.05 and ID 920.39 for ash and EE, respectively).

Table 2. Nutritional characteristics of the diets, ingredient inclusion, and synchronize index per treatment.

Item	Treatment	
	CPC	MPD
Synchrony index (SI) †	0.80	0.83
Ingredient inclusion in the diet		
Pasture haylage, %	50.0	58. 0
Canola meal, %	15.0	4.0
Sugar beet pulp, %	20.0	0.0
Triticale grain, %	15.0	38.0
Dry Matter, %	64.1	61.3
Crude Protein, %	15.1	12.7
Soluble Protein, %	5.6	5.7
Neutral Detergent Fiber, %	44.8	40.1
Acid Detergent Fiber, %	24.0	21.5
Digestibility value, %	81.2	74.4
TDN, %	66.2	68.8
Organic Matter, %	94.3	95.1
NFC, %	30.5	39.1
Ether Extract, %	3.9	3.2
ME, Mcal kg^{-1} DM	2.9	3.0
NEm, Mcal kg^{-1} DM	2.0	2.0
NEg, Mcal kg^{-1} DM	1.3	1.4
RDP, (% CP) ‡	86.0	83.3
RUP, (% CP) ‡	14.0	16.8
MPB, (g/d) ‡	5.8	0.1
RDPB, (g/d) ‡	240.0	−10.0

CPD: Crude protein diet; MPD: metabolizable protein diet; TDN = Total Digestible Nutrients estimated as ((ME/0.82)/4.4) × 100; NFC = Non-fiber carbohydrates estimated as 100—(% NDF + % CP + % Ash + % EE); ME = Metabolizable energy; NEm = Net energy of maintenance, and NEg = Net energy of gain; RDP = Ruminal degradable protein; RUP = Ruminal undegradable protein; MPB: Metabolizable protein balance; RDPB = Ruminal degradable protein balance. † Calculated from in situ degradation parameters and Verbič et al. [23] equations. ‡ Data obtained from the BCNRM software v1.0.37.14. [24]. Both diets were supplemented with 80 g d^{-1} per heifer of mineral salts.

2.3. *Organic Matter and Crude Protein in Situ Degradation Parameters of Feed Ingredients*

Food samples were lyophilized and grounded through a 5 mm sieve. Polyester bags of 10 × 20 cm (40–60 μm porosity) contained 4 g of sample achieving a ratio of 16 mg of sample per cm^2. Two replicates (bags) by incubation period were introduced in a fistulated cow (Holstein breed in a state of maintenance) in reverse order to remove all the bags from the rumen at the same time. Incubation times corresponded to 0, 2, 4, 8, 10, 14, 24, and 48, and 72 h were used. The cow was fed a ration formulated according to its nutritional requirements and managed in the EEAA housing yard. Prior to ruminal incubation, the bags were placed in 20 × 30 cm porous laundry bags and soaked in warm water (30 °C) for 20 min and then introduced into the donor's cow rumen. Once removed from the rumen, bags were washed with running water until the water was clear and then frozen at −20 °C for 24 h to stop any fermentation activity. Thereafter, samples were thawed and washed in a conventional washing machine for 10 min and dried in a forced-air oven at 60 °C for 48 h. The time 0 h was not incubated in the rumen and was used to determine the soluble fraction of the DM and CP. Residues after ruminal incubation were weighed to determine the amount of feed degraded at a certain incubation time.

Ruminal degradation parameters of OM and CP were determined using the model proposed by Ørskov and McDonald [28], through the nonlinear procedure of the GraphPad Prism v6.0 software, using the exponential model without a lag phase [28]:

$$PD = A + B\,(1 - e^{-kt}) \tag{1}$$

where A is the soluble fraction (g kg^{-1} fraction of bags washed at time 0 h), B is the insoluble but potentially degradable fraction (g kg^{-1}), k is the degradation rate constant (% h^{-1}), and t is the incubation time (h).

Effective degradability (ED) was calculated from the aforementioned parameters assuming fractional passage rates (kp) of 5% h^{-1} according to McDonald [29]:

$$ED\ (g\ kg^{-1}) = A + B[k\,/\,(k + kp)]e^{-kp} \tag{2}$$

where A (g kg^{-1}) is the soluble fraction; B (g kg^{-1}) is the insoluble but potentially degradable fraction; k and kp (h^{-1}) are the ruminal degradation rate constant and passage rate constant (5%), respectively, according to AFRC [30].

A correction for small particle losses was made according to Hvelplund and Weisbjerg [31]. Thus, samples of each food (1.5 g) were placed in containers to which 40 mL of water was added and stored at room temperature (20 °C) for an hour. Then they were filtered and washed eight times with 20 mL of water; the residue was dried in a forced-air oven at 60 °C for 48 h, and then weighed, calculating the water soluble (SOL) fraction. Assuming that, losses of small particles are degraded similarly to the particles left in the bag, corrections can be made for the loss:

$$PDcor\ (ti) = PD\ (ti) - P\,[1 - ((PD\ (ti) - (P + SOL))\,/\,(1 - (P + SOL)))] \tag{3}$$

$$EDcor = SOL + [((1 - SOL)\,/\,(1 - (P + SOL))) \times (ED - (P + SOL))] \tag{4}$$

$$acor = A - P \tag{5}$$

$$bcor = B + P\,[B\,/\,(1 - (P + SOL))] \tag{6}$$

$$ccor = c \tag{7}$$

where: *PDcor (ti)* is the degradability corrected at the time of incubation *ti*, *EDcor* is the corrected effective degradation, *P* are the losses of small particles, and *SOL* is the water solubility.

The values of the in-situ degradation characteristics of OM and CP are shown in Table S1, which was mainly used for the formulation of treatments and to determine the SI. The adjusted degradation curves of the OM and CP are shown in Figure S1.

2.4. In Vitro Fermentation

In vitro gas production was evaluated according to Theodorou et al. [32]. Three replicates (120 mL bottles) were used, including two "control" bottles (a standard concentrate with a known gas production pattern to compare and ensure the proper fermentation process) and two blanks (bottles without substrate) in three incubation runs. One gram of dried lyophilized substrate was used in each bottle. Then, they were filled with 85 mL of Goering–Van Soest medium, gasified with CO_2, and closed with rubber stoppers, and left at 4 °C overnight. The next day, 4 mL of reducing agent (Distilled water, sodium sulfate, cysteine HCl, and 1 mol/L NaOH) were added; each bottle was gasified again, completely closed with the rubber stopper, and sealed with aluminum. Then they were placed in a water bath at 39 °C. Rumen fluid was obtained directly from heifers in the experiment, randomly selected from the two treatments of each block. The extraction of ruminal liquor was conducted in the morning before heifers were fed, using an oro-ruminal probe (FLORA Ruminator; Profs-Products, Guelph, ON, Canada), stored immediately in thermos flasks and transported to the laboratory. The ruminal liquid was filtered through a cheesecloth, mixed and placed under constant CO_2 gasification. Later, ruminal liquor from heifers fed their corresponding diet was added (10 mL) to bottles containing the same diet. Once the rumen fluid was inoculated, the initial gas was extracted from the bottles. After inoculation, the bottles were placed in a water bath at 39 °C under continuous horizontal movement at 50 rpm. The gas pressure in the headspace of the bottles, above atmospheric pressure, was measured manually with a pressure transducer (PCE Instruments, Tobarra, Albacete, Spain) at 1, 2, 3, 4, 5, 6, 8, 10, 12, 15, 18, 24, 36, and 48 h, and the volume of gas produced was measured by extraction, using syringes connected through a three-way Luer valve from the bottles until the visual display of the transducer read zero. Once the volume of gas produced was recorded, it was eliminated. Fermentations were stopped after 48 h by placing the bottles on ice. For the in vitro gas production kinetics, after correcting for the white gas production, it was adjusted by the Michaelis–Menten model without lag phase [33] to obtain the corresponding fermentation parameters of the treatments and each of the feeds.

$$GP = A \times [T^n / (T^n + K^n)] \quad (8)$$

where GP is the production of gas at time T; A is the volume of asymptotic gas (mL g^{-1} MS); n is the coefficient that determines the shape of the curve in the function, and K is the time in which half of A (h^{-1}) occurs. The other parameters were calculated according to Groot et al. [34] and France et al. [33]:

$$C = n / (2 \times K) \quad (9)$$

$$MDR = (n - 1)\ ((n - 1) / n) / K \quad (10)$$

$$ta,\ tb \text{ and } tc = K \times (((X / (1 - X))\ (1 / n)) \quad (11)$$

where C is the degradation rate in the middle of the asymptote; MDR is the maximum degradation rate; ta, tb, and tc (h^{-1}) correspond to the time at which 25, 75, and 90% of A occurs; and X varies between 0.25, 0.75, and 0.90.

Volatile fatty acids (VFA) and ammonium (NH_3) concentrations were measured only for diets. For this, samples were obtained from the bottles at 4 and 48 h post-incubation, plus the two samples of inoculum to calculate net VFA and NH_3 production. Then, 5 mL was extracted from the supernatant, and 0.1 mL of concentrated hydrochloric acid (37%) were added. These were stored frozen (−20 °C) and were subsequently analyzed to determine the concentration (mmol L^{-1}) and proportion (%) of VFA (acetic acid (C2), propionic acid (C3), butyric acid (C4), isobutyric, valeric and isovaleric [VFAscr]), and the concentration of NH_3 (mg L^{-1}).

2.5. *The Concentration of N in Feces, Purine and N Derivatives in Urine*

A total of 12 heifers (three per pen) were randomly selected to collect fecal samples during the experiment, but one of them was dismissed from the study because of pregnancy. Samples were obtained three times during the experimental period (one day after the 1st, 3rd, and 4th weighing) and always from the same animals. The fecal samples were collected manually from the rectum, then frozen at −20 °C and then lyophilized. The determination of the concentration of N was made with the LECO FP528 nitrogen analyzer described above.

Additionally, urine samples were also obtained from the same 11 heifers by manual stimulation of the vulva. Urine was collected in a 1.0 L container covered with a mesh of lingerie to avoid solid remains coming from around the vulva. A minimum volume of 60 mL of urine was collected for the analysis of the N concentration, plus another 20 mL to determine purine derivatives in the urine. Previously, each bottle was filled with H_2SO_4 (10% v v^{-1}) of the collected volume. Samples were stored in a cooled container, then frozen at −20 °C. Samples destined to obtain N were lyophilized and subsequently analyzed through the LECO FP528 nitrogen analyzer. Samples for purine derivatives analyses (n = 10) were thawed and analyzed by using the HPLC-UV technique described by Vlassa et al. [35]. Urine volume was estimated according to Al-Khalidi [36], whereas daily total creatinine excretion was calculated according to Chizzotti [37]. Daily excretion of purine derivatives was estimated with the equation of Faichney et al. [38]. The creatinine daily excretion coefficient (mg d^{-1} K) = 113.12 × $LW^{-0.25}$ used was based on Ørskov et al. [39]. Daily purine absorption and the contribution of microbial N were calculated according to Chen and Gomes [40]. Finally, the total microbial protein (TMP) was obtained by the following equation:

$$TMP = MN \times 6.25 \quad (12)$$

where TMP: Total microbial protein (g d^{-1}); MN: Microbial Nitrogen (g d^{-1}); and 6.25 is the N content in proteins. These results were utilized to estimate the daily synthesis of microbial protein.

2.6. Nitrogen Use Efficiency (NUE)

Calculation of NUE was made based on the results of the N balance according to the BCNRM software v1.0.37.14. [24] and the equations proposed by Cole et al. [41] to estimate the retained N based on ADG and LW. The software estimates the amount of N excreted in the urine and feces separately, in addition to the N retained by the animal. Likewise, it estimates the N intake from the diet CP, considering 16% of N. The absorbed N was estimated according to the amount of MP, and the N retained according to the net growth protein. In addition, the software considers both excretions and endogenous N from the animal. Estimations of NUE were made for two periods (Period 1 = first 43 d, and Period 2 = the last 21 d) since consumption was different during these periods. Therefore, the nitrogen consumed was also different at each stage.

$$NUE = (N\ retained\ /\ N\ intake) \times 100 \quad (13)$$

2.7. Carcass Characteristics

Heifers were transported into a slaughterhouse for processing, where the following data were collected: cold carcass weight, carcass pH (measured directly 24 h postmortem), backfat thickness, ribeye area (measured in the cross-section of the *Longissimus dorsi* muscle, between the 9th and 10th ribs), and marbling (on the surface of the 9th rib muscle, determined by using USDA standards.

2.8. Data Analysis

The experimental model corresponded to a randomized complete block design (live weight as a block factor) with a univariate treatment structure with two levels (diets). For performance and carcass quality, each animal was considered as an experimental and observational unit. For variables derived from purine in urine, there were 10 observational units (four PCD and six MPD) for the variables of N in urine and feces n = 11 (five CPD and six MPD). The ANOVA model was

$$Y_{ijk} = \mu + \beta_j + \alpha_j + \varepsilon_{ijk} \quad (14)$$

where μ corresponds to the general mean; βi represents the effect of the ith block; α_j represents the effect of the jth diet, and ε_{ijk} the experimental error associated to the kth animal of the jth diet in the ith block. For multiple comparisons, the Tukey test was conducted with a significance = 0.05 when corresponded. In the case of the microbial protein synthesis, N in urine and feces, and results of in vitro fermentation, a repeated measure in the time model was used ($Y_{ijk} = \mu + \alpha_i + S_{k(i)} + \beta_j + [\alpha\beta_{ij}] + [\beta \times S_{jk(i)}] + \varepsilon_{ijk}$, where Y_{ijk} is the observation of the ith level diet and the jth level of time for subject k; μ is the general mean, α_i is the fixed effect level i of diet, $S_{k(i)}$ is the effect of subject nested within a block, β_j is the fixed effect of level j of factor time, $\alpha\beta_{ij}$ is the interaction effect between diet and time, ε_{ijk} is the random error effect). Since there was no replication for each combination of subject and factor time, the [$\beta \times S$] interaction effect cannot be separated from the error term and must be assumed to be zero. Animal performance and carcass data were analyzed with the statistical package JMP v14.0 (SAS Institute Inc.,

Cary, NC, USA). Repeated measures variables were analyzed with the SAS statistical package version 9.4 PROC MIXED (SAS Institute Inc.).

3. Results

3.1. Animal Performance and Carcass Characteristics

No diet effects for final LW, DMI, ADG, and Feed Conversion (Table 3) were observed. Likewise, no diet × time interaction was observed for ADG ($p = 0.89$), but there was an effect of the time on ADG ($p \leq 0.001$). During the first 21 d, ADG was as expected (near 1.0 kg/d), while in the second period (from 21 to 43 d), there was a slight increase for both diets. Finally, in the last period (43 to 64 d), there was a steep decrease in ADG in both treatments. In addition, no differences ($p > 0.05$) were observed for the weights recorded in the second and third weightings. However, in the last period (43 to 64 d), a decrease in the growth response for both treatments was observed for both the ADG and LW variables.

Table 3. Least square means for live weight, dry matter consumption, feed conversion, and daily weight gain per diet.

Variable	CPD ± SEM	MPD ± SEM	*p*-Value
Initial LW, kg	390.7 ± 6.09	386.6 ± 5.51	0.302
Final LW, kg	445.9 ± 4.72	437.5 ± 4.33	0.207
ADG, kg d^{-1}			
d 0–21	1.01 ± 0.15	0.90 ± 0.14	0.604
d 21–43	1.24 ± 0.15	1.02 ± 0.14	0.308
d 43–64	0.36 ± 0.15	0.30 ± 0.14	0.774
d 0–43	1.11 ± 0.10	0.95 ± 0.09	0.234
d 0–64	0.87 ± 0.07	0.74 ± 0.06	0.200
DMI, kg d^{-1}	8.56 ± 0.04	8.49 ± 0.03	0.192
Feed conversion rate, kg DM kg LW gain $^{-1}$	10.7 ± 1.42	12.3 ± 1.59	0.381

CPD: Crude Protein Diet; MPD: Metabolizable Protein Diet; ADG = Average daily gain; DMI = Dry matter intake; LW = Live body weight.

Heifers fed with the CPD showed greater cold carcass weight (Table 4), approximately 12 kg more than those fed with the MPD diet ($p = 0.04$). However, no differences were observed in the other carcass characteristics ($p > 0.10$), even when the rib eye area was numerically higher in heifers fed with the MPD.

Table 4. Least square means for carcass characteristics variables measured in heifers fed with two different diets.

Variable	CPD	MPD	*p*-Value
Cold carcass weight, kg †	240.2 ± 4.12 ^{a}	228.3 ± 3.55 ^{b}	0.041
Ribeye area, cm^2	88.1 ± 3.10	92.6 ± 2.68	0.281
Back fat thickness, mm	7.0 ± 1.10	6.5 ± 0.91	0.704
Marbling ‡	1.9 ± 0.20	1.8 ± 0.17	0.728
Muscle pH	5.0 ± 0.10	5.1 ± 0.05	0.135
Dressing, %	54.0	53.0	0.150

CPD: Crude Protein Diet; MPD: Metabolizable Protein Diet; † Different letters between columns indicate statistical difference, Tukey test; ‡ Marbling values scale from 1 to 3 (1 = Choice +; 2 = Choice; and 3 = Choice −).

3.2. In Vitro Fermentation Products

Parameters of in vitro gas production kinetics for both diets are shown in Table 5. No differences in asymptotic gas and 48 h GP were observed. The MPD exceeded the CPD in the MDR and C, whereas the time to produce 25, 50, 75, and 90% of asymptotic production was greater for CPD than MPD.

Table 5. Least square means of the parameters of the in vitro gas production kinetics of diets.

Variable	Diets		SEM	*p*-Value
	CPD	MPD		
GP48	248.6	257.4	9.33	0.125
A	266.7	269.1	10.84	0.715
n	1.8	1.8	0.06	0.128
K	12.2	11.1	0.42	0.008
C	0.1	0.1	0	0.003
MDR	0.1	0.1	0	0.004
ta	6.6	6.1	0.36	0.039
tb	22.6	20.2	0.44	0.004
tc	42.1	36.6	1.22	0.004

CPD = Crude Protein Diet; MPD = Metabolizable Protein Diet; GP48: Gas volume produced at 48 h of incubation (mL g^{-1} DM); A = Asymptote gas volume (mL g^{-1} DM); n = coefficient that determines the curve in the function; K = time to produce 50% of A (h), C: fractional rate of gas production in the middle of the asymptote (h^{-1}), MDR = maximum degradation rate (h^{-1}); ta, tb y tc = time to produce 0.25, 0.75 y 0.90 of A (h^{-1}).

The production and proportion of VFAs and NH_3 concentrations are presented in Table 6. There was a diet by time interaction for the concentration of NH_3, being similar at 4 h after incubation but at 48 h, the concentrations were 72% higher in the CPD. The total VFA production was greater for CPD when compared to MPD. However, the proportions of propionic acid and VFAscr were higher for MPD, whereas no differences were observed for the proportions of acetic and butyric acids, nor for the acetic: propionic and (acetic + butyric): propionic ratios.

Table 6. Least square means for the production and concentration of volatile fatty acids (VFAs) and the concentration of ammonium (NH_3) for both diets.

Variable	Diets			Time			*p*-Values		
	CPD	MPD	SEM	4 h	48 h	SEM	Diet	Time	D × T
NH_3, mg L^{-1}	89.9	66.5	4.6	67.8	88.6	4.48	<0.05	<0.01	<0.01
Total VFA, mmol L^{-1}	21.0	15.8	0.9	4.1	32.7	1.28	<0.01	<0.01	NS
Acetate, mol 100 mol^{-1}	58.0	52.4	2.5	63.6	46.8	2.83	NS	<0.05	NS
Propionate, mol 100 mol^{-1}	29.8	32.9	1.9	27.0	35.7	2.27	<0.05	<0.05	NS
Butyrate, mol 100 mol^{-1}	9.9	11.3	2.1	8.9	12.3	1.84	NS	NS	NS
VFAscr, mol 100 mol^{-1}	2.3	3.5	0.1	0.6	5.2	0.25	<0.01	<0.01	NS
C2:C3	2.1	1.7	0.2	2.5	1.3	0.29	NS	<0.05	NS
(C2 + C4):C3	2.4	2.1	0.3	2.8	1.7	0.33	NS	<0.05	NS

CPD = Crude Protein Diet; MPD = Metabolizable Protein Diet; SEM = Standard error of the mean; NH_3 = ammonia concentration; VFAscr = Branched chain volatile fatty acids production (isobutiric + valeric + isovaleric); C2:C3 = ratio acetic: propionic acid; (C2 + C4)/C3 = ratio ketogenic acids: glucogenic acid.

3.3. *Nitrogen Balance*

There was an increase in TMP synthesis over time for both diets (Table 7; $p < 0.01$). In addition, D1, D43, and D64 showed no differences between treatments in the daily amount of TMP produced. Likewise, there was no diet effect on the synthesis of TMP ($p > 0.05$), and the interaction between diet and time was also not significant ($p > 0.05$). There were no differences in time ($p > 0.05$) in the three dates when urinary N was evaluated. The N concentration in the urine was higher in the CPD ($p < 0.01$). Therefore, heifers that received the MPD had a lower concentration of nitrogen in the urine during the fattening period. In addition, there was no interaction between treatments and time ($p > 0.05$).

Table 7. Least square means for microbial protein, purine derivatives, creatinine, urine, and feces nitrogen concentration for both diets.

Variable	Diets				Time				*p*-Values		
	CPD	SEM	MPD	SEM	Day 1	Day 43	Day 64	SEM	Diet	Time	D × T
TMP, g d^{-1}	581.2	47.8	643.4	52.3	420.4	670.3	746.2	44.6	NS	<0.01	NS
N_{Mic}, g d^{-1}	93	7.6	102.9	8.4	67.3	197.2	119.4	7.1	NS	<0.01	NS
A, mmol L^{-1}	11.4	2.2	13.9	2.4	14.2	11.2	12.6	2.2	NS	NS	NS
UA, mmol L^{-1}	1.7	0.3	2.0	0.3	2.6	2.3	0.5	0.2	NS	<0.01	NS
C, mmol L^{-1}	11.5	2.1	14.7	2.3	20.1	9.9	9.2	2.1	NS	<0.01	NS
TPD, mmol L^{-1}	13.0	2.4	15.9	2.6	16.8	13.5	13.1	2.3	NS	NS	NS
N in urine, %	7.4	0.4	4.9	0.3	6.1	6.5	6.3	0.5	<0.01	NS	NS
N in feces, %	3.3	0.1	2.7	0.1	2.9	2.8	3.0	0.1	<0.01	NS	<0.05

CPD = Crude Protein Diet; MPD = Metabolizable Protein Diet; SEM = Standard error of the mean; TMP = Total microbial protein; N_{Mic} = Microbial nitrogen; A = Allantoin; UA = Uric acid; C = Creatinine; TDP = Total purine derivatives (allantoin + uric acid); N urine = Concentration of nitrogen in urine; N feces = Nitrogen concentration in feces; NS = Not significant.

A significant interaction between diet and time ($p < 0.05$) was observed for the N concentration in the feces, with CPD showing a higher percentage of N over time. Therefore, total N concentrations (urine + feces) of heifers fed a balanced diet in terms of ruminal degradable protein was lower than in those who received an unbalanced ruminal degradable protein diet. Finally, estimation of NUE using the BCNRM software was higher in the MPD (17.44%) when compared to CPD (14.27%) for the whole period. However, when NUE was estimated according to Cole et al. [41], lower values were obtained (8.36% and 9.19%, respectively) decreasing even more in the second period (43 to 64 d), but a without difference between diets ($p > 0.05$).

4. Discussion

Nutritional management of ruminants is a relevant issue to a better understanding of the use of nutrients and for optimization purposes, especially in the framework of where environmental, economic, and social matters acquire more relevance [42–46]. The present experiment was conducted to assess the hypothesis that balancing the diet by the MP system would improve N metabolism, rumen fermentation, and microbial protein synthesis in heifers. To test this hypothesis, experimental diets were formulated to have similar SI but differences in RUP and MP balances. To our knowledge, this is the first study that jointly evaluate the SI with the MP system in beef production systems in South America.

In vitro, GP is mainly used to analyze the fermentation kinetics in feed for ruminants [32]. In addition, it has been shown that there is a positive relationship between in vitro GP and digestibility [47]. Although we found no differences in GP, there were differences in VFA production between the two diets ($p < 0.01$). When comparing diets, we observed that CPD produced more total VFA, but that did not reflect a greater production of microbial protein. A possible explanation is that the greater amount of protein was transformed into NH_3, leaving the carbonated chains that make it up; these would be fermented, producing a greater amount of VFA compared to MPD. However, MPD presented a lower value of the K parameter when compared to CPD, which implies faster fermentation, is in accordance with the MDR being higher for MPD.

On the other hand, increasing CP content in the diet (or N intake) increased the amount of NH_3 exponentially [3,48]. In addition, the saturation of microbial ammonia uptake has been reported to occur between 10 and 14% CP [49]. Our results agree with these authors, since, an interaction between diet and time was found, with CPD achieving a higher concentration of NH_3 at 48 h, which would be explained by the higher consumption of N (15.13% vs. 12.73% of CP in CPD and MPD, respectively). Consequently, more N was lost in CPD, which was not used to synthesize microbial protein. In this study, a similar amount of microbial protein was synthesized in both diets, but N in urine (%) was higher in CPD compared to MPD. In this context, Firkins et al. [50] pointed out that there is a higher microbial protein synthesis efficiency when the NH_3 concentration in the rumen is lower. However, in our study, no differences were observed in the synthesis of microbial protein between diets that showed different concentrations of NH_3 in the rumen.

Chumpawadee et al. [13] showed that as the SI increases, the greater the synthesis of microbial protein that reaches the small intestine since it seeks to synchronize the source of carbohydrates with N to maximize the efficiency of microbial growth [51]. In our trial, both diets had a similar rate of synchrony, both close to 1.0. This would explain that the synthesis of the microbial protein has not been different between both dietary strategies. In a similar way, the increase in the synthesis of the microbial protein over time could be explained by the same concept. In fact, in the first measurement (day 1), heifers were not yet consuming a synchronized or balanced diet. Thus, as time progressed, the microorganisms adapted to the simultaneous availability of carbohydrates and N sources, which was reflected on measurements done on days 43 and 64. Therefore, microbial growth for both diets was more efficient.

The MP system allows to better adjust the nutrient needs by separating the requirements of the animal from the microbes of the rumen. Consequently, an improvement of some productive parameters and a significant decrease in the excretions of N into the environment is achieved [52]. In this same sense, Henning et al. [9] concluded that the synchrony between energy and the availability of N might be of less importance for microbial growth. However, the energy supply can improve the efficiency of microbial growth. Other authors support this idea [10], since the animal does not have an endogenous system to ensure the supply of energy, but in the case of N, the animal would adapt to ensure the availability of the nutrient through recycling, particularly when it is low in the diet. It has also been reported that at a higher percentage of TDN, which is an indicator of energy in the diet, there is a greater amount of synthesized microbial protein [6]. In this experiment, TDN energy values of both diets were similar (difference of 2.64%), which would confirm the similarity in the synthesis of microbial protein in both dietary strategies.

Seo et al. [53] reported a greater daily microbial growth with higher SI (0.81 and 0.83) when compared with an SI of 0.77. These authors attributed to the fact that their diet with the lower SI had greater amounts of non-structural carbohydrates; therefore, greater production of lactic acid, and consequently, an effect on rumen pH that affects the synthesis of microbial protein [54]. However, in their results, Seo et al. [53] did not report differences in rumen pH. In our study, no differences in pH (data not shown) were observed either, so it follows that it was not a determining factor affecting microbial protein synthesis. Seo et al. [55] suggest that the lower ruminal NH_3-N concentration correlates with higher utilization of NH_3-N for microbial protein synthesis. This is in accordance with our study, as MPD had a lower CP concentration and N intake showed a lower concentration of NH_3 and resulted in similar microbial protein synthesis.

The amount of microbial N produced was within the values described by the literature that used similar breeds [56–58], but when compared with breeds destined for milk production, these values were much lower, mainly due to the passage rates compared to those seen in dairy cattle, which were higher and also had higher consumption than beef production cattle [59]. The optimal range of NH_3 for microbial production is 5 to 8 mg 100 mL^{-1}. In our study, NH_3 was measured in vitro, and for none of the diets, it was limiting microbial growth.

The lowest concentration of N on the three sampling dates was observed in MPD. This would be mainly explained by the balance of the protein through the MP system established by the BCNRM software. The foregoing given that both treatments had a similar SI (0.80 and 0.83), and the difference between them was that it was balanced to adjust the availability of the rumen degradable protein and MP. Meanwhile, CPD was calculated considering the total demand for CP. In our study, a decrease in the total CP concentration of the diet was achieved when it was balanced by the MP system (MPD), reaching 12.73% CP, while the conventional CPD diet was 15.13% CP. Therefore, MPD resulted in a lower excretion of N, as observed in the N concentrations in the urine and feces. Several authors have reported similar results, that is, the higher the content of N consumed, the greater the excretions of N, both for urine and feces [60–62]. This is mainly because, by increasing the consumption of CP or rather the RDP, an excess of NH_3 is produced in the rumen, which is converted to urea in the liver and subsequently eliminated in the urine [63].

On the other hand, Erickson and Klopfenstein [52] found differences in the excretions of N for both one-year-old calves and calves when using a nutritional strategy that incorporates the feeding in phases balanced by the MP system. The authors reported that lower excretions of N were observed in treatments balanced with the MP system. Likewise, as occurred in our study, they pointed out that the contents of CP in the diet balanced by MP were lower. Finally, they reported that adopting this method could reduce excretions in steers in the evaluated fattening period (137 d) by up to 6.1 kg of N. Cole et al. [64] reported similar results regarding the concentration of N in feces and urine, evaluating diets with different CP contents (11.5%, 13.0%, and 14.5%). They observed that the diet with the highest concentration of CP was higher in the N concentration in feces and urine, coinciding with our results.

Carcass characteristics were not affected by the treatments, with the exception of cold carcass weight. This was not expected due to the similarity in the final LW of both dietary treatments. Nevertheless, another study reported that the weight of the hot carcass responded to the increase in the concentration of PC in the diet [65]. On the other hand, Cole et al. [64] and Gleghorn et al. [66] reported a lack of differences in the degree of marbling and fat coverage regarding the change of protein offered in the diet. In contrast, there was a trend for a smaller ribeye area in diets with higher CP, that is, 14% vs. 12% (Cole et al. 2003). Nevertheless, our results must be analyzed with caution because of the low number of replicates in the study.

The values of the NUE herein reported are within the ranges reported in the literature [21,41,45]. The N consumed in MPD was lower compared to CPD, because CP content was lower, and the RUP balance was better. The latter resulted in a lower concentration of NH_3 in MPD, and when there is less ruminal NH_3, NUE use to be higher [54]. In addition, the concentration of N in the feces and urine was lower in heifers fed with MPD. Therefore, they consumed less N and utilized it more efficiently than heifers fed the CPD. Although both, MP system and the SI, seek to improve NUE [12,44], using them together can produce better results than using each one separately. Galyean [67] promotes the use of the MP system since the protein intake is calculated according to the requirements of the microorganisms, thus avoiding excesses of N in the diets and fewer excretions of N into the environment. Therefore, using the MP system can reduce N excretions and therefore improve NUE [44,45].

By using the equations of Cole et al. [41], a numerically lower NUE was obtained in both periods but showing the same behavior as the values obtained from the BCNRM software. The lower NUE observed in the second period was due to animals decreasing their ADG caused by the decrease in DMI. Therefore, they were less efficient in retaining the ingested N. This scenario was not reflected with the values obtained using the BCNRM software.

Although there is no official data in Chile, some reports [68], as well as the empirical experience, indicate values of CP ranging from 10.5 to 20.0%, whereas energy metabolizable energy (ME) range between 2.60 to 2.90 Mcal/kg DM. This variation is explained in part because many production systems, particularly in Southern Chile, are grass-fed and use strategic supplementation with concentrates (usually small cereal grains and byproducts). Meanwhile, feed yards are closer to 13.5% CP and 2.75 Mcal ME/kg DM, but still showing great variability among them. Ruminal protein balance (RPB) has been proposed as a new trait to be used in diet formulation by Sauvant and Nozière [69] in the INRA (Institute National de Recherche Agronomique) 2018 system because it induces several interesting responses of protein and energy efficiency. These authors also indicated a negative relation of RPB with microbial growth efficiency and highlighted the relevance of degradable CP and energy in the rumen. Also, it has been demonstrated a decrease of OM digestibility when RPB decrease under zero [69,70].

Most of the actuals systems (NASEM, INRA 2018, DVE/OEB, NorFor, AFRC, CSIRO) have a mechanistic approach that looks at improving animal performance, product quality, the efficiency of feed use, and at the same time considers the environmental impacts by reducing N excretion and CH_4 emissions. The last release of INRA system (2018), has considered as a relevant criterion not only to evaluate the balance between degradable N and the energy available in the rumen but also to integrate the quantitative effects of the energy × nitrogen interactions in digestive processes, as previously described by Sauvant and Nozière [69]. A major focus in the new model was fractional passage rates of liquids, forages, and concentrates (proportion of concentrate in the diet), as well as the level of feeding. In our case, we use a fractional passage rate constant of 0.05%/h, but based on equations of the INRA 2018, it should be 0.04%/h. This, as mentioned before, affected the digestibility of the OM and the results associated with the NUE.

5. Conclusions

A dietary strategy that combines a high synchrony index with the MP system increases the NUE without negatively affecting animal performance or carcass characteristics compared to heifers fed a diet formulated based on their CP requirements. The lower N excretions show that the MPD is more environmentally friendly. In addition, the lower CP content in the MP diet may increase the beef production system's profitability because protein is one of the highest costs in the ration. Finally, we accepted the hypothesis, since the combination of synchrony and the MP system achieve greater efficiency in the use of nitrogen, without negatively affecting animals' performance neither the quality of the carcass.

Author Contributions: R.A.A. conceived and supervised the study and wrote the first draft. S.K. and G.G. executed the experiment, collected and analyzed the data. J.P.K. supervised in vitro and in situ incubations, collaborated in the statistical analysis, and contributed to editing the manuscript. C.A.-G. was the Laboratory responsible and contributed the final review of the manuscript. All authors have read and agreed to the published version of the manuscript.

Acknowledgments: The authors thank all the members of the Animal Nutrition Laboratory at Universidad Austral de Chile for their generous collaboration in this research.

References

1. Arias, R.A. Intensificación de los sistemas ganaderos y la contaminación ambiental, el caso de los gases efecto invernadero y el nitrógeno. In Proceedings of the XXII Jornadas de Extensión Agrícola "Producción de carne: Aspectos técnicos para enfrentar las demandas de calidad y sustentabilidad en un clima cambiante", Temuco, Chile, 31 May 2011; p. 11. (In Spanish).
2. Alfaro, M.; Salazar, F. Ganadería y contaminación difusa, implicancias para el sur de Chile. *Agricultura Técnica* **2005**, *65*, 330–340. (In Spanish) [CrossRef]
3. Vasconcelos, J.T.; Tedeschi, L.O.; Fox, D.G.; Galyean, M.L.; Greene, L.W. Feeding nitrogen and phosphorus in beef cattle feedlot production to mitigate environmental impacts. *Prof. Anim. Sci.* **2007**, *23*, 8–17. [CrossRef]
4. Klemesrud, M.J.; Klopfenstein, T.J.; Lewis, A.J. Metabolizable methionine and lysine requirements of growing cattle. *J. Anim. Sci.* **2000**, *78*, 199–206. [CrossRef] [PubMed]
5. NRC. *Nutrient Requirement of Beef Cattle, update 2000*; National Academy Press: Washington DC, USA, 2000; p. 232.
6. Owens, F.N.; Qi, S.; Sapienza, D.A. Invited Review: Applied protein nutrition of ruminants—Current status and future directions. *PAS* **2014**, *30*, 150–179. [CrossRef]
7. Herrera-Saldana, R.; Gomez-Alarcon, R.; Torabi, M.; Huber, J.T. Influence of Synchronizing Protein and Starch Degradation in the Rumen on Nutrient Utilization and Microbial Protein Synthesis. *J. Dairy Sci.* **1990**, *73*, 142–148. [CrossRef]
8. Clark, J.H.; Klusmeyer, T.H.; Cameron, M.R. Microbial Protein-Synthesis and Flows of Nitrogen Fractions to the Duodenum of Dairy-Cows. *J. Dairy Sci.* **1992**, *75*, 2304–2323. [CrossRef]
9. Henning, P.H.; Steyn, D.G.; Meissner, H.H. The effect of energy and nitrogen supply pattern on rumen bacterial growth in vitro. *Anim. Sci. J.* **1991**, *53*, 165–175. [CrossRef]
10. Hall, M.B.; Huntington, G.B. Nutrient synchrony: Sound in theory, elusive in practice. *J. Anim. Sci.* **2008**, *86*, E287–E292. [CrossRef]

11. Keim, J.P.; Anrique, R. Nutritional strategies to improve nitrogen use efficiency by grazing dairy cows. *Chil. J. Agric. Res.* **2011**, *71*, 623–633. [CrossRef]
12. Sinclair, L.A.; Garnsworth, P.C.; Newbold, J.R.; Buttery, P.J. Effect of synchronizing the rate of dietary energy and nitrogen release on rumen fermentation and microbial protein synthesis in sheep. *J. Agric. Sci.* **1993**, *120*, 251–263. [CrossRef]
13. Chumpawadee, S.; Sommart, K.; Vongpralub, T.; Pattarajinda, V. Effects of synchronizing the rate of dietary energy and nitrogen release on ruminal fermentation, microbial protein synthesis, blood urea nitrogen and nutrient digestibility in beef cattle. *Asian Australas. J. Anim. Sci.* **2006**, *19*, 181–188. [CrossRef]
14. Yang, J.Y.; Seo, J.; Kim, H.J.; Seo, S.; Ha, J.K. Nutrient Synchrony: Is it a Suitable Strategy to Improve Nitrogen Utilization and Animal Performance? *Asian Australas J. Anim. Sci.* **2010**, *23*, 972–979. [CrossRef]
15. Berthiaume, R.; Benchaar, C.; Chaves, A.V.; Tremblay, G.F.; Castonguay, Y.; Bertrand, A.; Bélanger, G.; Michaud, R.; Lafrenière, C.; McAllister, T.A.; et al. Effects of nonstructural carbohydrate concentration in alfalfa on fermentation and microbial protein synthesis in continuous culture1. *J. Dairy Sci.* **2010**, *93*, 693–700. [CrossRef] [PubMed]
16. Dewhurst, R.J.; Davies, D.R.; Merry, R.J. Microbial protein supply from the rumen. *Anim. Feed Sci. Technol.* **2000**, *85*, 1–21. [CrossRef]
17. Huntington, G.B. Starch utilization by ruminants: From basics to the bunk. *J. Anim. Sci.* **1997**, *75*, 852–867. [CrossRef]
18. Chumpawadee, S.; Sommart, K.; Vongpralub, T.; Pattarajinda, V. Effect of synchronizing the rate of degradation of dietary energy and nitrogen release on growth performance in Brahman beef cattle. *Suranaree J. Sci. Technol.* **2006**, *28*, 59–70.
19. Duarte, M.d.S.; Paulino, P.V.R.; Filho, S.d.C.V.; Paulino, M.F.; Detmann, E.; Zervoudakis, J.T.; Monnerat, J.P.I.d.S.; Viana, G.d.S.; Silva, L.H.P.; Serão, N.V.L. Performance and meat quality traits of beef heifers fed with two levels of concentrate and ruminally undegradable protein. *Trop. Anim. Health Prod.* **2011**, *43*, 877–886. [CrossRef]
20. Cooper, R.J.; Milton, T.; Klopfenstein, T.J. *Phase-Feeding Metabolizable Protein for Finishing Steers*; University of Nebraska: Lincoln, NE, USA, 2000; pp. 63–65.
21. Erickson, G.E.; Klopfenstein, T.J.; Milton, T.; Herold, D. *Effects of Matching Protein Requirements on Performance and Waste Management in the Feedlot*; Nebraska Beef Reports; University of Nebraska: Lincoln, NE, USA, 1 January 1999 ; pp. 60–63.
22. Arias, R.A.; Keim, J.P.; Gandarillas, M.; Velasquez, A.; Alvarado-Gilis, C.; Mader, T.L. Performance and carcass characteristics of steers fed with two levels of metabolizable energy intake during summer and winter season. *Animal* **2019**, *13*, 221–230. [CrossRef]
23. Verbič, J.; Ørskov, E.R.; Žgajnar, J.; Chen, X.B.; Žnidaršič-Pongrac, V. The effect of method of forage preservation on the protein degradability and microbial protein synthesis in the rumen. *Anim. Feed Sci. Technol.* **1999**, *82*, 195–212. [CrossRef]
24. National Academies of Sciences Engineering and Medicine (NASEM). *Nutrient Requirements of Beef Cattle: Eighth Revised Edition*; The National Academies Press: Washington, DC, USA, 2016; p. 494. [CrossRef]
25. Tilley, J.M.A.; Terry, R.A. A two-stage technique for the in vitro digestion of forage crops. *Grass Forage Sci.* **1963**, *18*, 104–111. [CrossRef]
26. Van Soest, P.J.; Robertson, J.B.; Lewis, B.A. Methods for Dietary Fiber, Neutral Detergent Fiber, and Nonstarch Polysaccharides in Relation to Animal Nutrition. *J. Dairy Sci.* **1991**, *74*, 3583–3597. [CrossRef]
27. Association of Official Analytical Chemists (AOAC). *Official Methods of Analysis*, 16th ed.; Association of Official Analytical Chemists: Washington, DC, USA, 1996.
28. Ørskov, E.R.; McDonald, I. The estimation of protein degradability in the rumen from incubation measurements weighted according to rate of passage. *J. Agr. Sci. Cambridge* **1979**, *92*, 499–503. [CrossRef]
29. Mcdonald, I. A Revised Model for the Estimation of Protein Degradability in the Rumen. *J. Agric. Sci.* **1981**, *96*, 251–252. [CrossRef]
30. AFRC. *Energy and Protein Requirments of Ruminants*; CAB International: Wallingford, UK, 1993.
31. Hvelplund, T.; Weisbjerg, M.R. In situ techniques for the estimation of protein degradability and postrumen availability. In *Forage Evaluation in Ruminant Nutrition*; Givens, D.I., O'wen, E., Axford, R.F.E., Omed, H.M., Eds.; CAB international: Wallingford, UK, 2000; pp. 233–258. [CrossRef]

32. Theodorou, M.K.; Williams, B.A.; Dhanoa, M.S.; McAllan, A.B.; France, J. A simple gas production method using a pressure transducer to determine the fermentation kinetics of ruminant feeds. *Anim. Feed Sci. Technol.* **1994**, *48*, 185–197. [CrossRef]
33. France, J.; Dijkstra, J.; Dhanoa, M.S.; López, S.; Bannink, A. Estimating the extent of degradation of ruminant feeds from a description of their gas production profiles observed in vitro: Derivation of models and other mathematical considerations. *Br. J. Nutr.* **2000**, *83*, 143–150. [CrossRef]
34. Groot, J.C.J.; Cone, J.W.; Williams, B.A.; Debersaques, F.M.A.; Lantinga, E.A. Multiphasic analysis of gas production kinetics for in vitro fermentation of ruminant feeds. *Anim. Feed Sci. Technol.* **1996**, *64*, 77–89. [CrossRef]
35. Vlassa, M.; Filip, M. Determination of purine derivatives in bovine urine using rapid chromatographic techniques. *Arch. Zootech.* **2009**, *12*, 59–70.
36. Al-Khalidi, U.A.S.; Chaglassian, T.H. The species distribution of xanthine oxidase. *Biochem. J.* **1965**, *97*, 318–320. [CrossRef]
37. Chizzotti, M.L.; Valadares Filho, S.C.; Valadares, R.F.D.; Chizzotti, F.H.M.; Tedeschi, L.O. Determination of creatinine excretion and evaluation of spot urine sampling in Holstein cattle. *Livest. Sci.* **2008**, *113*, 218–225. [CrossRef]
38. Faichney, G.J.; Welch, R.J.; Brown, G.H. Prediction of the excretion of allantoin and total purine derivatives by sheep from the 'creatinine coefficient'. *J. Agric. Sci.* **1995**, *125*, 425–428. [CrossRef]
39. Ørskov, E.R.; MacLeod, N.A.; Nakashima, Y. Effect of different volatile fatty acids mixtures on energy metabolism in cattle. *J. Anim. Sci.* **1991**, *69*, 3389–3397. [CrossRef]
40. Chen, X.B.; Gomes, M.J. *Estimation of Microbial Protein Supply to Sheep and Cattle based on Urinary Excretion of Purine Derivatives—An Overview of the Technical Details*; International Feed Resources Unit: Aberdeen, UK, September 1995; pp. 21–25.
41. Cole, N.A.; Defoor, P.J.; Galyean, M.L.; Duff, G.C.; Gleghorn, J.F. Effects of phase-feeding of crude protein on performance, carcass characteristics, serum urea nitrogen concentrations, and manure nitrogen of finishing beef steers. *J. Anim. Sci.* **2006**, *84*, 3421–3432. [CrossRef]
42. Erickson, G.E.; Klopfenstein, T.J.; Milton, T.C. Dietary protein effects on nitrogen excretion and volatilization in open-dirt feedlots. In Proceedings of the 8th International Symposium on Animals, Agriculture, and Food Processing Wastes, St. Joseph, MO, USA, 9–11 October 2000; pp. 297–304.
43. Hersom, M.J. Opportunities to enhance performance and efficiency through nutrient synchrony in forage-fed ruminants. *J. Anim. Sci.* **2008**, *86*, E306–E317. [CrossRef]
44. Erickson, G.; Klopfenstein, T. Nutritional and management methods to decrease nitrogen losses from beef feedlots. *J. Anim. Sci.* **2010**, *88*, E172–E180. [CrossRef]
45. Arias, R.A.; Velasquez, A.; Toneatti, M. Simulation of the nitrogen use efficiency in pasture-finished steers in southern Chile. *Arch. Med. Vet.* **2013**, *45*, 125–134. [CrossRef]
46. Waldrip, H.M.; Cole, N.A.; Todd, R.W. Nitrogen sustainability and beef-cattle feedyards: I. Introduction and influence of pen surface conditions and diet. *PAS* **2015**, *31*, 89–100. [CrossRef]
47. Xu, M.; Rinker, M.; McLeod, K.R.; Harmon, D.L. *Yucca schidigera* extract decreases in vitro methane production in a variety of forages and diets. *Anim. Feed Sci. Technol.* **2010**, *159*, 18–26. [CrossRef]
48. Detmann, E.; Valente, É.E.L.; Batista, E.D.; Huhtanen, P. An evaluation of the performance and efficiency of nitrogen utilization in cattle fed tropical grass pastures with supplementation. *Livest. Sci.* **2014**, *162*, 141–153. [CrossRef]
49. Detmann, E.; Paulino, M.F.; Mantovani, H.C.; Valadares, S.D.; Sampaio, C.B.; de Souza, M.A.; Lazzarini, I.; Detmann, K.S.C. Parameterization of ruminal fibre degradation in low-quality tropical forage using Michaelis-Menten kinetics. *Livest. Sci.* **2009**, *126*, 136–146. [CrossRef]
50. Firkins, J.L.; Berger, L.L.; Merchen, N.R.; Fahey, G.C., Jr. Effects of forage particle size, level of feed intake and supplemental protein degradability on microbial protein synthesis and site of nutrient digestion in steers. *J. Anim. Sci.* **1986**, *62*, 1081–1094. [CrossRef]
51. Reynolds, C.K.; Kristensen, N.B. Nitrogen recycling through the gut and the nitrogen economy of ruminants: An asynchronous symbiosis. *J. Anim. Sci.* **2008**, *86*, E293–E305. [CrossRef]
52. Erickson, G.E.; Klopfenstein, T.J. Managing N inputs and the effect on N losses following excretion in open-dirt feedlots in Nebraska. *Sci. World J.* **2001**, *2*, 830–835. [CrossRef]

53. Seo, J.K.; Yang, J.; Kim, H.J.; Upadhaya, S.D.; Cho, W.M.; Ha, J.K. Effects of Synchronization of Carbohydrate and Protein Supply on Ruminal Fermentation, Nitrogen Metabolism and Microbial Protein Synthesis in Holstein Steers. *Asian Australas. J. Anim. Sci.* **2010**, *23*, 1455–1461. [CrossRef]
54. Bach, A.; Calsamiglia, S.; Stern, M.D. Nitrogen Metabolism in the Rumen. *J. Dairy Sci.* **2005**, *88*, E9–E21. [CrossRef]
55. Seo, J.K.; Kim, M.H.; Yang, J.Y.; Kim, H.J.; Lee, C.H.; Kim, K.H.; Ha, J.K. Effects of Synchronicity of Carbohydrate and Protein Degradation on Rumen Fermentation Characteristics and Microbial Protein Synthesis. *Asian Australas. J. Anim. Sci.* **2013**, *26*, 358–365. [CrossRef]
56. Prendergast, S.L.; Gibbs, S.J. A comparison of microbial protein synthesis in beef steers fed ad libitum winter ryegrass or fodder beet. *Proc. New. Zeal. Soc. An.* **2015**, *75*, 251–256.
57. Koenig, K.M.; Beauchemin, K.A. Nitrogen metabolism and route of excretion in beef feedlot cattle fed barley-based finishing diets varying in protein concentration and rumen degradability. *J. Anim. Sci.* **2013**, *91*, 2310–2320. [CrossRef]
58. Davies, K.L.; McKinnon, J.J.; Mutsvangwa, T. Effects of dietary ruminally degradable starch and ruminally degradable protein levels on urea recycling, microbial protein production, nitrogen balance, and duodenal nutrient flow in beef heifers fed low crude protein diets. *Can. J. Anim. Sci.* **2013**, *93*, 123–136. [CrossRef]
59. Gehman, A.M.; Kononoff, P.J. Nitrogen utilization, nutrient digestibility, and excretion of purine derivatives in dairy cattle consuming rations containing corn milling co-products. *J. Dairy Sci.* **2010**, *93*, 3641–3651. [CrossRef]
60. Yan, T.; Frost, J.P.; Keady, T.W.J.; Agnew, R.E.; Mayne, C.S. Prediction of nitrogen excretion in feces and urine of beef cattle offered diets containing grass silage. *J. Anim. Sci.* **2007**, *85*, 1982–1989. [CrossRef]
61. Dijkstra, J.; Reynolds, C.K.; Kebreab, E.; Bannink, A.; Ellis, J.L.; France, J.; van Vuuren, A.M. Challenges in ruminant nutrition: Towards minimal nitrogen losses in cattle. In *Energy and Protein Metabolism and Nutrition in Sustainable Animal Production*; Oltjen, J., Kebreab, E., Lapierre, H., Eds.; Wageningen Academic Publishers: Wageningen, The Netherlands, 2013; pp. 47–58.
62. Cantalapiedra-Hijar, G.; Peyraud, J.L.; Lemosquet, S.; Molina-Alcaide, E.; Boudra, H.; Noziere, P.; Ortigues-Marty, I. Dietary carbohydrate composition modifies the milk N efficiency in late lactation cows fed low crude protein diets. *Animal* **2014**, *8*, 275–285. [CrossRef]
63. Erickson, G.; Watson, A.K.; MacDonald, J.C.; Klopfenstein, T.J. Protein nutrition evaluation and application to growing and finishing cattle. In Proceedings of the 27th annual Florida Ruminant Nutrition Symposium, Gainesville, FL, USA, 15–17 February 2016; pp. 91–102.
64. Cole, N.A.; Greene, L.W.; McCollum, F.T.; Montgomery, T.; McBride, K. Influence of oscillating dietary crude protein concentration on performance, acid-base balance, and nitrogen excretion of steers. *J. Anim. Sci.* **2003**, *81*, 2660–2668. [CrossRef]
65. Boonsaen, P.; Soe, N.W.; Maitreejet, W.; Majarune, S.; Reungprim, T.; Sawanon, S. Effects of protein levels and energy sources in total mixed ration on feedlot performance and carcass quality of Kamphaeng Saen steers. *Agric. Nat. Resour.* **2017**, *51*, 57–61. [CrossRef]
66. Gleghorn, J.F.; Elam, N.A.; Galyean, M.L.; Duff, G.C.; Cole, N.A.; Rivera, J.D. Effects of crude protein concentration and degradability on performance, carcass characteristics, and serum urea nitrogen concentrations in finishing beef steers. *J. Anim. Sci.* **2004**, *82*, 2705–2717. [CrossRef]
67. Galyean, M.L. Protein levels in beef cattle finishing diets: Industry application, university research, and systems results. *J. Anim. Sci.* **1996**, *74*, 2860–2870. [CrossRef]
68. Morales, R.; Folch, C.; Iraira, S.; Teuber, N.; Realini, C.E. Nutritional quality of beef produced in Chile from different production systems. *Chil. J. Agric. Res.* **2012**, *72*, 80–86. [CrossRef]
69. Sauvant, D.; Nozière, P. Quantification of the main digestive processes in ruminants: The equations involved in the renewed energy and protein feed evaluation systems. *Animal* **2016**, *10*, 755–770. [CrossRef]
70. Sauvant, D.; Nozière, P. The Rumen Protein Balance as a key trait to model ruminant responses to dietary proteins. In Proceedings of the 6th EAAP International Symposium on Energy and Protein Metabolism and Nutrition, Belo Horizonte, Brazil, 9–12 September 2019; p. 4.

Nutrigenomic Effects of Long-Term Grape Pomace Supplementation in Dairy Cows

Marianna Pauletto [1,†], Ramy Elgendy [1,2,†], Andrea Ianni [3], Elettra Marone [3], Mery Giantin [1], Lisa Grotta [3], Solange Ramazzotti [3], Francesca Bennato [3], Mauro Dacasto [1,*] and Giuseppe Martino [3]

1 Department of Comparative Biomedicine and Food Science, University of Padua, Viale dell'Università 16, 35020 Legnaro (PD), Italy; marianna.pauletto@unipd.it (M.P.); ramy.elgendy@igp.uu.se (R.E.); mery.giantin@unipd.it (M.G.)
2 Department of Pharmacology, Faculty of Veterinary Medicine, Suez Canal University, Ismailia 41522, Egypt
3 Faculty of Bioscience and Technology for Food, Agriculture and Environment, University of Teramo, 64100 Teramo (TE), Italy; aianni@unite.it (A.I.); emarone@unite.it (E.M.); lgrotta@unite.it (L.G.); sramazzotti@unite.it (S.R.); fbennato@unite.it (F.B.); gmartino@unite.it (G.M.)
* Correspondence: mauro.dacasto@unipd.it.
† These Authors contributed equally to this work.

Simple Summary: The aim of this study was to evaluate the effect of grape pomace (GP), the polyphenol-rich agricultural by-product, on dairy cows' whole-blood transcriptome, milk production and composition. Twelve lactating Holstein-Friesian cows were randomly assigned to two groups; the first received a GP-supplemented diet for 60 days (group GP), whereas the second was given only a basal diet (CTR). The results reveal 40 protein-coding genes differentially expressed in the GP group when compared with the CTR group, but no effects were noticed on milk production, concentrations of crude protein, fat, casein, lactose and urea, or somatic cell count. Compared to CTR, GP had a transcriptomic signature mainly reflecting a reinforced immunogenic response.

Abstract: The increasing demand for more animal products put pressure on improving livestock production efficiency and sustainability. In this context, advanced animal nutrition studies appear indispensable. Here, the effect of grape pomace (GP), the polyphenol-rich agricultural by-product, was evaluated on Holstein-Friesian cows' whole-blood transcriptome, milk production and composition. Two experimental groups were set up. The first one received a basal diet and served as a control, while the second one received a 7.5% GP-supplemented diet for a total of 60 days. Milk production and composition were not different between the group; however, the transcriptome analysis revealed a total of 40 genes significantly affected by GP supplementation. Among the most interesting down-regulated genes, we found the DnaJ heat-shock protein family member A1 (DNAJA1), the mitochondrial fission factor (MFF), and the impact RWD domain protein (IMPACT) genes. The gene set enrichment analysis evidenced the positive enrichment of 'interferon alpha (IFN-α) and IFN-γ response', 'IL6-JAK-STAT3 signaling' and 'complement' genes. Moreover, the functional analysis denoted positive enrichment of the 'response to protozoan' and 'negative regulation of viral genome replication' biological processes. Our data provide an overall view of the blood transcriptomic signature after a 60-day GP supplementation in dairy cows which mainly reflects a GP-induced immunomodulatory effect.

Keywords: nutrigenomics; grape pomace; polyphenols; dairy cows; RNA-sequencing

1. Introduction

Historically, human populations largely depend on domestic farm animals for the production of animal food-products (e.g., meat, fat, milk and other dairy products). The recent and increasing demand for more animal products, especially in developing countries with rising living standards, puts pressure on global livestock productivity. With a rapidly growing world population, this increasing demand will put a strain on our natural resources and the whole environment as livestock directly compete with humans for the available arable land [1]. Nonetheless, the high demand for these animal proteins has to deal with other challenges such as climate change. To improve further animal production efficiency and limit the global land use for livestock production, advanced animal nutrition studies are today indispensable.

In this context, a thorough understanding of animal gene expression driven by dietary nutrients can be regarded as the bottom line of advanced animal nutrition research. Nutrigenomics (including transcriptomics) is a new branch of animal nutrition at the basic molecular biological level. In other words, it studies the effects of dietary nutrients on cellular gene expression and ultimately, phenotypic changes in living animals. Among the transcriptomic approaches, RNA-sequencing (RNA-seq) simultaneously measures the differential expression of thousands of genes in any biological matrix. The resulting information can be used to determine the effects of a certain condition (e.g., a supplemented diet) from the perspective of the biological processes or molecular functions involved rather than from the expression levels of individual genes [2].

In the European Union (EU), agro-industrial by-products are being considered a key element in the EU's 2020 Environment Action Programme concerning waste management, recycling, and reuse [3]. The reuse of agricultural by-products to produce bio-liquids, oils [4] and bioactive compounds such as carotenoids and polyphenols [5,6] or as animal feed and land fertilizers [5–7], has valorized them as sustainable resources. Indeed, the use of several agricultural by-products in animal nutrition has been largely explored [7]. In addition to the 'traditional' agricultural residues such as oil meals, bran, brewers' grains, beet pulp and molasses, other 'alternative' by-product residues, resulting from fruit and vegetable processing, have become attractive as animal feedstuffs [8].

Grape pomace (GP) is the residue (by-product) that remains after crushing the grapes to collect its juice. It makes about one third of the total volume of the grapes used in wine production [9]. Although biodegradable, GP requires time to mineralize; thus, if left unexploited, it might represent a potential source of pollution to the environment [10]. However, GP is rich in phenolic compounds [9] that hold substantial antioxidant properties [11,12]. Moreover, due to its low cost and high fiber content, GP was proposed earlier as an alternative feed ingredient to 'partially' replace the forage portion in the diet of ruminants [9].

Recently, there has been growing interest in studying the effects of the dietary inclusion of GP on the overall productive performances and production aspects of food-producing animals. For instance, GP increased the biodiversity degree of intestinal microflora in broiler chicks [13], and improved the gain-to-feed ratio and overall performance in pigs [14,15]. Further, GP altered the nitrogen metabolism and decreased the ruminal ammonia production in male sheep [16], was shown to modify the rumen microbial population involved in methane metabolism [17], enhance the growth of facultative probiotic bacteria and inhibit the growth of pathogenic ones in lambs [18], but had no effect on milk yield and composition of dairy ewes [19]. In dairy cows, few studies, with varying results, have investigated the effects of GP dietary inclusion on milk production, composition and total polyphenols content as well as on the overall performance of animals [10,20–23]. We have previously described the transcriptomic signature of veal calves fed with a GP-supplemented diet [24]. Nevertheless, to our knowledge, few papers concerning the nutrigenomic effect of long-term dietary GP supplementation in dairy cows are available. It is conceivable that the dietary inclusion of GP (5% of total diet) would have a noticeable transcriptomic signature that could reflect its claimed antioxidant effects in dairy cows. Therefore, in the present study, a whole-transcriptome profiling, using the RNA-seq technology, was performed in dairy cows fed with a GP-supplemented diet and the obtained results are described herein.

2. Materials and Methods

2.1. Animals and Study Design

Animals were managed as per the Directive 2010/63 [25] of the European Parliament regarding the protection of animals used for experimentation or other scientific purposes. Blood sampling was performed concurrently with planned blood withdrawal for prophylaxis. Dairy cows had not been subjected to breeding practices different than those normally envisaged; thus, ethical declaration was not necessary.

Twelve lactating Holstein-Friesian cows (~163 days in milk), with an average calving number of 2.55 and an average body condition score of 2.97 were used as experimental animals for this study. The cows had an acclimation period of 2 weeks; then, they were randomly allocated into 2 experimental groups of 6 animals each. The first group received a basal diet and served as a control (CTR), while the other group received a diet supplemented with 7.5% GP-supplemented diet on a dry matter basis (GP group). The experimentation was conducted for a period of 60 days, in which all animals were housed in 2 separate areas of free housing with access to an identical feeding area in which each animal had an individual feeding bin, with water freely available all throughout the study. The GP, obtained from different wineries and distilleries in the region of Abruzzo (Italy), was prepared according to the procedure described in [26], and analyzed for its chemical composition before being added to the animals' diet. Weekly from the start of the trial, the daily feed intake (dry matter) has been calculated for each animal by subtracting the refused meal to the administered diet.

2.2. Extraction and Determination of GP Total Polyphenols Content (TPC)

In order to extract the GP polyphenols, 1 g of pomace was added to 5 mL of methanol and the mixture was stirred for 4 h at room temperature. The solution was then transferred in a 15 mL test tube and placed in an ultrasonic bath for 10 min. At the end of this procedure, centrifugation (4000 rpm) was carried out for 20 min. The supernatant was recovered and filtered with paper filters (Whatman 41) in order to remove the interfering substances; this operation was repeated three times. Total Phenolic Content (TPC) was estimated by using the Folin-Ciocalteu colorimetric method, according to the procedure reported by [27]. The chemical composition and total polyphenols content of the GP extract are reported in Table 1.

Table 1. Chemical composition and main total polyphenols content of the GP extract.

Chemical Composition of GP	
Dry Matter, %	92.20
Crude protein [1], %	12.40
Ether extract [1], %	4.20
Crude fibre [1], %	31.80
Neutral-detergent fibre [1], %	42.20
Acid detergent fibre [1], %	41.30
Acid-detergent lignin [1], %	28.80
Starch [1], %	-
Ash [1], %	8.50
Total Polyphenols Content (TCP) in GP	
TPC (GAE [2] mg/g)	16.07 ± 1.41

[1] On a dry matter (DM) basis; [2] Gallic Acid Equivalents.

2.3. Sampling

Milk, feed, and whole-blood (WB) samples were individually collected from each group. For the RNA-seq analysis, 2.5 mL of jugular venous blood were collected at T60 (2 groups × 6 animals each). Duplicate WB samples were collected in PAXgene™ tubes (Qiagen SpA, Milan, Italy), stored at room temperature overnight and, then, at −20 °C until RNA isolation. Milk yield (morning milking) was recorded for each cow and individual milk samples were collected from all animals and stored at −20 °C for further analysis.

2.4. Milk and Feed Composition Analysis

Milk samples were analyzed for their protein, casein, lactose, fat, urea, pH, and total solids content (%) using Fourier Transformed InfraRed (FTIR) spectrophotometry (MilkoScanTM FT 6000; Foss, Hillerød, Denmark) as per the manufacturer's instructions. The milk somatic cell count (SCC) was analyzed by a fluoro-opto-electronic counter (FossomaticTM FC; Foss, Hillerød, Denmark). The method is based on flow cytometry technology that counts somatic cells, following the mixing of milk samples (1 to 10 μL, at 30–42 °C) with a DNA molecule-binding staining solution. The administered diets were analyzed according to the methods described by AOAC International [28] in order to obtain information about dry matter (method 930.15), crude protein (method 954.01), fat content (method 920.39), crude fibre (method 962.09) and ash (method 942.05). The procedure reported by Goering and Van Soest [29] was instead used for the evaluation of Neutral Detergent Fibre (NDF) and Acid Detergent Fibre (ADF). The composition of the basal and supplemental diets is reported in Table 2.

Table 2. Ingredients and chemical composition of the two diets.

	Diets	
Ingredients of the diets	CTR/basal	GP
Corn silage, %	45.70	45.50
Second-cut alfalfa hay, %	20.30	20.20
Wheat straw, %	3.80	1.30
Corn grain meal, %	13.50	13.10
Soybean meal, %	8.40	7.60
Barley meal, %	2.50	2.50
Cotton seed, %	3.80	2.80
Dried grape pomace, %	-	5.00
Vitamins and minerals, %	2.00	2.00
Nutrient composition		
Dry Matter, %	61.50	61.70
Crude protein [1], %	15.30	15.90
Ether extract [1], %	2.30	2.20
Crude fibre [1], %	17.80	18.90
Neutral-detergent fibre [1], %	34.40	36.30
Acid detergent fibre [1], %	20.30	22.50
Acid-detergent lignin [1], %	4.60	5.90
Starch [1], %	25.20	24.80
Ash [1], %	4.50	5.10
Net energy (Milk UF/kg DM)	0.89	0.91

[1] On a dry matter (DM) basis.

2.5. Library Preparation and RNA-Seq Analysis

Total RNA was isolated from blood samples by using the PAXgene blood RNA kit (Qiagen, Milan, Italy) as per the manufacturer's instructions. To reduce the possible presence of genomic DNA contamination, a 15-minutes on-column DNase digestion step was included in the RNA isolation protocol. Total RNA concentration was determined by the NanoDrop ND-1000 spectrophotometer (NanoDrop Technologies Inc., Wilmington, DE, USA), and its quality was measured by the 2100 Bioanalyzer and RNA 6000 Nano kit (Agilent Technologies, Santa Clara, CA, USA). Strand-specific RNA-seq libraries were prepared using the SureSelect strand-specific mRNA library preparation kit (Agilent Technologies, Santa Clara, CA, USA) as per the manufacturer's protocol. In brief, poly(A) RNA was purified from 1 μg of total RNA using two serial rounds of binding to oligo(dT) magnetic particles;

then, the nucleic acid was fragmented and reverse transcribed to generate cDNA. Illumina-specific adaptor was sequentially ligated to the 3′ end of cDNA fragments, purified using the AMPure XP beads (Beckman Coulter, Brea, CA, USA) and, finally, PCR-amplified (13 cycles) using an appropriate indexing primer to allow further samples multiplexing. The PCR-amplified libraries were purified with the AMPure XP beads (Beckman Coulter, Brea, CA, USA) and then assessed for their quality and fragments distribution using the 2100 Bioanalyzer DNA 1000 assay (Agilent Technologies, Santa Clara, CA, USA). In the presence of adaptor-dimers (electropherogram's peak at 100 to 150-bp), another round of magnetic beads purification was performed. Libraries were quantified by both the Qubit®Fluorometer (Life Technologies, Monza, Italy) and the qPCR-based NEBNext library quantification kit (New England BioLabs, Hitchin, UK). Finally, equimolar amounts of each 6 index-tagged libraries were multiplexed together in one pool (total of 4 pools, 24 single libraries) and then sequenced by an Illumina HiSeq 2500 for 50 sequencing cycles.

The raw 50 bp single-end sequences (Sanger/Illumina 1.9 encoding) were quality-controlled using FastQC v.0.11.4 [30], and the low-quality bases (Phred quality score <30) and adaptor contamination (if present) were removed by Trimmomatic v.0.36 [31], using the parameters 'LEADING:3 SLIDINGWINDOW:4:20 MINLEN:25'. The high-quality reads were mapped by HISAT v. 2.0.4 [32] against the *Bos taurus* reference genome (Ensembl UMD 3.1). The uniquely-mapped reads aligned to exons were counted with HTSeq v.0.6.1 [33], then tested—by the DESeq2 R package v.1.14.1 [34]—for the presence of differentially expressed genes (DEGs) between CTR and GP groups after the 60-day supplementation period. All genes with a false discovery rate (FDR) less than 0.1 were considered DEGs regardless of their fold-change (FC) value. The sequencing data (FASTQ files) associated with this project are deposited in the GenBank's Sequence Read Archive (SRA) under the accession number SRP105401.

2.6. Functional Analysis and Statistics

The Gene Set Enrichment Analysis (GSEA) was used to examine the significantly enriched pathways in the GP group—in comparison with the CTR group. GSEA is a computational method that identifies shared differential gene expression of predefined, functionally related gene sets representing biological pathways. This is quantified by using a different type of Enrichment Score (ES), a weighted Kolmogorov–Smirnov-like statistic that evaluates if the members of the pathway are randomly distributed or found at the extremes (top or bottom) of the list [35]. In the present study, the GSEA pre-ranked option was used to analyze the deregulated pathways in the GP-supplemented group (GP T60 vs. CTR T60). All the Ensembl gene IDs were collapsed to their corresponding HUGO gene symbols; then, the entire normalized transcriptome dataset was ranked by the logarithm transformed (base 2) FC, where the up- and down-regulated genes were assigned positive and negative values, respectively. The pre-ranked dataset was analyzed (1000 permutations) against the Hallmarks (h.all.v7.0), i.e., the curated canonical KEGG pathways (c2.cp.kegg.v7.0.) and the gene ontology biological process (c5.bp.v7.0) catalogs from the Molecular Signatures Database (MsigDB) [35]. The GSEA output reported in the present study represents the most negatively- or positively-enriched hallmarks, pathways or biological processes (highest normalized enrichment score; NES) in the GP group compared with CTR. To visualize the gene expression data of GP and CTR samples, principal component analysis (PCA) and hierarchical clustering plots were generated using the ClustVis R package [36].

3. Results

The health status of all the experimental animals was satisfactory all throughout the experiment, with no visible signs of illness or supplementation-related stress. The dairy cows produced an average daily amount of 17 ± 2.4 liters of milk. The effect of the GP supplementation on the daily amount of milk produced as well as on the milk biochemical profile is shown in Supplementary Figure S1. Long-term GP supplementation did not result in differences between the GP and CTR groups either in the daily amount of milk produced than in milk's biochemical profile, i.e., the percentage of fat, protein, casein, lactose, urea and solids. In addition to this, and in strict accordance with what was observed for the milk yield, no significant variations were highlighted in feed intake for the entire duration of the trial, with a daily dry matter intake (DMI) equal to 19.3 ± 1.2 and 19.5 ± 0.9 kg in CTR and GP group respectively.

Regarding RNA-seq, the sequencing of 12 samples (2 groups × 6 animals/group) resulted in an average of~21 million reads per sample. A summary of the sequencing output is reported in Supplementary Table S1. After the assumption of a GP-supplemented diet for 60 days, 40 protein-coding genes were found to be differentially expressed in the GP group, of which 10 and 30 genes were down- and up-regulated, respectively (Supplementary Table S2). The most down-regulated gene in the GP group (FDR < 0.05) was the DnaJ heat shock protein family (HSP40) member A1. Additional interesting genes down-regulated by GP supplementation were the mitochondrial fission factor (MFF), the impact RWD domain protein (IMPACT), the ATP binding cassette subfamily A member 13 (ABCA13), the regulator of G-protein signaling 2 (RGS2), and the integrin subunit alpha 3 (ITGA3). The top 100 deregulated genes (using their normalized RNA-seq counts) were able to discriminate the two experimental groups either through a principal component analysis (PCA) plot (with the first 2 components accounting for > 60% of the total variability; Figure 1a) or by a hierarchical clustering heatmap (Figure 1b).

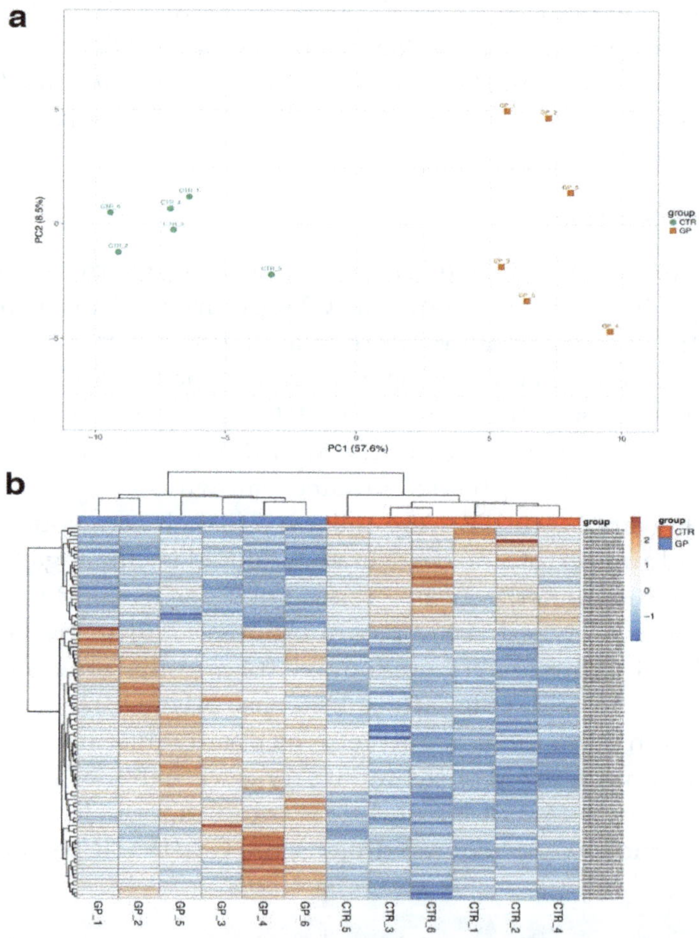

Figure 1. Principal Component Analysis (PCA) and hierarchical clustering heatmap. PCA and heatmap of the top 100 deregulated genes in dairy cows fed for 60 days with a GP-supplemented diet (GP T60 vs. CTR T60). (**a**) PCA plot of the GP (orange squares, right side) and CTR individuals (green circles, left side) after 60 days of GP dietary supplementation. (**b**) Heatmap generated from the DEGs using the hierarchical clustering (Euclidean distance clustering algorithm) of the top 100 deregulated genes of the GP group.

To further pinpoint the collective effect of the transcriptomic changes, we used GSEA to identify long-term GP-supplementation-associated pathways and biological processes. The GSEA analysis revealed that most of the core-enriched genes in the GP group, such as 'interferon alpha (IFN-α) response' (NES = 2.02), 'IFN-γ response' (NES = 1.94), 'IL6-JAK-STAT3 signaling' (NES = 1.74) or 'complement' hallmark genes (NES = 1.57), fell into the immune system-related hallmark gene sets (Figure 2). Moreover, some biological processes related with immune defense mechanisms were positively enriched in the GP group. Among these ones, the most enriched processes were the 'response to protozoan' (NES = 1.99) and the 'negative regulation of viral replication' (NES = 1.96; Figure 3). A summary of the GSEA analysis output is reported in Supplementary Table S3.

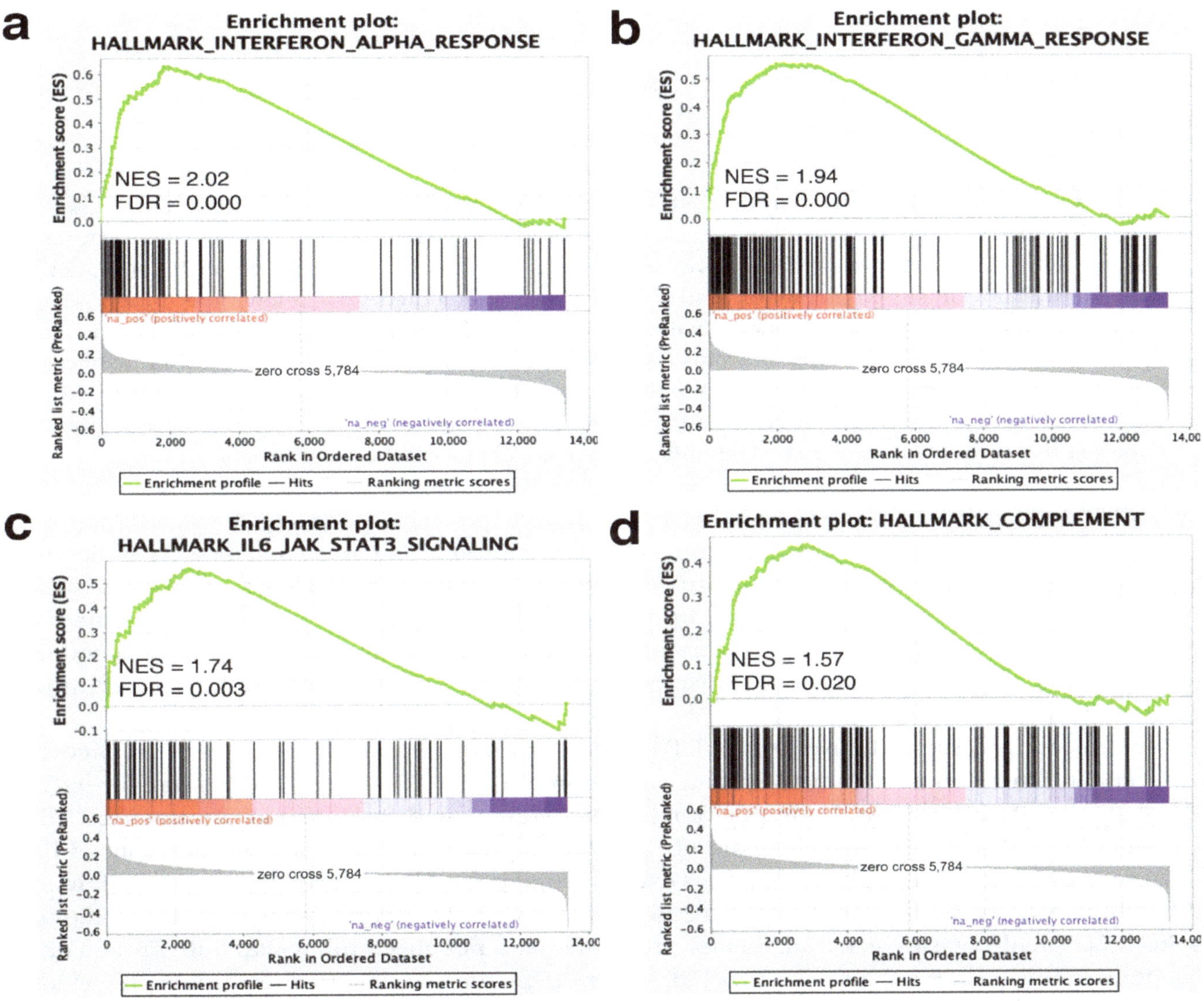

Figure 2. Gene set enrichment analysis (GSEA) plot (score curves) for enriched hallmarks. The GSEA analysis was performed with the hallmarks gene sets of GSEA Molecular Signatures Database. The "Signal-to-Noise" ratio (SNR) statistic was used to rank the genes per their correlation with the dietary GP supplementation (red) or the un-supplemented CTR phenotype (blue). The heatmap on the right side of each panel visualize the genes mostly contributing to the enriched gene set. For the fully detailed list, see Supplementary Table S3. The green curve corresponds to the Enrichment Score (ES) curve, which is the running sum of the weighed ES obtained from the GSEA software, while the Normalized Enrichment Score (NES) and the corresponding p-value are reported within each graph. Panels (**a**–**d**) denote the most enriched (significant) pathways (i.e., gene sets).

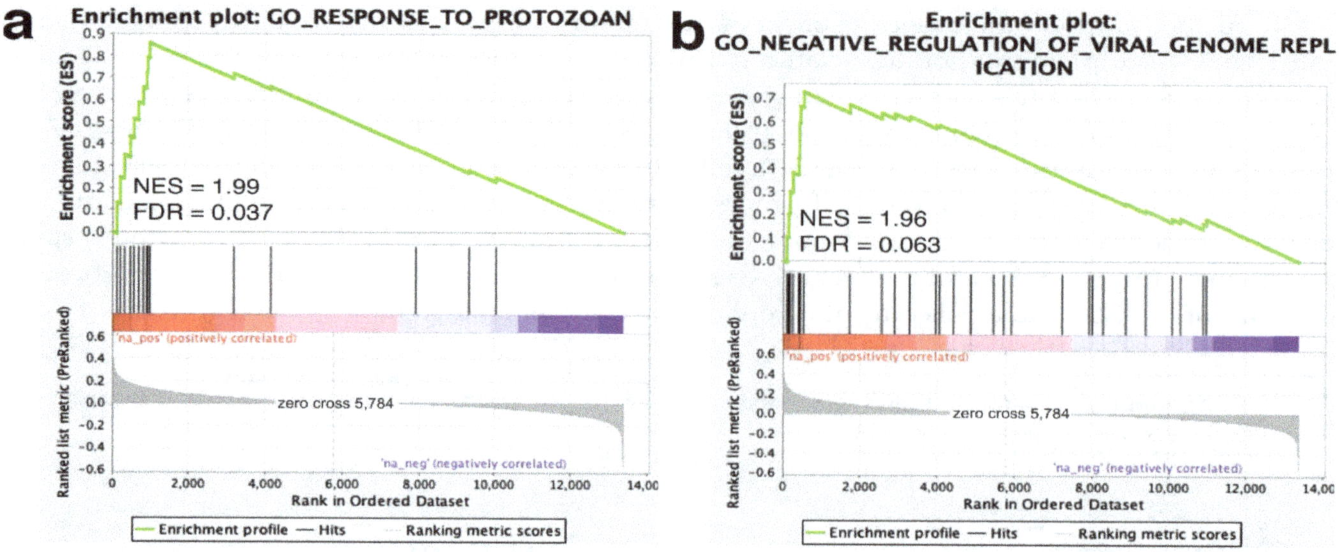

Figure 3. Gene set enrichment analysis (GSEA) plot (score curves) for enriched biological processes. The GSEA analysis was performed with the biological process gene sets of GSEA Molecular Signatures Database. The "Signal-to-Noise" ratio (SNR) statistic was used to rank the genes per their correlation with the dietary GP supplementation (red) or the un-supplemented CTR phenotype (blue). The heatmap on the right side of each panel visualize the genes mostly contributing to the enriched gene set. For the fully detailed list, see Supplementary Table S3. The green curve corresponds to the Enrichment Score (ES) curve, which is the running sum of the weighed ES obtained from the GSEA software, while the Normalized Enrichment Score (NES) and the corresponding *p*-value are reported within each graph. Panels (**a**,**b**) denote the most enriched (significant) biological processes (i.e., gene sets).

4. Discussion

The objective of this study was to examine the transcriptomic signature of a long-term dietary GP supplementation in dairy cows and to evaluate whether this signature reflects known GP-associated potential health benefits [37]. The main finding was that a 60-day dietary GP supplementation can modulate the expression of a considerable, albeit less than expected, number of genes in dairy cows and induce a noticeable transcriptomic signature that mainly reflects a GP-induced immunomodulatory effect.

Several studies assessed the effects of GP on milk production and composition in dairy cows. In two recent and independent studies, a dietary addition of ~10% GP (~2 kg GP/head/day for~60 days) did not show differences in milk composition [38,39]. Moreover, different concentrations (50, 75 and 100 g/kg of DM) of grape residue silage did not affect milk production nor the concentrations of CP, fat or lactose in dairy cows throughout a 21 days experimental period [10]. Further, the milk of dairy cows fed for three months with a diet containing 15% GP preserved the normal levels of fat, protein and caseins [22]. Finally, in a 4-week trial, dietary grape marc did not affect milk yield, milk protein or milk fat content of mid-lactation Holstein cows [40]. Overall, the present data further support the evidence that dietary GP does not alter milk yield or composition even after long-term supplementation.

Grape pomace accounts for ~25%–35% of the total weight of processed grapes; it contains significant amounts of dietary fiber, for which it may partially replace the forage portion in ruminants' diet [22,37]. Far more interesting, GP is enriched with potent bioactive compounds, especially polyphenols (~2%–6.5%). Grape polyphenolics, possessing different chemical structures and activities, are essentially categorized into two major classes of compounds: flavonoids (the most abundant polyphenols) and non-flavonoids [37,39,41]. Even though the biological activity of GP extracts mostly relate to their antioxidant properties, the dietary intake of these bioactive compounds results in pleiotropic health-promoting responses, e.g., cardio- and neuroprotective, anti-diabetic, anti-cancer, anti-aging, anti-adipogenetic as well as prebiotic and anti-microbial beneficial effects [39,41–43]. Overall, these evidences would suggest a link between polyphenol-rich food consumption and reduction in the incidence of numerous chronic disorders effect [42]. A separate discussion deserves the effects of polyphenols on the immune system, which is the second most important mechanism of action (after the antioxidant activity). Several studies have shown how polyphenols target multiple inflammatory components and lead to anti-inflammatory mechanisms and/or inflammation antagonism. Further, polyphenols today represent also promising candidates for the therapy of autoimmune diseases [43–46].

In this study, the change in expression of more than 40 protein-coding genes indicates that dietary GP has a traceable transcriptomic signature. Grape pomace contains lipid, proteins, fiber, minerals and large amounts of polyphenols (e.g., 10.4–64.8 g gallic acid equivalents/kg); moreover, polyphenolics show a linear correlation with in vitro antioxidant activity [47]. Therefore, one can safely presume that present transcriptional changes are due to GP polyphenols content. However, in the present experimental condition, only a few DEGs were significantly modulated by the dietary GP supplementation, and no genes were markedly over-expressed (in terms of FC). This was not surprising as in transcriptomics (e.g., microarray, RNA-seq), a large part of biologically meaningful DEGs possess FC less than 2. As a whole, these DEGs are not a minor population; rather, a significant fraction of transcriptional changes occurring in that sample (tissue, blood, etc.) [48]. This begs the question of whether such small changes reflect biologically meaningful events. In nutrigenomic studies, nutritionally relevant concentrations of bioactive compounds may elicit subtle changes in gene expression, resulting in critically important biological insights even though difficult to be detected reliably [24,49–51]. Moreover, small transcriptional variations are somehow expected when investigating long-term effects of a supplemented diet. In our opinion, such an event should not be ignored. Noteworthy, some interesting genes such as DNAJA1, MFF, IMPACT, ABCA13, RGS2, and ITGA were significantly downregulated in animals fed with a GP-supplemented diet (FDR < 0.05).

The top downregulated gene, DNAJA1, also known as heat shock protein (HSP) 40, plays a fundamental role during both physiological and stress conditions as it functions in mediating inappropriate proteins aggregation and folding [52]. Several studies have suggested that some heat shock proteins (both HSP70 and HSP40) are induced by wide ranges of stressful and pathological conditions [53]. Intriguingly, the natural flavonoid quercetin, containing a polyphenolic chemical structure and possessing healthy antioxidant properties, was shown to inhibit the expression of some HSPs at basal levels in human tissues [54–56]. Therefore, we might suggest that a lower expression of DNAJA1 in the GP group of animals compared to controls could be associated to healthier conditions of cows fed a GP-supplemented diet.

In living organisms, mitochondria continually change their morphology to maintain cell homeostasis. This process (mitochondrial dynamics) consists of two interconnected events, i.e., fission and fusion [57,58]. Mitochondrial fission is mediated by dynamin-related protein 1 (Drp1), a cytosolic highly conserved guanosine triphosphatase of the dynamin superfamily of proteins [57,59,60]. In mammals, the recruitment of Drp1 to mitochondria is mediated by MFF, a member of an array of Drp1-binding receptor factors localized at the mitochondrial outer membrane [57,58,60,61]. Fission is a key mechanistic step in apoptosis induction. Furthermore, Drp1 and MFF may contribute to apoptotic phenomena resulting from oxidative stress and reactive oxygen species (ROS) overproduction [57,58,60,62,63]. Some xenobiotics trigger oxidative stress, mitochondrial disfunction and ultimately apoptosis by increasing the expression of Drp1 (mostly) and MFF; among these ones ammonia [63], T2-toxin [64], cadmium [65], and microcystin [66]. However, curcumin, resveratrol, anthocyanins, quercetin and polyphenols inhibit mitochondrial fission through transcriptional and post-translational modifications of Drp-1, MFF as well as of other factors contributing to mitochondrial dynamics, thus preventing cells from oxidative stress [67–69]. Therefore, present MFF downregulation might be viewed as a protective effect provoked by GP supplementation against cellular oxidative stress, resulting in minor recruitment and transfer of Drp-1 inside mitochondria and, consequently, no mitochondrial fission/fusion imbalance and apoptosis. Concomitant inhibition of Drp-1 (i.e., DNM1L) was expected but no effects were observed for this gene.

IMPACT is a protein-coding gene highly expressed in neuronal tissue of mammals, where it promotes neuronal development [70–72]. Noteworthily, it substantially contributes to the adaptive response to nutritional stress and dietary restriction (i.e., a reduction in food intake without malnutrition) [70]. Under these stress conditions (e.g., amino acids depletion, proteasome inhibition, lack of glucose), IMPACT can inhibit a serine/threonine kinase coded by the General Control Non-derepressible 2 (GCN2) gene [70–74]. This latter gene, which is not listed among the DEGs, confers an adaptive response through distinct biochemical reactions (e.g., phosphorylation) and stress response genes activation. This cascade of events constitutes the Integrated Stress Response (ISR) pathway [70]. In addition to GCN2 inhibition, IMPACT overexpression also abrogates the phosphorylation of another GCN2 molecular target, i.e., the eukaryotic initiation factor 2, α subunit (eIF2α) [70,73]. Interestingly, IMPACT, GCN2 and ISR "molecular sensors" contribute substantially to immune system regulation [71,72,74,75]. This crosstalk between nutrient metabolism, gut immune system and host microbiota potentially shapes the overall gut responses to nutrients, and today represents a "hot" research topic [76,77]. As to IMPACT, KO mice are leaner than wild-type ones, especially when fed a high-fat diet. Furthermore, they show a defect in the control body temperature in response to starvation. This evidence, albeit not completely independent from GCN2, would suggest that IMPACT protein is involved in the maintenance of energy homeostasis [72]. The IMPACT gene downregulation we observed in the present study might be hypothetically interpreted as the result of a positive effect of GP on cattle nutritional status and immune system.

RGS2 is a conserved regulator of G protein-coupled receptors (GPCRs) signaling pathway. It belongs to the RGS superfamily of proteins, whose expression is regulated by epigenetic, transcriptional and post-translational mechanisms. These proteins determine the intensity and duration of GPCR-mediated effects through binding to the active Gα subunit of G proteins and activating GTPase. This protein-coding gene is expressed in several tissues such as CNS, heart, vasculature, kidney, immune system, lungs, bone and ovaries [78,79]. Interestingly, oxidative stress conditions increase RGS2 mRNA levels [79,80]; moreover, RGS2 contributes to the regulation of some enzymes involved in antioxidant defense, namely glyoxalase-1 and glutathione reductase-1 [81]. Finally, RGS2 has been shown to be involved in the pathophysiology of gastrointestinal inflammation and visceral pain, mostly regulating T-cell immunity [79,82,83]. In our opinion, the observed RGS2 gene downregulation can be viewed as the consequence of an indirect positive effect of GP on the overall cattle gastrointestinal tract homeostasis.

Limited information is currently available for the remaining two DEGs, i.e., ABCA13 and ITGA3. The former is a member of the ATP-binding cassette (ABC) superfamily of efflux drug transporters, encoding for the largest protein of the same family (5,058 residues) [84]. This gene is expressed in trachea, testis, bone marrow and blood-derived cells [85,86]. Interestingly, broilers fed diets supplemented with trace minerals (Zn, Cu, and Mn) showed reduced footpad lesions and improved wound healing process. At the molecular level, upregulation of genes involved in collagen synthesis, deposition and organization, cell migration, matrix remodelling, and angiogenesis, including ITGA2 and ITGA3, was noticed [87]. Despite this, no information about a possible modulatory effect of polyphenols on ABCA13 and ITGA3 are actually available.

In the present study, most of the core-enriched genes in the GP group (e.g., IFN-α, IFN-γ, and IL6-JAK-STAT3) belong to hallmarks and biological processes related to the immune response. All this was expected. Widespread evidence shows that polyphenols possess predominantly anti-oxidant, anti-inflammatory and immunomodulatory properties in both humans and farm animal species including cattle; hence, they are beneficial all-purpose nutraceuticals or supplements [45,46,88]. Polyphenols inhibit NF-κB, PI3K/Akt, MAPKs, and arachidonic acid-dependent signaling pathways. Regarding their immunosuppressant properties, polyphenols modulate T-cell function (e.g., suppressing T helper 2 activation and promoting the development of regulatory T cells), inhibit mast cell degranulation, and downregulate inflammatory cytokine responses [45,46,76]. Owing to these immunomodulatory properties, polyphenols have emerged as potential tools for the treatment of autoimmune disorders [46,89].

Additional biological processes positively enriched in GP group were the 'response to protozoan' and the 'negative regulation of viral replication'. Again, our results corroborate recent evidence about the use of polyphenolic compounds as potential therapeutic tools against protozoa and viruses, albeit scarce information is available on farm animals. Some dietary polyphenols (e.g., quercetin, resveratrol, rutin) show a certain efficacy against leishmaniosis [90], and giardiasis [91]; furthermore, resveratrol itself has been suggested as a potential restorative agent against toxoplasmosis brain disorders [92] and *Schistosoma mansoni* infections [93]. As far as the effects on viruses, human and animal evidence suggest polyphenolic compounds (e.g., resveratrol, tea polyphenols) possess potential preventive and therapeutic effects, and might be used as adjuvant therapy for the management of viral infection, e.g., influenza, herpes simplex, pseudorabies, hepatitis C, porcine reproductive and respiratory syndrome, and even Ebola [88,94–97].

5. Conclusions

This study provides evidence on the transcriptomic signature of long-term dietary GP supplementation in dairy cows. The present results strengthen the concept of using a nutrigenomic approach to assess the beneficial nor detrimental effects of farm animals' diets supplemented with grape polyphenolics and, in general, with undervalued food-related by-products and sources such as aromatic plants, culinary herbs, spices, vegetables and fruit trimmings. In our experimental conditions, the GP dietary supplementation resulted in a differential transcriptomic signature mostly reflecting an immunomodulatory effect. Overall, these data contribute to the growing body of nutrigenomics research in dairy cows and, to a wider extent, in ruminants.

Although additional studies should be performed to ascertain benefits and drawbacks of GP supplementation in dairy cows, and to define the optimal feeding schemes, the results here obtained let us suggest that this agricultural by-product has an overall positive impact on immune status of cows, without affecting milk quality.

The present study comes with some shortcomings. First, the number of DEGs was lower than expected; moreover, most of them were downregulated and showed lower FC. Nevertheless, we were expecting to find a higher number of DEGs, possibly confirming (e.g., gene upregulation) the antioxidant and anti-inflammatory properties of grape polyphenolics. Actually, the transcriptomic signature of GP-supplemented animals (namely, the list of DEGs and GSEA outputs) is suggestive of improved dairy cows' general health conditions, thereby confirming the GP wide range of health-promoting activities, e.g., anti-inflammatory, anti-aging, anti-oxidant, anti-microbial and antiviral properties. Moreover, it is important to highlight that feed might exert its biological effect at different levels. Thus, it would be possible that GP targets mechanisms other than transcription, such as mRNA silencing, mRNA translation, and enzyme activity/functionality. To this respect, the previous literature demonstrated that GP supplementation affects catalase activity in lambs [18], superoxide dismutase and glutathione peroxidase activity in piglets [98]. To our knowledge, proteomic studies investigating the effect of GP have not been so far performed in farmed animals. However, it is likely that protein expression may be a target of GP. Indeed, dietary inclusion of other bioactive components have been demonstrated to significantly affect the intestinal mucosa proteome of growing pigs [99] and the skeletal muscle proteins of pre-weaning calves [100].

Another possible limitation of the study is having measured the GP-dependent transcriptomic changes in dairy cows WB. Many experiments aiming at the characterization of the beneficial effect of nutraceuticals on the individual's transcriptome involve the use of gastrointestinal biopsies (e.g., in rodents, chicken, fish). Apart from the ethical considerations, there is evidence that the WB is a suitable surrogate tissue for transcriptomics studies. The final potential bias is intrinsic to grape polyphenolics composition and the corresponding biological activity. In general, the varying composition of GP originating from different grapes and locations are likely to show remarkable differences in their biological activity. As a consequence, this "biodiversity" impacts GP bioavailability, metabolism, bioactivity and, ultimately the animal's transcriptomics signature. This is still a major criticism of the use of grape polyphenols in food-producing animals as well as in other species including humans. In perspective, crude or highly purified forms of GP extract might be used to characterize the most of beneficial effect of a GP-supplemented diet. Despite these potential biases, the present results corroborate the health benefits, sustainability, and environmental impact of GP utilization in dairy animals and in general in food-producing species. Clearly, further investigations are recommended to improve our knowledge on the best use of polyphenolic compounds as nutraceuticals, e.g., evaluating different feeding schemes and/or percentage of GP supplementation, exploring the now in fashion "host-drug-nutraceutical-microbiota interactions".

Author Contributions: Conceptualization, M.G., M.D. and G.M.; Data curation, R.E., A.I. and E.M.; Formal analysis, R.E. and E.M.; Funding acquisition, G.M.; Investigation, M.P., A.I. and S.R.; Methodology, R.E., A.I., L.G., S.R. and F.B.; Project administration, G.M.; Resources, G.M.; Supervision, M.D. and G.M.; Writing—original draft, M.P., R.E. and M.D.; Writing—review and editing, A.I. and M.G. All authors have read and agreed to the published version of the manuscript.

References

1. Van Kernebeek, H.R.J.; Oosting, S.J.; Van Ittersum, M.K.; Bikker, P.; De Boer, I.J.M. Saving land to feed a growing population: Consequences for consumption of crop and livestock products. *Int. J. Life Cycle Assess.* **2016**, *21*, 677–687. [CrossRef]
2. Hasan, M.S.; Feugang, J.M.; Liao, S.F. A Nutrigenomics Approach Using RNA Sequencing Technology to Study Nutrient-Gene Interactions in Agricultural Animals. *Curr. Dev. Nutr.* **2019**, *3*, nzz082. [CrossRef]
3. EU. A General Union Environment Action Programme to 2020 "Living well, within the limits of our planet". *Off. J. Eur. Union* **2013**, *L354*, 171–200.
4. Leiva-Candia, D.E.; Pinzi, S.; Redel-Macías, M.D.; Koutinas, A.; Webb, C.; Dorado, M.P. The potential for agro-industrial waste utilization using oleaginous yeast for the production of biodiesel. *Fuel* **2014**, *123*, 33–42. [CrossRef]
5. Schieber, A.; Stintzing, F.C.; Carle, R. By-products of plant food processing as a source of functional compounds—Recent developments. *Trends Food Sci. Technol.* **2001**, *12*, 401–413. [CrossRef]
6. Vauzour, D.; Rodriguez-Mateos, A.; Corona, G.; Oruna-Concha, M.J.; Spencer, J.P.E. Polyphenols and human health: Prevention of disease and mechanisms of action. *Nutrients* **2010**, *2*, 1106–1131. [CrossRef]
7. Federici, F.; Fava, F.; Kalogerakis, N.; Mantzavinos, D. Valorisation of agro-industrial by-products, effluents and waste: Concept, opportunities and the case of olive mill waste waters. *J. Chem. Technol. Biotechnol.* **2009**, *84*, 895–900. [CrossRef]
8. Mirzaei-Aghsaghali, A.; Maheri-Sis, N. Nutritive value of some agro-industrial by-products for ruminants—A review. *World J. Zool.* **2008**, *3*, 40–46.
9. Makris, D.P.; Boskou, G.; Andrikopoulos, N.K. Polyphenolic content and in vitro antioxidant characteristics of wine industry and other agri-food solid waste extracts. *J. Food Compos. Anal.* **2007**, *20*, 125–132. [CrossRef]
10. Santos, N.W.; Santos, G.T.D.; Silva-Kazama, D.C.; Grande, P.A.; Pintro, P.M.; de Marchi, F.E.; Jobim, C.C.; Petit, H.V. Production, composition and antioxidants in milk of dairy cows fed diets containing soybean oil and grape residue silage. *Livest. Sci.* **2014**, *159*, 37–45. [CrossRef]
11. Llobera, A.; Cañellas, J. Antioxidant activity and dietary fibre of Prensal Blanc white grape (Vitis vinifera) by-products. *Int. J. Food Sci. Technol.* **2008**, *43*, 1953–1959. [CrossRef]
12. Negro, C.; Tommasi, L.; Miceli, A. Phenolic compounds and antioxidant activity from red grape marc extracts. *Bioresour. Technol.* **2003**, *87*, 41–44. [CrossRef]
13. Viveros, A.; Chamorro, S.; Pizarro, M.; Arija, I.; Centeno, C.; Brenes, A. Effects of dietary polyphenol-rich grape products on intestinal microflora and gut morphology in broiler chicks. *Poult. Sci.* **2011**, *90*, 566–578. [CrossRef] [PubMed]

14. Gessner, D.K.; Fiesel, A.; Most, E.; Dinges, J.; Wen, G.; Ringseis, R.; Eder, K. Supplementation of a grape seed and grape marc meal extract decreases activities of the oxidative stress-responsive transcription factors NF-κB and Nrf2 in the duodenal mucosa of pigs. *Acta Vet. Scand.* **2013**, *55*, 18. [CrossRef]
15. Fiesel, A.; Gessner, D.K.; Most, E.; Eder, K. Effects of dietary polyphenol-rich plant products from grape or hop on pro-inflammatory gene expression in the intestine, nutrient digestibility and faecal microbiota of weaned pigs. *BMC Vet. Res.* **2014**, *10*, 196. [CrossRef]
16. Ishida, K.; Kishi, Y.; Oishi, K.; Hirooka, H.; Kumagai, H. Effects of feeding polyphenol-rich winery wastes on digestibility, nitrogen utilization, ruminal fermentation, antioxidant status and oxidative stress in wethers. *Anim. Sci. J.* **2015**, *86*, 260–269. [CrossRef] [PubMed]
17. Biscarini, F.; Palazzo, F.; Castellani, F.; Masetti, G.; Grotta, L.; Cichelli, A.; Martino, G. Rumen microbiome in dairy calves fed copper and grape-pomace dietary supplementations: Composition and predicted functional profile. *PLoS ONE* **2018**, *13*, e0205670. [CrossRef]
18. Kafantaris, I.; Kotsampasi, B.; Christodoulou, V.; Kokka, E.; Kouka, P.; Terzopoulou, Z.; Gerasopoulos, K.; Stagos, D.; Mitsagga, C.; Giavasis, I.; et al. Grape pomace improves antioxidant capacity and faecal microflora of lambs. *J. Anim. Physiol. Anim. Nutr.* **2017**, *101*, e108–e121. [CrossRef] [PubMed]
19. Nudda, A.; Correddu, F.; Marzano, A.; Battacone, G.; Nicolussi, P.; Bonelli, P.; Pulina, G. Effects of diets containing grape seed, linseed, or both on milk production traits, liver and kidney activities, and immunity of lactating dairy ewes. *J. Dairy Sci.* **2015**, *98*, 1157–1166. [CrossRef]
20. Gessner, D.K.; Koch, C.; Romberg, F.J.; Winkler, A.; Dusel, G.; Herzog, E.; Most, E.; Eder, K. The effect of grape seed and grape marc meal extract on milk performance and the expression of genes of endoplasmic reticulum stress and inflammation in the liver of dairy cows in early lactation. *J. Dairy Sci.* **2015**, *98*, 8856–8868. [CrossRef]
21. Manso, T.; Gallardo, B.; Salvá, A.; Guerra-Rivas, C.; Mantecón, A.R.; Lavín, P.; de la Fuente, M.A. Influence of dietary grape pomace combined with linseed oil on fatty acid profile and milk composition. *J. Dairy Sci.* **2016**, *99*, 1111–1120. [CrossRef]
22. Chedea, V.S.; Pelmus, R.S.; Lazar, C.; Pistol, G.C.; Calin, L.G.; Toma, S.M.; Dragomir, C.; Taranu, I. Effects of a diet containing dried grape pomace on blood metabolites and milk composition of dairy cows. *J. Sci. Food Agric.* **2017**, *97*, 2516–2523. [CrossRef] [PubMed]
23. Ianni, A.; Martino, G. Dietary Grape Pomace Supplementation in Dairy Cows: Effect on Nutritional Quality of Milk and Its Derived Dairy Products. *Foods* **2020**, *9*, 168. [CrossRef]
24. Iannaccone, M.; Elgendy, R.; Giantin, M.; Martino, C.; Giansante, D.; Ianni, A.; Dacasto, M.; Martino, G. RNA sequencing-based whole-transcriptome analysis of friesian cattle fed with grape pomace-supplemented diet. *Animals* **2018**, *8*, 188. [CrossRef]
25. European Union. Directive 2010/63/EU of the European Parliament and of the Council of 22 September 2010 on the protection of animals used for scientific purposes. *Off. J. Eur. Union* **2010**, *53*, 33–79.
26. Ianni, A.; Di Luca, A.; Martino, C.; Bennato, F.; Marone, E.; Grotta, L.; Cichelli, A.; Martino, G. Dietary supplementation of dried grape pomace increases the amount of linoleic acid in beef, reduces the lipid oxidation and modifies the volatile profile. *Animals* **2019**, *9*, 578. [CrossRef] [PubMed]
27. Singleton, V.L.; Rossi, J. Colorimetry of total phenolics with phosphomolybdic. *Am. J. Enol. Vitic.* **1965**, *16*, 144–158.
28. Helrick, K. *Official Methods of Analysis (AOAC)*; Association of Official Analytical Chemist: Washington, DC, USA, 1990; ISBN 8589439488.
29. Goering, H.K.; Van Soest, P.J. *Forage Fiber Analyses (Apparatus, Reagent, Procedures and Some Applications): Agriculture Handbook No. 379*; Agricultural Research Service, US Department of Agriculture: Washington, DC, USA, 1970.
30. Andrews, S.; Krueger, F.; Seconds-Pichon, A.; Biggins, F.; Wingett, S. FastQC. A Quality Control Tool for High Throughput Sequence Data. Babraham Bioinformatics. Available online: http://www.bioinformatics.babraham.ac.uk/projects/fastqc/ (accessed on 14 April 2020).
31. Bolger, A.M.; Lohse, M.; Usadel, B. Trimmomatic: A flexible trimmer for Illumina sequence data. *Bioinformatics* **2014**, *30*, 2114–2120. [CrossRef]
32. Kim, D.; Langmead, B.; Salzberg, S.L. HISAT: A fast spliced aligner with low memory requirements. *Nat. Methods* **2015**, *12*, 357–360. [CrossRef]

33. Anders, S.; Pyl, P.T.; Huber, W. HTSeq-A Python framework to work with high-throughput sequencing data. *Bioinformatics* **2015**, *31*, 166–169. [CrossRef] [PubMed]
34. Love, M.I.; Huber, W.; Anders, S. Moderated estimation of fold change and dispersion for RNA-seq data with DESeq2. *Genome Biol.* **2014**, *15*, 550. [CrossRef] [PubMed]
35. Subramanian, A.; Tamayo, P.; Mootha, V.K.; Mukherjee, S.; Ebert, B.L.; Gillette, M.A.; Paulovich, A.; Pomeroy, S.L.; Golub, T.R.; Lander, E.S.; et al. Gene set enrichment analysis: A knowledge-based approach for interpreting genome-wide expression profiles. *Proc. Natl. Acad. Sci. USA* **2005**, *102*, 15545–15550. [CrossRef]
36. Metsalu, T.; Vilo, J. ClustVis: A web tool for visualizing clustering of multivariate data using Principal Component Analysis and heatmap. *Nucleic Acids Res.* **2015**, *43*, W566–W570. [CrossRef] [PubMed]
37. Averilla, J.N.; Oh, J.; Kim, H.J.; Kim, J.S.; Kim, J.S. Potential health benefits of phenolic compounds in grape processing by-products. *Food Sci. Biotechnol.* **2019**, *28*, 1607–1615. [CrossRef]
38. Ianni, A.; Di Maio, G.; Pittia, P.; Grotta, L.; Perpetuini, G.; Tofalo, R.; Cichelli, A.; Martino, G. Chemical–nutritional quality and oxidative stability of milk and dairy products obtained from Friesian cows fed with a dietary supplementation of dried grape pomace. *J. Sci. Food Agric.* **2019**, *99*, 3635–3643. [CrossRef] [PubMed]
39. Ianni, A.; Innosa, D.; Martino, C.; Bennato, F.; Martino, G. Short communication: Compositional characteristics and aromatic profile of caciotta cheese obtained from Friesian cows fed with a dietary supplementation of dried grape pomace. *J. Dairy Sci.* **2019**, *102*, 1025–1032. [CrossRef] [PubMed]
40. Scuderi, R.A.; Ebenstein, D.B.; Lam, Y.W.; Kraft, J.; Greenwood, S.L. Inclusion of grape marc in dairy cattle rations alters the bovine milk proteome. *J. Dairy Res.* **2019**, *86*, 154–161. [CrossRef]
41. Cotoras, M.; Vivanco, H.; Melo, R.; Aguirre, M.; Silva, E.; Mendoza, L. In vitro and in vivo evaluation of the antioxidant and prooxidant activity of phenolic compounds obtained from grape (*Vitis vinifera*) pomace. *Molecules* **2014**, *19*, 21154–21167. [CrossRef]
42. Singh, A.K.; Cabral, C.; Kumar, R.; Ganguly, R.; Rana, H.K.; Gupta, A.; Lauro, M.R.; Carbone, C.; Reis, F.; Pandey, A.K. Beneficial effects of dietary polyphenols on gut microbiota and strategies to improve delivery efficiency. *Nutrients* **2019**, *11*, 2216. [CrossRef]
43. Zhang, H.; Tsao, R. Dietary polyphenols, oxidative stress and antioxidant and anti-inflammatory effects. *Curr. Opin. Food Sci.* **2016**, *8*, 33–42. [CrossRef]
44. Yahfoufi, N.; Alsadi, N.; Jambi, M.; Matar, C. The immunomodulatory and anti-inflammatory role of polyphenols. *Nutrients* **2018**, *10*, 1618. [CrossRef] [PubMed]
45. Focaccetti, C.; Izzi, V.; Benvenuto, M.; Fazi, S.; Ciuffa, S.; Giganti, M.G.; Potenza, V.; Manzari, V.; Modesti, A.; Bei, R. Polyphenols as immunomodulatory compounds in the tumor microenvironment: Friends or foes? *Int. J. Mol. Sci.* **2019**, *20*, 1714. [CrossRef] [PubMed]
46. Khan, H.; Sureda, A.; Belwal, T.; Çetinkaya, S.; Süntar, İ.; Tejada, S.; Devkota, H.P.; Ullah, H.; Aschner, M. Polyphenols in the treatment of autoimmune diseases. *Autoimmun. Rev.* **2019**, *18*, 647–657. [CrossRef]
47. Jin, Q.; O'Hair, J.; Stewart, A.C.; O'Keefe, S.F.; Neilson, A.P.; Kim, Y.T.; McGuire, M.; Lee, A.; Wilder, G.; Huang, H. Compositional characterization of different industrial white and red grape pomaces in Virginia and the potential valorization of the major components. *Foods* **2019**, *8*, 667. [CrossRef] [PubMed]
48. St. Laurent, G.; Shtokalo, D.; Tackett, M.R.; Yang, Z.; Vyatkin, Y.; Milos, P.M.; Seilheimer, B.; McCaffrey, T.A.; Kapranov, P. On the importance of small changes in RNA expression. *Methods* **2013**, *63*, 18–24.
49. Garosi, P.; De Filippo, C.; van Erk, M.; Rocca-Serra, P.; Sansone, S.-A.; Elliott, R. Defining best practice for microarray analyses in nutrigenomic studies. *Br. J. Nutr.* **2005**, *93*, 425–432. [CrossRef]
50. Pagmantidis, V.; Méplan, C.; Van Schothorst, E.M.; Keijer, J.; Hesketh, J.E. Supplementation of healthy volunteers with nutritionally relevant amounts of selenium increases the expression of lymphocyte protein biosynthesis genes. *Am. J. Clin. Nutr.* **2008**, *87*, 181–189. [CrossRef]
51. Elgendy, R.; Giantin, M.; Castellani, F.; Grotta, L.; Palazzo, F.; Dacasto, M.; Martino, G. Transcriptomic signature of high dietary organic selenium supplementation in sheep: A nutrigenomic insight using a custom microarray platform and gene set enrichment analysis. *J. Anim. Sci.* **2016**, *94*, 3169–3184. [CrossRef]
52. Kim, Y.E.; Hipp, M.S.; Bracher, A.; Hayer-Hartl, M.; Ulrich Hartl, F. Molecular Chaperone Functions in Protein Folding and Proteostasis. *Annu. Rev. Biochem.* **2013**, *82*, 323–355. [CrossRef]
53. Liu, Q.; Liang, C.; Zhou, L. Structural and functional analysis of the Hsp70/Hsp40 chaperone system. *Protein Sci.* **2020**, *29*, 378–390. [CrossRef]

54. Elia, G.; Santoro, M.G. Regulation of heat shock protein synthesis by quercetin in human erythroleukaemia cells. *Biochem. J.* **1994**, *300*, 201–209. [CrossRef] [PubMed]
55. Nagai, N.; Nakai, A.; Nagata, K. Quercetin suppresses heat shock response by down regulation of HSF1. *Biochem. Biophys. Res. Commun.* **1995**, *208*, 1099–1105. [CrossRef] [PubMed]
56. Aalinkeel, R.; Bindukumar, B.; Reynolds, J.L.; Sykes, D.E.; Mahajan, S.D.; Chadha, K.C.; Schwartz, S.A. The dietary bioflavonoid, quercetin, selectively induces apoptosis of prostate cancer cells by down-regulating the expression of heat shock protein 90. *Prostate* **2008**, *68*, 1773–1789. [CrossRef] [PubMed]
57. Youle, R.J.; Van Der Bliek, A.M. Mitochondrial fission, fusion, and stress. *Science* **2012**, *337*, 1062–1065. [CrossRef] [PubMed]
58. Gilkerson, R. A disturbance in the force: Cellular stress sensing by the mitochondrial network. *Antioxidants* **2018**, *7*, 126. [CrossRef] [PubMed]
59. Inoue-Yamauchi, A.; Oda, H. Depletion of mitochondrial fission factor DRP1 causes increased apoptosis in human colon cancer cells. *Biochem. Biophys. Res. Commun.* **2012**, *421*, 81–85. [CrossRef]
60. Tak, H.; Eun, J.W.; Kim, J.; Park, S.J.; Kim, C.; Ji, E.; Lee, H.; Kang, H.; Cho, D.H.; Lee, K.; et al. T-cell-restricted intracellular antigen 1 facilitates mitochondrial fragmentation by enhancing the expression of mitochondrial fission factor. *Cell Death Differ.* **2017**, *24*, 49–58. [CrossRef]
61. Prieto, J.; León, M.; Ponsoda, X.; García-García, F.; Bort, R.; Serna, E.; Barneo-Muñoz, M.; Palau, F.; Dopazo, J.; López-García, C.; et al. Dysfunctional mitochondrial fission impairs cell reprogramming. *Cell Cycle* **2016**, *15*, 3240–3250. [CrossRef]
62. Toyama, E.Q.; Herzig, S.; Courchet, J.; Lewis, T.L., Jr.; Losón, O.C.; Hellberg, K.; Young, N.P.; Chen, H.; Polleux, F.; Chan, D.C.; et al. AMP-activated protein kinase mediates mitochondrial fission in response to energy stress. *Science* **2016**, *351*, 275–281. [CrossRef]
63. Shah, S.W.A.; Chen, J.; Han, Q.; Xu, Y.; Ishfaq, M.; Teng, X. Ammonia inhalation impaired immune function and mitochondrial integrity in the broilers bursa of fabricius: Implication of oxidative stress and apoptosis. *Ecotoxicol. Environ. Saf.* **2020**, *190*, 110078. [CrossRef]
64. Yang, J.; Guo, W.; Wang, J.; Yang, X.; Zhang, Z.; Zhao, Z. T-2 toxin-induced oxidative stress leads to imbalance of mitochondrial fission and fusion to activate cellular apoptosis in the human liver 7702 cell line. *Toxins* **2020**, *12*, 43. [CrossRef] [PubMed]
65. Ge, J.; Zhang, C.; Sun, Y.C.; Zhang, Q.; Lv, M.W.; Guo, K.; Li, J.L. Cadmium exposure triggers mitochondrial dysfunction and oxidative stress in chicken (*Gallus gallus*) kidney via mitochondrial UPR inhibition and Nrf2-mediated antioxidant defense activation. *Sci. Total Environ.* **2019**, *689*, 1160–1171. [CrossRef] [PubMed]
66. Zhang, C.; Wang, J.; Zhu, J.; Chen, Y.; Han, X. Microcystin-leucine-arginine induced neurotoxicity by initiating mitochondrial fission in hippocampal neurons. *Sci. Total Environ.* **2020**, *703*, 134702. [CrossRef] [PubMed]
67. Chen, C.; Huang, J.; Shen, J.; Bai, Q. Quercetin improves endothelial insulin sensitivity in obese mice by inhibiting Drp1 phosphorylation at serine 616 and mitochondrial fragmentation. *Acta Biochim. Biophys. Sin.* **2019**, *51*, 1250–1257. [CrossRef] [PubMed]
68. Li, A.; Zhang, S.; Li, J.; Liu, K.; Huang, F.; Liu, B. Metformin and resveratrol inhibit Drp1-mediated mitochondrial fission and prevent ER stress-associated NLRP3 inflammasome activation in the adipose tissue of diabetic mice. *Mol. Cell. Endocrinol.* **2016**, *434*, 36–47. [CrossRef] [PubMed]
69. Naoi, M.; Wu, Y.; Shamoto-Nagai, M.; Maruyama, W. Mitochondria in Neuroprotection by Phytochemicals: Bioactive Polyphenols Modulate Mitochondrial Apoptosis System, Function and Structure. *Int. J. Mol. Sci.* **2019**, *20*, 2451. [CrossRef]
70. Ferraz, R.C.; Camara, H.; De-Souza, E.A.; Pinto, S.; Pinca, A.P.F.; Silva, R.C.; Sato, V.N.; Castilho, B.A.; Mori, M.A. IMPACT is a GCN2 inhibitor that limits lifespan in Caenorhabditis elegans. *BMC Biol.* **2016**, *14*, 87. [CrossRef]
71. Cambiaghi, T.D.; Pereira, C.M.; Shanmugam, R.; Bolech, M.; Wek, R.C.; Sattlegger, E.; Castilho, B.A. Evolutionarily conserved IMPACT impairs various stress responses that require GCN1 for activating the eIF2 kinase GCN2. *Biochem. Biophys. Res. Commun.* **2014**, *443*, 592–597. [CrossRef]
72. Pereira, C.M.; Filev, R.; Dubiela, F.P.; Brandão, B.B.; Queiroz, C.M.; Ludwig, R.G.; Hipolide, D.; Longo, B.M.; Mello, L.E.; Mori, M.A.; et al. The GCN2 inhibitor IMPACT contributes to diet-induced obesity and body temperature control. *PLoS ONE* **2019**, *14*, e0217287. [CrossRef]
73. Castilho, B.A.; Shanmugam, R.; Silva, R.C.; Ramesh, R.; Himme, B.M.; Sattlegger, E. Keeping the eIF2 alpha kinase Gcn2 in check. *Biochim. Biophys. Acta Mol. Cell Res.* **2014**, *1843*, 1948–1968. [CrossRef]

74. Ravishankara, B.; Liua, H.; Shindea, R.; Chaudharya, K.; Xiaoa, W.; Bradleya, J.; Koritzinsky, M.; Madaio, M.P.; McGaha, T.L. The amino acid sensor GCN2 inhibits inflammatory responses to apoptotic cells promoting tolerance and suppressing systemic autoimmunity. *Proc. Natl. Acad. Sci. USA* **2015**, *112*, 10774–10779. [CrossRef] [PubMed]
75. Murguía, J.R.; Serrano, R. New functions of protein kinase Gcn2 in yeast and mammals. *IUBMB Life* **2012**, *64*, 971–974. [CrossRef] [PubMed]
76. Hachimura, S.; Totsuka, M.; Hosono, A. Immunomodulation by food: Impact on gut immunity and immune cell function. *Biosci. Biotechnol. Biochem.* **2018**, *82*, 584–599. [CrossRef] [PubMed]
77. Revelo, X.S.; Winer, S.; Winer, D.A. Starving Intestinal Inflammation with the Amino Acid Sensor GCN2. *Cell Metab.* **2016**, *23*, 763–765. [CrossRef]
78. Alqinyah, M.; Hooks, S.B. Regulating the regulators: Epigenetic, transcriptional, and post-translational regulation of RGS proteins. *Cell. Signal.* **2018**, *42*, 77–87. [CrossRef]
79. Ota, A.; Sawai, M.; Sakurai, H. Stress-induced transcription of regulator of G protein signaling 2 (RGS2) by heat shock transcription factor HSF1. *Biochimie* **2013**, *95*, 1432–1436. [CrossRef]
80. Zmijewski, J.W.; Song, L.; Jope, R.S.; Harkins, L.; Cobbs, C.S. Oxidative stress and heat shock stimulate RGS2 expression in 1321N1 astrocytoma cells. *Arch. Biochem. Biophys.* **2001**, *392*, 192–196. [CrossRef]
81. Salim, S.; Asghar, M.; Taneja, M.; Hovatta, I.; Wu, Y.L.; Saha, K.; Sarraj, N.; Hite, B. Novel role of RGS2 in regulation of antioxidant homeostasis in neuronal cells. *FEBS Lett.* **2011**, *585*, 1375–1381. [CrossRef] [PubMed]
82. Salaga, M.; Storr, M.; Martemyanov, K.A.; Fichna, J. RGS proteins as targets in the treatment of intestinal inflammation and visceral pain: New insights and future perspectives. *BioEssays* **2016**, *38*, 344–354. [CrossRef]
83. Wang, D. The essential role of G protein-coupled receptor (GPCR) signaling in regulating T cell immunity. *Immunopharmacol. Immunotoxicol.* **2018**, *40*, 187–192. [CrossRef]
84. Vasiliou, V.; Vasiliou, K.; Nebert, D.W. Human ATP-binding cassette (ABC) transporter family. *Hum. Genomics* **2009**, *3*, 281. [CrossRef]
85. Chaiworapongsa, T.; Romero, R.; Whitten, A.; Tarca, A.L.; Bhatti, G.; Draghici, S.; Chaemsaithong, P.; Miranda, J.; Hassan, S.S. Differences and similarities in the transcriptional profile of peripheral whole blood in early and lateonset preeclampsia: Insights into the molecular basis of the phenotype of preeclampsia. *J. Perinat. Med.* **2013**, *41*, 485–504. [CrossRef] [PubMed]
86. Dréan, A.; Rosenberg, S.; Lejeune, F.X.; Goli, L.; Nadaradjane, A.A.; Guehennec, J.; Schmitt, C.; Verreault, M.; Bielle, F.; Mokhtari, K.; et al. ATP binding cassette (ABC) transporters: Expression and clinical value in glioblastoma. *J. Neurooncol.* **2018**, *138*, 479–486. [CrossRef] [PubMed]
87. Chen, J.; Tellez, G.; Escobar, J.; Vazquez-Anon, M. Impact of Trace Minerals on Wound Healing of Footpad Dermatitis in Broilers. *Sci. Rep.* **2017**, *7*, 1–9. [CrossRef] [PubMed]
88. Gessner, D.K.; Ringseis, R.; Eder, K. Potential of plant polyphenols to combat oxidative stress and inflammatory processes in farm animals. *J. Anim. Physiol. Anim. Nutr.* **2017**, *101*, 605–628. [CrossRef]
89. Ding, S.; Jiang, H.; Fang, J. Regulation of immune function by polyphenols. *J. Immunol. Res.* **2018**. [CrossRef]
90. Da Silva, E.R.; Brogi, S.; Lucon-Júnior, J.F.; Campiani, G.; Gemma, S.; Maquiaveli, C.D.C. Dietary polyphenols rutin, taxifolin and quercetin related compounds target: Leishmania amazonensis arginase. *Food Funct.* **2019**, *10*, 3172–3180. [CrossRef]
91. Anthony, J.P.; Fyfe, L.; Stewart, D.; McDougall, G.J. Differential effectiveness of berry polyphenols as anti-giardial agents. *Parasitology* **2011**, *138*, 1110–1116. [CrossRef]
92. Bottari, N.B.; Schetinger, M.R.C.; Pillat, M.M.; Palma, T.V.; Ulrich, H.; Alves, M.S.; Morsch, V.M.; Melazzo, C.; de Barros, L.D.; Garcia, J.L.; et al. Resveratrol as a Therapy to Restore Neurogliogenesis of Neural Progenitor Cells Infected by Toxoplasma gondii. *Mol. Neurobiol.* **2019**, *56*, 2328–2338. [CrossRef]
93. Soliman, R.H.; Ismail, O.A.; Badr, M.S.; Nasr, S.M. Resveratrol ameliorates oxidative stress and organ dysfunction in Schistosoma mansoni infected mice. *Exp. Parasitol.* **2017**, *174*, 52–58. [CrossRef]
94. Bahramsoltani, R.; Sodagari, H.R.; Farzaei, M.H.; Abdolghaffari, A.H.; Gooshe, M.; Rezaei, N. The preventive and therapeutic potential of natural polyphenols on influenza. *Expert Rev. Anti. Infect. Ther.* **2016**, *14*, 57–80. [CrossRef] [PubMed]
95. Chowdhury, P.; Sahuc, M.E.; Rouillé, Y.; Rivière, C.; Bonneau, N.; Vandeputte, A.; Brodin, P.; Goswami, M.; Bandyopadhyay, T.; Dubuisson, J.; et al. Theaflavins, polyphenols of black tea, inhibit entry of hepatitis C virus in cell culture. *PLoS ONE* **2018**, *13*, e0198226. [CrossRef] [PubMed]

96. Zhang, M.; Wu, Q.; Chen, Y.; Duan, M.; Tian, G.; Deng, X.; Sun, Y.; Zhou, T.; Zhang, G.; Chen, W.; et al. Inhibition of proanthocyanidin A2 on porcine reproductive and respiratory syndrome virus replication in vitro. *PLoS ONE* **2018**, *13*, e0193309. [CrossRef] [PubMed]
97. Zhao, X.; Cui, Q.; Fu, Q.; Song, X.; Jia, R.; Yang, Y.; Zou, Y.; Li, L.; He, C.; Liang, X.; et al. Antiviral properties of resveratrol against pseudorabies virus are associated with the inhibition of IκB kinase activation. *Sci. Rep.* **2017**, *7*, 1–11. [CrossRef]
98. Chedea, V.S.; Palade, L.M.; Pelmus, R.S.; Dragomir, C.; Taranu, I. Red grape pomace rich in polyphenols diet increases the antioxidant status in key organs—Kidneys, liver, and spleen of piglets. *Animals* **2019**, *9*, 149. [CrossRef] [PubMed]
99. Herosimczyk, A.; Lepczyński, A.; Ożgo, M.; Tuśnio, A.; Taciak, M.; Barszcz, M. Effect of dietary inclusion of 1% or 3% of native chicory inulin on the large intestinal mucosa proteome of growing pigs. *Animal* **2020**. [CrossRef] [PubMed]
100. Yu, K.; Matzapetakis, M.; Horvatić, A.; Terré, M.; Bach, A.; Kuleš, J.; Yeste, N.; Gómez, N.; Arroyo, L.; Rodríguez-Tomàs, E.; et al. Metabolome and proteome changes in skeletal muscle and blood of pre-weaning calves fed leucine and threonine supplemented diets. *J. Proteomics* **2020**, *216*, 103677. [CrossRef] [PubMed]

12

Effects of Mannan Oligosaccharides on Gas Emission, Protein and Energy Utilization, and Fasting Metabolism in Sheep

Chen Zheng [1], Junjun Ma [1], Ting Liu [1,*], Bingdong Wei [2] and Huaming Yang [2]

[1] College of Animal Science and Technology, Gansu Agricultural University, No. 1 Yingmen Village, Anning District, Lanzhou 730070, China; zhengc@gsau.edu.cn (C.Z.); majunjungsau@gmail.com (J.M.)
[2] Branch of Animal Husbandry, Jilin Academy of Agricultural Sciences, No.186 East Xinhua Road, Gongzhuling 136100, China; weibingdong@dlut.edu.cn (B.W.); yhmjl@163.com (H.Y.)
* Correspondence: liuting@gsau.edu.cn.

Simple Summary: Mannan oligosaccharides (MOS) are a promising feed additive to improve animal health, immune capacity, and antioxidation. Based on the previous studies, we carried out three experiments to investigate the effects of MOS on the gas emission, protein and energy utilization, and fasting metabolism of sheep. The results showed that 2.0% MOS supplementation led to the lowest *in vitro* CO_2 production and lower CH_4 production and decreased *in vivo* intake. However, it also decreased urine nitrogen excretion and energy released as CH_4, and then improved the utilization of crude protein and energy of sheep. There were no differences in the parameters of respiration and energy metabolism of sheep under the fasting condition. The findings indicated that MOS slightly affected the gas emission and nutrients and energy utilization of sheep.

Abstract: This study investigated the effects of mannan oligosaccharides (MOS) on *in vitro* and *in vivo* gas emission, utilization of crude protein (CP) and energy, and relative parameters of sheep under fasting metabolism conditions. *In vitro* gas productions were evaluated over 12 h in sheep diets containing different amounts of MOS (from 0% to 6.0%/kg, the increment was 0.5%). A control experiment was used to assess the gas emission, utilization of CP and energy, and fasting metabolism in control sheep and sheep treated with 2.0% MOS over 24 days (d). The results showed that 2.0% MOS supplementation led to the lowest *in vitro* CO_2 production and less CH_4 production, while also leading to decrease *in vivo* nutrients intake, CP and energy excretion, digested and retained CP, and energy released as CH_4 ($p < 0.05$). Furthermore, 2.0% MOS supplementation appeared to decrease *in vivo* O_2 consumption and CH_4 production per metabolic body weight ($BW^{0.75}$), and increase the CP retention rate of sheep ($p < 0.074$). MOS did not affect other parameters, along with the same parameters of sheep under fasting metabolism conditions ($p > 0.05$). The findings indicate MOS has only slight effects on the gas emission and nutrients and energy metabolism of sheep.

Keywords: energy; gas; mannan oligosaccharides; protein; sheep

---

1. Introduction

Prebiotics are non-digestible food ingredients that benefit the host by selectively stimulating the activity of one or a limited number of bacteria in the intestine [1,2]. The predominant prebiotics include oligosaccharides such as fructooligosaccharides (FOS), inulin, mannan oligosaccharides (MOS), and xylooligosaccharides [1]. They have been termed a nutricine, meaning that they have no direct nutritive value, but maintain intestinal digestive and absorptive functions, thus improving the health and performance of farmed animals [3].

MOS are found in the outer layer of yeast, *Saccharomyces cerevisiae*, in the cell wall. As the most bioactive compounds, they contain both mannan proteins and complex carbohydrates, including β-glucans [4], and are widely used as a dietary supplement to boost the immune system and eliminate pathogens from the intestinal tract [5,6]. For instance, MOS bind to the mannose-specific lectin of Gram-negative pathogens that express Type-1 fimbriae (e.g., *Escherichia coli*), resulting in their excretion from the intestine [7]. The limited research focusing on MOS in ruminants found that MOS improved the immunoglobulin G (IgG) concentration of lamb's blood [8], enhanced the health of the ruminal epithelium of sheep [9], improved the colostrum quantity of cows [4], and increased the antioxidant capacity of sheep [10]. In previous study, β 1-4 galacto-oligosaccharides (GOS) reduced the CH_4 production in sheep [11]; however, very few scientists have studied the effect of MOS on the energy metabolism and gas emission of ruminants. So, we hypothesized that MOS could adjust the CO_2 and CH_4 emission in the rumen and improve the energy utilization of sheep. The purpose of this study was to investigate the effects of MOS on *in vitro* and *in vivo* gas production, respiration, protein digestion, energy metabolism, and fasting metabolism in sheep.

2. Materials and Methods

2.1. Experimental Design, Animals, and Housing

All experiments in this study were carried out in accordance with the approved guidelines of the Regulation of the Standing Committee of Gansu People's Congress. All experimental protocols and the collection of samples were approved by the Ethics Committee of Gansu Agriculture University under permission no. DK-005.

A single-factor experimental design was used for the *in vitro* gas production experiments. Thirteen different doses of MOS (Bio-Mos®, Alltech, Nicholasville, KY, USA) were tested: 0, 0.5, 1.0, 1.5, 2.0, 2.5, 3.0, 3.5, 4.0, 4.5, 5.0, 5.5, and 6.0%/kg of basal diet (as fed basis). Each treatment was repeated three times. Four Chinese Northeast Merino rams with an eternal ruminal fistula and similar body weights (64.82 ± 3.17 kg) were used as ruminal fluid donors, and were fed basal diet and water ad libitum. An automatic recording device of trace gas production with six channels (Branch of Animal Husbandry, Jilin Academy of Agricultural Sciences, Gongzhuling, Jilin Province, China) equipped with gas flowmeters (CH_4 and CO_2, Sensors, Inc., Saline, MI, USA; H_2, CITY Technology, Inc., Hampshire, UK) was used to measure the gas flow. The *in vitro* tests lasted for 12 h.

A control experiment was used to assess *in vivo* CO_2, CH_4, and NH_3 production, O_2 consumption, urine nitrogen excretion, total heat output, and the apparent digestibility and retention rates of crude protein (CP) and energy in control sheep and sheep treated with 2.0% MOS (because adding 2.0% MOS to the basal diet led to the lowest CO_2 and lower CH_4 production in an *in vitro* experiment; then, this dose was chosen to apply in an *in vivo* experiment). The experimental group consisted of eight healthy rams (*Dorper* ♂× *Small tail Han-yang* ♀; four rams per group) with similar body weights (42.50 ± 3.24 kg). The test period included a 14-day (d) acclimation period, a 6-d digestive and metabolic experiment, a 60-hour (h) fasting acclimation period (the respiratory quotient (RQ) approximately was below 0.71 when sheep fasted for over 60 h), and a 24-h fasting metabolism experiment. All rams were housed individually in a respiratory chamber (Branch of Animal Husbandry, Jilin Academy of Agricultural Sciences, Gongzhuling, Jilin Province, China). Each chamber was equipped with a feeder and a drinker that provided ad libitum access to feed and water. Each chamber was also equipped with automatic gas flowmeters (CH_4 and CO_2, Sensors, Inc., Saline, MI, USA; O_2, Advanced Micro Instruments, Inc., Costa Mesa, CA, USA; NH_3, Industrial Health & Safety Instrumentation, Inc., St. Petersburg, FL, USA). Before rams went into the respiratory chamber, the 6.14-L calibrating CO_2 was pumped into the respiratory chamber to test the recovery rate, and repeated three times; the result of the gas recovery tests was 98.22% (Table S1). The data on gas production and O_2 consumption were calibrated by the gas recovery rate of each chamber.

2.2. Experimental Diets

The basal diet was formulated to meet or exceed the recommendations for all nutrients of ram sheep under China Agricultural Industry Standard NY/T816-2004 (Table 1).

Table 1. Ingredient and chemical composition of the basal diet.

Items	Concentrate
Ingredient (%)	
Corn	24.90
Soybean meal	12.72
Chinese wildrye	60.00
Calcium hydrophosphate	0.78
Limestone	0.80
Salt	0.40
Additive premix [1]	0.40
Chemical composition (%) [2]	
DM	85.71
DE (MJ/kg) [3]	9.25
CP	9.46
DE/CP (MJ/g)	0.10
NDF	46.06
ADF	23.75
Ca	0.65
P	0.40

[1] Additive premix includes mineral elements (mg/kg): S, 200; Fe, 25; Zn, 40; Cu, 8; I, 0.3; Mn, 40; Se, 0.2; Co, 0.1; vitamins (IU/kg): vitamin A, 940; vitamin E, 20. [2] Concentrations of dry matter (DM), crude protein (CP), Ca, and P were measured in accordance with AOAC (DM: method 930.15, CP: method 990.03, Ca: method 978.02, P: method 946.06) [12], and concentrations of neutral detergent fiber (NDF) and acid detergent fiber (ADF) were measured in accordance with Goering and Soest [13]. [3] DE = digestible energy.

2.3. In Vitro Procedures

Rumen contents from each donor sheep were obtained immediately before morning feeding by soft tube and syringe via ruminal fistula and strained through four layers of cheesecloth. Then, the fluid from each sheep was pooled and mixed with culture medium solution in a 1:2 ratio (vol/vol) at 39 °C in accordance with the description by Menke et al. [14], and the mixed solution was also constantly filled with N_2 to maintain anaerobic condition. Before incubation, in order to eliminate the metering error, 0.2843% calibrating CO_2 was pumped into the channel, and then the gas flowmeter was calibrated to 0.2843%. All *in vitro* cultures contained 2 g of experimental diets, which were carefully weighed into the fermentation channel of the automatic recording device of trace gas production, while the control channel did not receive any diet in order to calibrate the gas production from the ruminal fluid. Additionally, different doses of MOS were added into the channel in terms of experimental treatment. Then, each channel received 150 mL of mixed solution (including 50 mL of ruminal fluid and 100 mL of culture medium solution), and was maintained at 39 °C for 12 h of incubating. Since there were six channels for incubation, the culture was divided into nine periods, which meant that control, 0%, 0.5%, 1.0%, 1.5%, and 2.0% MOS treatments were carried out during the first period; control, 2.5%, 3.0%, 3.5%, 4.0%, and 4.5% MOS treatments were carried out during the second period; control, 5.0%, 5.5%, and 6.0% MOS treatments were carried out during the third period; and then, this procedure was repeated two times during the fourth to ninth periods. The productions of CO_2, CH_4, and H_2 were recorded by a gas flowmeter automatically per every 6 min. Then, the gas production was calculated per min from a total of 12 h of incubated gas production, after which the gas production per every 24 h was calculated.

2.4. In Vivo Procedures

In the acclimation period and digestive and metabolic experiments, all sheep received a basal diet or 2.0% MOS-treated diet randomly. Since they were housed in the respiratory chamber, the productions of CO_2, CH_4, NH_3, and H_2 along with the consumption of O_2 were measured using the gas flowmeter automatically per every 27 min.

The weight of feed intake and residues were recorded daily carefully. During the digestive and metabolic experiments, 10% of total diet, 10% of total feces output, and 5% of total urine output (5 mL of sulfuric acid added into total urine daily before collection to prevent nitrogen losses) were sampled for 6 sequential d and stored at −20 °C. At the end of the data collection period, the diets, fecal samples, and urine samples were thawed and pooled for each sheep. For nitrogen analysis, 3% of total feces output was sampled daily and stored in wide-mouth bottles containing 20 mL of 10% sulfuric acid for nitrogen fixation, and these samples were also pooled for 6 d individually.

After the digestive and metabolic experiments, sheep fasted approximately over 60 h until the RQ was below 0.71. Then, the sheep fasting metabolism experiment was carried out. The procedures of gas production and O_2 consumption recording, feces output sampling, and urine output sampling was as previously stated, but the recording and sampling times were 24 h.

2.5. Chemical Analysis

The diets and fecal samples were dried at 65 °C for 72 h in a forced-air oven, and then ground through a 1-mm screen using a Wiley Mill (Ogaw Seiki Co., Ltd., Tokyo, Japan). The gross energy of diet, feces, and urine samples were measured using a calorimeter (C2000, IKA®-Werke GmbH & Co. KG, Staufen, Germany). The nitrogen contents of the diet, feces containing 10% sulfuric acid, and urine samples were determined by the Kjeldahl method (AOAC, method 990.03) [12].

2.6. Calculations and Statistical Analysis

A gas flowmeter records the dynamic gas production or consumption; therefore, the total gas production or consumption is calculated using the following equations:

$$\textit{In vitro}\text{ gas production (mL/24 h)} = \text{average production (mL/min)} \times 60 \times 24;$$

$$\textit{In vivo}\text{ gas production/consumption (L/24 h)} = \text{average production/consumption (L/min)} \times 60 \times 24.$$

Parameters related to energy metabolism are calculated using the following equations [15]:

$$\text{Total heat output (J)} = 16.175 \times O_2\text{ (L)} + 5.021 \times CO_2\text{ (L)} - 2.167 \times CH_4\text{ (L)} - 5.987 \times \text{Urine nitrogen (g)}.$$

$$\text{Oxidation of protein (g)} = \text{Urine nitrogen (g)} \times 6.25.$$

$$CO_2\text{ production from the oxidation of protein (L)} = \text{Urine nitrogen (g)} \times 4.754.$$

$$O_2\text{ consumption from the oxidation of protein (L)} = \text{Urine nitrogen (g)} \times 5.923.$$

$$\text{Heat output from the oxidation of protein (J)} = \text{Urine nitrogen (g)} \times 113.76.$$

Data from the *in vitro* experiment were analyzed by one-way analysis of variance (SPSS 19.0, IBM Co. Limited, Chicago, IL, USA) using the following model:

$$X_{ij} = \mu + \alpha_i + e_{ij} \tag{1}$$

where X_{ij} is the observation of the dependent variable ($i = 1$ to 13, $j = 1$ to 3), μ is the population mean, α_i is the random effect of treatment, and e_{ij} is the random error associated with the observation.

Data from the *in vivo* experiments were analyzed by an independent samples t test.

Significance was determined at $p \leq 0.05$ and tendency at $0.05 < p \leq 0.10$ using Tukey's multiple comparison tests.

3. Results

3.1. In Vitro CO_2, CH_4 and H_2 Production

The 24-h *in vitro* CO_2, CH_4, and H_2 production under different amounts of adding MOS are shown in Table 2. MOS influenced *in vitro* CO_2 and CH_4 productions significantly ($p < 0.05$). An MOS of 3.5% led to the highest production of CO_2 and CH_4, while 2.0% MOS resulted in the lowest amount of CO_2 emissions, and 0.5% and 5.0% MOS resulted in the lowest CH_4 emissions ($p < 0.05$). No differences in H_2 production were observed between different MOS doses ($p > 0.05$).

Table 2. *In vitro* CO_2, CH_4, and H_2 production with different amounts of adding mannan oligosaccharides (MOS).

MOS (%) [1]	CO_2 (mL/24 h)	CH_4 (mL/24 h)	H_2 (mL/24 h)
0	161.52 ^{cde}	27.87 ^{bc}	0.11
0.5	153.03 ^{cde}	26.21 ^{c}	0.06
1.0	155.61 ^{cde}	28.29 ^{bc}	0.04
1.5	145.64 ^{de}	28.50 ^{bc}	0.10
2.0	143.65 ^{e}	27.65 ^{bc}	0.10
2.5	169.92 ^{abc}	30.80 ^{abc}	0.05
3.0	169.16 ^{abc}	30.88 ^{abc}	0.06
3.5	189.27 ^{a}	36.58 ^{a}	0.08
4.0	186.53 ^{ab}	34.66 ^{ab}	0.09
4.5	172.96 ^{abc}	28.92 ^{abc}	0.11
5.0	162.40 ^{cde}	26.64 ^{c}	0.10
5.5	166.69 ^{bcd}	27.73 ^{bc}	0.09
6.0	162.60 ^{cde}	27.12 ^{bc}	0.10
SEM [2]	2.05	0.58	0.01
p-value	<0.001	0.001	0.841

[1] MOS doses were 0, 0.5, 1.0, 1.5, 2.0, 2.5, 3.0, 3.5, 4.0, 4.5, 5.0, 5.5, and 6.0% added to a 2-g basal diet. Mean results of *in vitro* gas production are shown for yield per 24 h ($n = 3$ per treatment, *in vitro* gas production (mL/24 h) = average production (mL/min) × 60 × 24). [2] SEM: standard error of mean. Values within a column with different superscripts differ significantly at $p < 0.05$.

3.2. In Vivo Gas Production, CP, and Energy Digestion and Retention of Sheep

The CO_2, CH_4, and NH_3 production, O_2 consumption, urine nitrogen excretion, total heat output, and relative parameters regarding the energy metabolism of the control and 2.0% MOS-treated sheep are shown in Table 3. MOS did not affect RQ, CO_2, CH_4, and NH_3 production, O_2 consumption, and total heat output ($p > 0.05$). However, MOS decreased the urine nitrogen excretion and parameters of protein oxidation related to the energy metabolism of sheep ($p < 0.05$). Furthermore, MOS tended to decrease O_2 consumption and CH_4 production per metabolic body weight ($BW^{0.75}$), along with the ratio of heat output from protein oxidation to total heat output ($p < 0.074$).

The effects of MOS on sheep dry matter intake (DMI), CP, and energy digestion and retention are shown in Table 4. The supplementation of MOS decreased dry matter (DM), CP, and energy intake, CP and energy excreted in feces and urine, digested and retained CP, and the energy released as CH_4 in sheep ($p < 0.05$). The addition of MOS did not affect the apparent digestibility of CP and energy, retention of energy, digestible energy (DE), and metabolizable energy (ME) ($p > 0.05$). However, MOS tended to increase the retention rate of CP ($p < 0.054$).

Table 3. *In vivo* gas production and energy metabolism of control and MOS-treated sheep.

Item [1]		Control	MOS-Treated	SEM [9]	*p*-Value
RQ [2]		0.86	0.87	0.002	0.328
CO_2 production	(L/24 h)	375.05	372.29	4.04	0.111
	(L/24 h/kg $BW^{0.75}$) [3]	20.77	20.58	0.20	0.641
CH_4 production	(L/24 h)	21.64	20.22	0.40	0.600
	(L/24 h/kg $BW^{0.75}$)	1.20	1.12	0.02	0.072
NH_3 production	(L/24 h)	32.58	33.43	0.61	0.923
	(L/24 h/kg $BW^{0.75}$)	1.80	1.85	0.03	0.507
O_2 consumption	(L/24 h)	438.37	430.69	4.96	0.074
	(L/24 h/kg $BW^{0.75}$)	24.27	23.81	0.25	0.358
Urine nitrogen excretion	(g/24 h)	14.87 ^{a}	13.56 ^{b}	0.30	0.028
	(g/24 h/kg $BW^{0.75}$)	0.82 ^{a}	0.75 ^{b}	0.02	0.035
Total heat output [4]	(MJ/24 h)	8.65	8.57	0.10	0.694
	(MJ/24 h/kg $BW^{0.75}$)	0.48	0.47	0.01	0.553
Oxidation of protein [5]	(g/24 h)	92.93 ^{s}	84.78 ^{b}	1.88	0.028
	(g/24 h/kg $BW^{0.75}$)	5.15 ^{a}	4.70 ^{b}	0.11	0.035
CO_2 production from the oxidation of protein [6]	(L/24 h)	70.69 ^{a}	64.48 ^{b}	1.43	0.028
	(L/24 h/kg $BW^{0.75}$)	3.92 ^{a}	3.57 ^{b}	0.08	0.035
O_2 consumption from the oxidation of protein [7]	(L/24 h)	88.07 ^{a}	80.34 ^{b}	1.78	0.028
	(L/24 h/kg $BW^{0.75}$)	4.88 ^{a}	4.45 ^{b}	0.10	0.035
Heat output form the oxidation of protein [8]	(MJ/24 h)	1.69 ^{a}	1.54 ^{b}	0.03	0.028
	(MJ/24 h/kg $BW^{0.75}$)	0.094 ^{a}	0.086 ^{b}	0.002	0.035
The ratio of heat output from the oxidation of protein to total heat output (%)		19.55	18.12	0.39	0.069

[1] Sheep were fed diets containing 0% and 2.0% MOS ($n = 4$ per treatment). Mean results of CO_2, CH_4, and NH_3 production, O_2 consumption, urine nitrogen excretion, total heat output, and relative parameters regarding energy metabolism are shown for the 6-d collection phase of the study for each treatment. The *in vivo* gas production or consumption was calculated using the following equation: *in vivo* gas production/consumption (L/24 h) = average production/consumption (L/min) × 60 × 24. [2] RQ = respiratory quotient. [3] $BW^{0.75}$ = Metabolic body weight. [4] Total heat output (J) = 16.175 × O_2 (L) + 5.021 × CO_2 (L) − 2.167 × CH_4 (L) − 5.987 × Urine nitrogen (g). [5] Oxidation of protein (g) = Urine nitrogen (g) × 6.25. [6] CO_2 production from the oxidation of protein (L) = Urine nitrogen (g) × 4.754. [7] O_2 consumption from the oxidation of protein (L) = Urine nitrogen (g) × 5.923. [8] Heat output from the oxidation of protein (J) = Urine nitrogen (g) × 113.76. [9] SEM: standard error of the mean. Values within a row with different superscripts differ significantly at $p < 0.05$.

Table 4. *In vivo* nutrients and energy digestion and retention of control and MOS-treated sheep.

Item [1]		Control	MOS-Treated	SEM [11]	*p*-Value
Dry matter intake (DMI, kg/24 h)		1.66 ^{a}	1.48 ^{b}	0.02	<0.001
Crude protein (CP)					
CP intake	(g/24 h)	182.96 ^{a}	163.56 ^{b}	2.27	<0.001
	(g/24 h/kg BW $^{0.75}$) [2]	10.07 ^{a}	9.22 ^{b}	0.13	<0.001
CP in feces	(g/24 h)	46.90 ^{a}	38.78 ^{b}	2.03	0.044
	(g/24 h/kg BW $^{0.75}$)	2.51	2.12	0.12	0.107
CP in urine	(g/24 h)	92.93 ^{a}	84.78 ^{b}	1.88	0.028
	(g/24 h/kg BW $^{0.75}$)	9.40	9.41	0.19	0.976
Digested CP [3]	(g/24 h)	136.05 ^{a}	124.78 ^{b}	2.14	0.007
	(g/24 h/kg BW $^{0.75}$)	7.56	7.09	0.12	0.058
Retained CP [4]	(g/24 h)	43.12 ^{a}	40.00 ^{b}	0.66	0.016
	(g/24 h/kg BW $^{0.75}$)	2.34	2.28	0.03	0.339
Apparent digestibility (%) [5]		74.41	76.40	1.02	0.336
Retention rate (%) [6]		23.56	24.42	0.22	0.054

Table 4. *Cont.*

Item [1]		Control	MOS-Treated	SEM [11]	p-Value
Energy					
Energy intake	(MJ/24 h)	35.39 ^{a}	31.64 ^{b}	0.44	< 0.001
	(MJ/d/kg BW $^{0.75}$)	1.96 ^{a}	1.73 ^{b}	0.03	< 0.001
Energy in feces	(MJ/24 h)	12.69 ^{a}	10.05 ^{b}	0.61	0.028
	(MJ/d/kg BW $^{0.75}$)	0.74 ^{a}	0.48 ^{b}	0.03	<0.001
Energy in urine	(MJ/24 h)	0.46 ^{a}	0.42 ^{b}	0.01	0.028
	(MJ/d/kg BW $^{0.75}$)	0.03	0.02	0.001	0.301
Energy in CH_4	(MJ/24 h)	0.86 ^{a}	0.80 ^{b}	0.01	0.015
	(MJ/d/kg BW $^{0.75}$)	0.05	0.04	0.001	0.094
Digestible energy (DE) [7]	(MJ/24 h)	22.70	21.59	0.46	0.231
	(MJ/d/kg BW $^{0.75}$)	1.22	1.25	0.02	0.549
Metabolizable energy (ME) [8]	(MJ/24 h)	21.40	20.39	0.45	0.270
	(MJ/d/kg BW $^{0.75}$)	1.15	1.18	0.02	0.470
Apparent digestibility (%) [9]		64.34	68.59	1.52	0.165
Retention rate (%) [10]		60.63	64.32	1.48	0.216

[1] Sheep were fed 0% and 2.0% MOS (n = 4 per treatment). Mean results of DMI, CP, and energy digestion and retention are shown for the 6-d collection phase of the study for each treatment. [2] BW $^{0.75}$ = Metabolic body weight. [3] Digested CP (g) = CP intake − CP in feces. [4] Retained CP (g) = CP intake − CP in feces − CP in urine. [5] CP apparent digestibility (%) = (CP intake − CP in feces)/(CP intake). [6] CP retention rate (%) = (CP intake − CP in feces − CP in urine)/(CP intake). [7] Digestible energy (MJ) = energy intake − energy in feces. [8] Metabolizable energy (MJ) = energy intake − energy in feces − energy in urine − energy in CH_4. [9] Energy apparent digestibility (%) = (energy intake − energy in feces)/(energy intake). [10] Energy retention rate (%) = (energy intake − energy in feces − energy in urine − energy in CH_4) / (energy intake). [11] SEM: standard error of the mean. Values within a row with different superscripts differ significantly at $p < 0.05$.

3.3. *In Vivo Gas Production and Energy Metabolism of Sheep under Fasting Metabolism Conditions*

There were no differences in RQ, CO_2 production, O_2 consumption, urine nitrogen excretion, total heat output, and relative parameters regarding energy metabolism between the control and MOS-treated sheep under fasting metabolism conditions ($p > 0.05$, Table 5).

Table 5. *In vivo* gas production and energy metabolism of control and MOS-treated sheep under fasting metabolism conditions.

Item [1]		Control	MOS-treated	SEM [9]	p-Value
RQ [2]		0.66	0.69	0.006	0.267
CO_2 production	(L/24 h)	69.00	75.03	3.93	0.849
	(L/24 h/kg BW$^{0.75}$) [3]	3.83	4.15	0.22	0.494
O_2 consumption	(L/24 h)	103.45	109.57	5.24	0.921
	(L/24 h/kg BW$^{0.75}$)	5.74	6.06	0.29	0.617
Urine nitrogen excretion	(g/24 h)	3.28	2.95	1.52	0.828
	(g/24 h/kg BW$^{0.75}$)	0.18	0.35	0.08	0.490
Total heat output [4]	(MJ/24 h)	2.03	1.99	0.18	0.783
	(MJ/24 h/kg BW$^{0.75}$)	0.11	0.11	0.004	0.788
Oxidation of protein [5]	(g/24 h)	20.53	18.42	3.87	0.828
	(g/24 h/kg BW$^{0.75}$)	1.26	1.11	0.26	0.824
CO_2 production from the oxidation of protein [6]	(L/24 h)	15.61	14.01	2.94	0.828
	(L/24 h/kg BW$^{0.75}$)	0.96	0.85	0.20	0.824
O_2 consumption from the oxidation of protein [7]	(L/24 h)	19.45	17.45	3.67	0.828
	(L/24 h/kg BW$^{0.75}$)	1.19	1.05	0.24	0.824
Heat output form the oxidation of protein [8]	(MJ/24 h)	0.37	0.34	0.07	0.828
	(MJ/24 h/kg BW$^{0.75}$)	0.02	0.02	0.005	0.824
The ratio of heat output from the oxidation of protein to total heat output (%)		19.84	19.09	4.95	0.953

[1] Sheep were under the fasting metabolism conditions (n = 4 per treatment). Mean results of CO_2 production, O_2 consumption, urine nitrogen excretion, total heat output, and relative parameters regarding the energy metabolism of control and MOS-treated sheep are shown for the 24-h collection phase of the study for each treatment. The *in vivo* gas production or consumption was calculated using following equation: *in vivo* gas production/consumption (L/24 h) = average production/consumption (L/min) × 60 × 24. [2] RQ = respiratory quotient. [3] BW$^{0.75}$ = Metabolic body weight. [4] Total heat (J) = 16.175 × O_2 (L) + 5.021 × CO_2 (L) − 5.987 × Urine N (g). [5] Oxidation of protein (g) = Urine nitrogen (g) × 6.25. [6] CO_2 production from the oxidation of protein (L) = Urine nitrogen (g) × 4.754. [7] O_2 consumption from the oxidation of protein (L) = Urine nitrogen (g) × 5.923. [8] Heat output from the oxidation of protein (J) = Urine nitrogen (g) × 113.76. [9] SEM: standard error of the mean.

4. Discussion

4.1. In Vitro CO_2, CH_4 and H_2 Production

In this study, the MOS addition range from 0.5% to 2.0% decreased *in vitro* CO_2 and CH_4 production. Basically, carbohydrate was degraded by microbes in the rumen to produce approximately 65.5% CO_2, 28.8% CH_4, and small quantities of N_2, O_2, and H_2 [16]. Although some bicarbonates from saliva through the ruminal wall produce CO_2, the degradation of carbohydrates by ruminal microbes is the major pathway for CO_2 production [16]. In the rumen, methanogenesis evolves from carbon oxide, acetic acids, and methanol by methanogens. For most methanogens, methanogenesis from CO_2 and H_2 is the sole energy source. CH_4 production is considerably enhanced by the CO_2 present in the headspace, and there is an equilibrium established between the dissolved CO_2 in the media and the partial pressure of CO_2 in headspace gas. Meanwhile, the higher CO_2 concentration in the headspace would result in a greater dissolved CO_2 concentration in the media [17,18]. The result of the current study was in accordance with the previous conclusion that a positive relationship existed between CO_2 and CH_4; however, some scientists believed that there was a negative relationship, because CO_2 and H_2 are in general the precursors for CH_4 formation in the rumen [19,20]. This experiment illustrated that in an *in vitro* condition, MOS as an additive had only slight effects on ruminal fermentation, the partial pressure of CO_2, and methanogens and methanogenesis, and adding 0.5–2.0% MOS decreased CO_2 production and led to lower CH_4 production. It might be that MOS doses ranging from 0.5% to 2.0% were appropriate doses for ruminal microbes' fermentation pathway, and most of the produced CO_2 was potentially absorbed by the rumen wall to synthesize some nutrients, such as aspartic acid and isoleucine, and a lower partial pressure of CO_2 and dissolved CO_2 led to lower methanogenesis. Furthermore, higher doses of MOS meant higher oligosaccharides transferred into rumen as a substrate for microbes' fermentation, so, more substrates produced more CO_2, and instead promoted methanogenesis. However, MOS doses ranging from 2.5% to 4.5% generated the highest CO_2 and CH_4 emissions, but MOS doses ranging 5.0% to 6.0% resulted in lower CO_2 and CH_4 production. Further investigation is needed to confirm whether moderate doses of MOS improve carbohydrate fermentation in the rumen and much higher doses of MOS inhibit ruminal fermentation because more soluble carbohydrates led to lower pH and were harmful to microbes [20]. In addition, the fate of CO_2 is more complicated because of the pooling and recycling of animal metabolic carbon as urea and bicarbonates in saliva with that produced by the rumen organisms [21]. Furthermore, the process of methanogenesis is affected by many environmental factors, including the carbohydrate type and digestion passage rate. Feed additives such as yeast have the ability to shift H_2 utilization from methanogenesis to reductive acetogenesis through the homoacetogenic bacteria that can produce acetate from CO_2 and H_2 [22], along with internal factors [23]. So, more research about MOS on ruminal fermentation and the microbial population need to be undertaken in the future.

4.2. In Vivo Gas Production, CP, and Energy Digestion and Retention of Sheep

In the present study, the supplementation of MOS decreased ingestion, CP, and energy excreted in feces and urine, digested and retained CP, energy released as CH_4, along with the same parameters per $BW^{0.75}$, but MOS did not affect the apparent digestibility of CP and energy. Surprisingly, MOS decreased the ingestion of DM in sheep; the reason may be that there were low replicates in each treatment, indicating that individual differences may be a key factor in the results. Additional experiments with more sheep are needed to verify these effects of MOS.

Although MOS decreased the DMI of sheep, it decreased the CP and energy excretion in feces and urine at the same time; as a result, no differences were observed on the digestion and retention of CP and energy, even though higher digestibility and retention rates occurred in MOS-treated sheep. It indicates that MOS negatively influences the nutrients and energy intake of sheep; however, it makes sheep improve the utilization of nutrients and energy. Accordingly, because of improving the utilization of energy both in the rumen and the whole body of sheep, the methanogenesis in rumen was restricted, and resulted in less energy released as CH_4. A previous study found that GOS significantly increased the DE of sheep fed with Italian ryegrass hay and concentrate (3:2, on a DM basis), but did not affect nitrogen digestion and retention [24]. Other researchers reported that MOS did not affect the apparent digestibility and retention rate of nutrients in cattle and sheep [4,10]; these findings are in accordance with that of Goiri et al. [25], who pointed out that chitosan did not influence the nutrients' apparent digestibility. However, another study indicated that MOS enhanced the health of the ruminal epithelium of sheep by reducing the thickness of the stratum corneum, and it might have increased the nutrients digestion [9]. In addition, similar to this study, Sharma et al. [26] also reported that MOS increased the apparent digestibility of CP in Murrah buffalo calves. These findings indicated that MOS improved the CP utilization via decreased urine nitrogen excretion. So, more research about the effects of MOS on nutrients and energy metabolism in rumen also need to be undertaken in the future.

In the current study, treatment with 2.0% MOS resulted in less DM ingestion compared to the control. This led to less O_2 consumption and CO_2 production, because the digestion of feed and metabolism of nutrients in MOS-treated sheep consumed less O_2 and produced less CO_2. In *in vivo* conditions, CO_2 production from respiration is much larger than ruminal fermentation, and MOS only adjusts ruminal fermentation slightly; so, the decreasing tendency of CH_4 is observed from sheep in the respiratory chamber, not CO_2 production. Thus, it is possible that the basic vital activities of sheep were not affected by MOS. Respiration is a kind of rhythmic vital activity; its frequency and amplitude are determined by the body condition, which depends on nervous humoral regulation. Normally, feed intake cannot change the basic respiration frequency and amplitude, not to mention that there was only a small quantity of MOS; the results of the current study are also in consonance with the physiological theory.

Food ingestion affects the production of body heat in animals; however, for example, animals continuously produce heat and lose it to their surroundings, either directly by radiation, conduction, and convection, or indirectly by the evaporation of water [27]. In the current study, MOS decreased the urine nitrogen excretion and relative protein oxidation in sheep; however, the ratio of heat output from protein oxidation in the body to total heat output was decreased at the same time, indicated that the basal metabolism process is hardly influenced by MOS.

4.3. In Vivo Gas Production and Energy Metabolism of Sheep under Fasting Metabolism Conditions

MOS did not affect the CO_2 production, O_2 consumption, urine nitrogen excretion, total heat output, and relative parameters regarding the energy metabolism of sheep under fasting metabolism conditions in the current study. In a fasting animal, the quantity of heat produced is equal to the energy of the tissue catabolized. When measured under specific conditions, the energy is known as the animal's basal metabolism. A fasting animal must oxidize reserves of nutrients to provide the energy needed for essential processes such as respiration and circulation of the blood [27]. MOS cannot modify primary vital activities; no differences were observed in CO_2 production, O_2 consumption, urine nitrogen excretion, total heat output, and relative parameters regarding energy metabolism in 2.0% MOS-treated sheep and non-MOS fed sheep under fasting metabolism conditions.

5. Conclusions

The supplementation of MOS did not affect *in vitro* H_2 production, *in vivo* CO_2, CH_4, and NH_3 production, O_2 consumption, total heat output, the apparent digestibility of CP and energy, the retention rates of energy of sheep, and the respiration and energy metabolism of sheep under fasting metabolism conditions. However, the addition of 2.0% MOS to the sheep diet led to the lowest *in vitro* CO_2 production and less CH_4 production. Furthermore, treatment with 2.0% MOS decreased the intake of DM, CP, and energy, along with the CP and energy in feces and urine, digested and retained CP, and released the energy as CH_4. Treatment with 2.0% MOS appeared to increase the retention rate of CP in sheep. The results suggest that MOS only slightly affects ruminal fermentation and metabolism in sheep, and the effects were not strong enough to lead to substantial changes.

Author Contributions: Conceptualization, C.Z., B.W., and H.Y.; methodology, C.Z. and B.W.; investigation, J.M.; data curation, C.Z.; writing—original draft preparation, C.Z.; writing—review and editing, T.L.; supervision, J.M., B.W. and H.Y.

Acknowledgments: We thank all the people who participated in the experiments. We also thank Zhiyuan Ma from the Institute of Subtropical Agriculture, Chinese Academy of Science for his comments on this paper along with the native English-speaking scientists of Elixigen Company (Huntington Beach, California) for editing this manuscript.

References

1. Adhikari, P.A.; Kim, W.K. Overview of prebiotics and probiotics: Focus on performance, gut health and immunity—a review. *Ann. Anim. Sci.* **2017**, *17*, 949–966. [CrossRef]
2. Gibson, G.R.; Roberfroid, M.B. Dietary modulation of the human colonic microbiota: Introducing the concept of prebiotics. *J. Nutr.* **1995**, *125*, 1401–1412. [CrossRef] [PubMed]
3. Halas, V.; Nochta, I. Mannan oligosaccharides in nursery pig nutrition and their potential mode of action. *Animals* **2012**, *2*, 261–274. [CrossRef] [PubMed]
4. Westland, A.; Martin, R.; White, R.; Martin, J.H. Mannan oligosaccharide prepartum supplementation: Effects on dairy cow colostrum quality and quantity. *Animal* **2017**, *11*, 1779–1782. [CrossRef] [PubMed]
5. Najdegerami, E.H.; Tokmachi, A.; Bakhshi, F. Evaluating the Effects of Dietary Prebiotic Mixture of Mannan Oligosaccharide and Poly-β-Hydroxybutyrate on the Growth Performance, Immunity, and Survival of Rainbow Trout, *Oncorhynchus mykiss* (Walbaum 1792), Fingerlings. *J. World Aquacult. Soc.* **2017**, *48*, 415–425. [CrossRef]
6. Fernandez, F.; Hinton, M.; van Gils, B. Dietary mannan-oligosaccharides and their effect on chicken caecal microflora in relation to Salmonella Enteritidis colonization. *Avian Pathol.* **2002**, *31*, 49–58. [CrossRef] [PubMed]
7. Thomas, W.E.; Nilsson, L.M.; Forero, M.; Sokurenko, E.V.; Vogel, V. Shear-dependent 'stick-and-roll' adhesion of type 1 fimbriated *Escherichia coli*. *Mol. Microbiol.* **2004**, *53*, 1545–1557. [CrossRef] [PubMed]
8. Demirel, G.; Turan, N.; Tanor, A.; Kocabagli, N.; Alp, M.; Hasoksuz, M.; Yilmaz, H. Effects of dietary mannanoligosaccharide on performance, some blood parameters, IgG levels and antibody response of lambs to parenterally administered *E. coli* O157:H7. *Arch. Anim. Nutr.* **2007**, *61*, 126–134. [CrossRef] [PubMed]
9. Diaz, T.G.; Branco, A.F.; Jacovaci, F.A.; Jobim, C.C.; Bolson, D.C.; Daniel, J.L.P. Inclusion of live yeast and mannan-oligosaccharides in high grain-based diets for sheep: Ruminal parameters, inflammatory response and rumen morphology. *PLoS ONE* **2018**, *13*, e0193313.
10. Zheng, C.; Li, F.D.; Hao, Z.L.; Liu, T. Effects of adding mannan oligosaccharides on digestibility and metabolism of nutrients, ruminal fermentation parameters, immunity, and antioxidant capacity of sheep. *J. Anim. Sci.* **2018**, *96*, 284–292. [CrossRef]
11. Mwenya, B.; Santoso, B.; Sar, C.; Gamo, Y.; Kobayashi, T.; Arai, I.; Takahashi, J. Effects of including β-1,4 galacto-oligosaccharides, lactic acid bacteria or yeast culture on methanogenesis as well as energy and nitrogen metabolism in sheep. *Anim. Feed Sci. Tech.* **2004**, *115*, 313–326. [CrossRef]
12. AOAC. *Official Methods of Analysis*, 17th ed.; Association of Offical Analytical Chemists: Arlington, VA, USA, 2000.
13. Goering, H.R.; Soest, P.J.V. *Forage Fiber Analysis. Agricultural Handbook No. 379*; Agricultural Research Service-US Department of Agriculture: Washington, DC, USA, 1970.
14. Menke, K.H.; Raab, L.; Salewski, A.; Steingass, H.; Fritz, D.; Schneider, W. The estimation of the digestibility and metabolizable energy content of ruminant feedingstuffs from the gas production when they are incubated with rumen liquor *in vitro*. *J. Agr. Sci.* **1979**, *93*, 217–222. [CrossRef]
15. Brouwer, E. Report of Subcommittee on Constants and Factors. In *Proceedings of the 3rd Symposium on Energy Metabolism, Troon, Scotland, May 1964*; EAAP Publ.: Academic Press: London, UK, 1965; pp. 441–443.
16. Aschenbach, J.R.; Penner, G.B.; Stumpff, F.; Gäbel, G. Ruminant Nutrition Symposium: Role of fermentation acid absorption in the regulation of ruminal pH. *J. Anim. Sci.* **2011**, *89*, 1092–1107. [CrossRef] [PubMed]
17. Patra, A.K.; Yu, Z.T. Effects of gas composition in headspace and bicarbonate concentrations in media on gas and methane production, degradability, and rumen fermentation using *in vitro* gas production techniques. *J. Dairy Sci.* **2013**, *96*, 4592–4600. [CrossRef] [PubMed]
18. Alford, J.S., Jr. Measurement of dissolved carbon dioxide. *Can. J. Microbiol.* **1976**, *22*, 52–56. [CrossRef]

19. Sirohi, S.K.; Pandey, N.; Singh, B.; Puniya, A.K. Rumen methanogens: A review. *Indian J. Microbiol.* **2010**, *50*, 253–262. [CrossRef] [PubMed]
20. Parra-Garcia, A.; Elghandour, M.M.M.Y.; Greiner, R.; Barbabosa-Pilego, A.; Camacho-Diaz, L.M.; Salem, A.Z.M. Effects of *Moringa oleifera* leaf extract on ruminal methane and carbon dioxide production and fermentation kinetics in a steer model. *Environ. Sci. Pollut. Res.* **2019**, *26*, 15333–15344. [CrossRef] [PubMed]
21. Soest, P.V. *Nutritional Ecology of the Ruminant*, 2nd ed.; Cornell University Press: New York, NY, USA, 1994.
22. Elghandour, M.M.Y.; Kholif, A.E.; López, S.; Mendoza, G.D.; Odongo, N.E.; Salem, A.Z.M. *In vitro* Gas, Methane, and Carbon Dioxide Productions of High Fibrous Diet Incubated With Fecal Inocula From Horses in Response to the Supplementation With Different Live Yeast Additives. *J. Equine Vet. Sci.* **2016**, *38*, 64–71. [CrossRef]
23. Zhao, L.P.; Meng, Q.X.; Li, Y.; Wu, H.; Huo, Y.L.; Zhang, X.Z.; Zhou, Z.M. Nitrate decreases ruminal methane production with slight changes to ruminal methanogen composition of nitrate-adapted steers. *BMC Microbiol.* **2018**, *18*, 21. [CrossRef]
24. Pen, B.; Takaura, K.; Yamaguchi, S.; Asa, R.; Takahashi, J. Effects of *Yucca schidigera* and *Quillaja saponaria* with or without β1–4 galacto-oligosaccharides on ruminal fermentation, methane production and nitrogen utilization in sheep. *Anim. Feed Sci. Tech.* **2007**, *138*, 75–88. [CrossRef]
25. Goiri, I.; Oregui, L.M.; Garcia-Rodriguez, A. Use of chitosans to modulate ruminal fermentation of a 50:50 forage-to-concentrate diet in sheep. *J. Anim. Sci.* **2010**, *88*, 749–755. [CrossRef] [PubMed]
26. Sharma, A.N.; Kumar, S.; Tyagi, A.K. Effects of mannan-oligosaccharides and *Lactobacillus acidophilus* supplementation on growth performance, nutrient utilization and faecal characteristics in Murrah buffalo calves. *J. Anim. Physiol. Anim. Nutr.* **2018**, *102*, 679–689. [CrossRef] [PubMed]
27. McDonald, P.; Edwards, R.A.; Greenhalgh, J.F.D.; Morgan, C.A. *Animal Nutrition*; Pearson Education Limited: Upper Saddle River, NJ, USA, 2002.

13

Evaluation of the Combined use of Saccharomyces Cerevisiae and Aspergillus Oryzae with Phytase Fermentation Products on Growth, Inflammatory and Intestinal Morphology in Broilers

Wen Yang. Chuang [1], Wei Chih. Lin [1], Yun Chen. Hsieh [1], Chung Ming. Huang [1], Shen Chang. Chang [2] and Tzu-Tai Lee [1,3,*]

1 Department of Animal Science, National Chung Hsing University, Taichung 402, Taiwan; xssaazxssaaz@yahoo.com.tw (W.Y.C.); waynezi2227738@gmail.com (W.C.L.); richard840909@gmail.com (Y.C.H.); wxads508@gmail.com (C.M.H.)
2 Kaohsiung Animal Propagation Station, Livestock Research Institute, Council of Agriculture, Pingtung 912, Taiwan; macawh@mail.tlri.gov.tw
3 The iEGG and Animal Biotechnology Center, National Chung Hsing University, Taichung 402, Taiwan
* Correspondence: ttlee@dragon.nchu.edu.tw.

Simple Summary: The stress and anti-nutrient effect caused by environmental problems and animal feed is an urgent problem in poultry production. As ancient probiotics, *Aspergillus oryzae* and *Saccharomyces cerevisiae* can effectively improve the immunity of animals. Furthermore, the anti-nutrient object, phytate, reduces nutrition absorption. Therefore, *S. cerevisiae* or *A. oryzae* with phytase co-fermentation may help solve these problems. Results show that the addition of a fermentation product can effectively reduce the inflammatory response and drop the number of harmful bacteria in the ileum of broilers. Among them, *A. oryzae* fermentation product has a better effect than *S. cerevisiae* fermentation product.

Abstract: *Saccharomyces cerevisiae* and *Aspergillus oryzae* are both ancient probiotic species traditionally used as microbes for brewing beer and soy sauce, respectively. This study investigated the effect of adding these two probiotics with phytase fermentation products to the broilers diet. Fermented products possess protease and cellulase, and the activities were 777.1 and 189.5 U/g dry matter (DM) on *S. cerevisiae* fermented products (SCFP) and 190 and 213.4 U/g DM on *A. oryzae* fermented products (AOFP), respectively. Liposaccharides stimulated PBMCs to produce nitric oxide to 120 μmol. Both SCFP and AOFP reduced lipopolysaccharides stimulated peripheral blood mononuclear cells (PBMCs) nitric oxide release to 40 and 60 μmol, respectively. Nevertheless, in an MTT (3-(4,5-dimethylthiazol-2-yl)-2,5-diphenyltetrazolium bromide) assay, SCFP and AOFP also increased the survival rate of lipopolysaccharides stimulated PBMCs by almost two-fold compared to the negative control. A total of 240 broilers were divided into four groups as Control, SCFP 0.1% (SCFP), SCFP 0.05% + AOFP 0.05% (SAFP), and AOFP 0.1% (AOFP) groups, respectively. Each group had 20 broilers, and three replicate pens. The results showed that the addition of SCFP, SAFP, and AOFP groups did not affect the growth performances, but increased the jejunum value of villus height and villus: crypt ratio on SAFP and AOFP groups compared to the control and SCFP groups. Furthermore, adding SCFP, SAFP, and AOFP significantly reduced the number of *Clostridium perfringens* in ileum chyme. SCFP, SAFP, and AOFP significantly reduced the amount of interleukin-1β, inducible nitric oxide synthases, interferon-γ, and nuclear factor kappa B mRNA expression in PBMCs, especially in the AOFP group. In summary, all the SCFP, SAFP, and AOFP groups can be suggested as a functional feed additive since they enhanced villus: crypt ratio and decreased inflammation-related mRNA expression, especially for AOFP group in broilers.

Keywords: *Saccharomyces cerevisiae*; *Aspergillus oryzae*; phytase; fermentation; broilers

1. Introduction

As the global climate changes, the price of commodity feed ingredients, such as corn and soybean, rise continuously. Therefore, it is imperative to find new agricultural by-products as substitutes. Every year, millions of tons of wheat are produced all over the world, and wheat bran (WB) is the main by-product [1]. WB is inexpensive and may become one of the replacement sources for animal feed ingredients. WB consists of about 15% crude protein, 4% crude fat, 28% carbohydrate, 42% insoluble fiber, and is low in energy content. WB contains many compounds, such as ferulic acid, tocopherol, and lutein [2], which are good antioxidants. However, it contains high fiber content, 1% phytate, about 84% phytate-p and the total-p ratio [3] and susceptibility to mycotoxin contaminations because of improper storage. These have negative impacts on the utilization of WB as a viable ingredient in animal feed.

It is well-known that yeast cell walls are rich in ß-glucan and mannan, which can regulate immunity and protect the host from mycotoxins damage [4]. Although past studies have indicated that the addition of *S. cerevisiae* powder does not improve the growth traits of animals, it can increase the villi height and reduce the number of *E. coli* in the gut [5]. *S. cerevisiae* also has considerable enzyme secretion capabilities, such as proteases, xylanases, and cellulases [6]. In addition to *S. cerevisiae*, a previous study also pointed out that *A. oryzae*, generally recognized as safe (GRAS) probiotic, can be used as a probiotic for fermenting; it also has good enzyme secretion ability and can increase the crude protein content of the fermented material [7]. Moreover, many studies indicated that WB could increase its utilization value after fermentation by probiotics such as *Bacillus amyloliquefaciens*, *S. cerevisiae*, or white rot fungi in broilers diet [5,8]. Furthermore, a previous study pointed out that probiotics can decrease the inflammation response in animals [9]. Inflammation can cause many chronic diseases that affect animal health and longevity [10]. The reduction of the inflammatory response can decrease energy loss and improve the cell survival rate [11].

It is well-known that phytate possess high content of phosphorus and can chelate sodium, calcium or amino acids, thus affecting the absorption and utilization of nutrients by animals; however, phytase addition could solve the above-mentioned problem, and increase the digestibility of amino acids such as Lys, Met, Cys, and Thr [12], and the weight of the tibia, as well as increase the retention of phosphorus [13].

Few studies have been carried out on co-fermenting WB with probiotics combined with phytase to examine inflammation in broilers. Therefore, this study aims to investigate the *S. cerevisiae* fermented products (SCFP) and *A. oryzae* fermented products (AOFP) additions in broiler diets on the growth characteristics, intestinal morphology, microbial morphology, and inflammation-related mRNA expression in broilers.

2. Materials and Methods

2.1. Probiotic Characteristics

S. cerevisiae was cultured in Yeast-Mald (YM) broth at 30 °C for 0, 6, 12, 18, and 24 h, and then the number of *S. cerevisiae* colony-forming unit (CFU) on YM agar (Sparks, MD 21152 USA) was counted after sequence dilution to measure the growth curve of *S. cerevisiae*. *A. oryzae* was cultured in potato dextrose broth (PDB, Neogen, Lansing, MI, USA) at 30 °C for 0, 6, 12, 18, and 24 h, and the content of dry mycelia weight of *A. oryzae* in PDB was measured for the growth curve of *A. oryzae*. Moreover, 10^8 CFU/mL *S. cerevisiae* or 10^6 spores/mL *A. oryzae* fluid were incubated in 85 °C hot water for 3 min for heat tolerance measurement. After heating, the surviving *S. cerevisiae* and *A. oryzae* were counted on the YM agar or potato dextrose agar (PDA, Neogen, Lansing, MI, USA) respectively.

Acid, base, and gastrointestinal fluid tolerance were measured by pH-adjusted phosphate buffered saline (PBS, pH 3.0, 4.0, or 13.0, adjusted by citric acid or NaOH) (Merck, Darmstadt, Germany), simulation intestinal fluid (pH 13.0, 0.3% bile acid (Merck, Darmstadt, Germany), and 0.1% trypsin (Merck, Darmstadt, Germany)), or gastric fluid (pH 3.0, 0.3% pepsin). Here, 10^8 CFU/mL *S. cerevisiae* or 10^6 spores/mL *A. oryzae* fluid were incubated in the pH adjusted fluid, simulation intestinal fluid, or gastric fluid for 0 or 3 h at 30 °C. After incubating, the surviving *S. cerevisiae* and *A. oryzae* were counted on the YM agar or PDA, respectively.

The adherence ability was measured by the method indicated by Annika et al. [14], briefly, sacrificing the broiler and removing its crop. After dissection of the crop, sterile pH 7.2 PBS was used for rinsing. After washing, the crop epithelial cells were scraped with a slide and *S. cerevisiae* or *A. oryzae* spore solution was added. The state of attachment was confirmed under an optical microscope.

2.2. *S. cerevisiae Fermented Products (SCFP) and A. oryzae Fermented Products (AOFP) Preparation and Characteristics*

Then 10^8 CFU/ mL of *S. cerevisiae* was added to dilute the fluid mentioned above into sterilized wheat bran at a ratio of 1 mL dilute fluid per 3 g of wheat bran. The moisture of *S. cerevisiae* pre-fermentation wheat bran was adjusted to 60%. After that, 500 phytase units (FTU) of phytase were added to 30 g *S. cerevisiae* pre-fermentation wheat bran. *S. cerevisiae* pre-fermentation wheat bran (contain phytase) was incubated at 30 °C for 5 days, and placed in a stove at 50 °C for 1 day. The *S. cerevisiae* fermented product (SCFP) was stored in a −20 °C refrigerator for the other test.

The spore counts of *A. oryzae* were diluted to 9×10^6 spores/mL with deionized water. The diluted fluid was added into sterilized WB at a ratio of 1 mL dilute fluid per 3 g wheat bran. The moisture of *A. oryzae* pre-fermentation wheat bran was adjusted to 50%. After that, 500 FTU was added to 30 g *A. oryzae* pre-fermentation wheat bran. *A. oryzae* pre-fermentation wheat bran (contain phytase) was incubated at 30 °C for 7 days, and placed in a stove at 50 °C for 1 day. The *A. oryzae* fermented product (AOFP) was stored in a −20 °C refrigerator for the other test. The xylanase [15], protease [16], cellulase, and ß-glucanase [17] activities in SCFP or AOFP were analyzed before the animal experiment.

2.3. *Peripheral Blood Mononuclear Cell Isolation*

The assay was conducted and modified according to Kaiser et al. [18]. Briefly, 5 mL whole blood was collected from a total of thirty-six 35-d-old broilers (3 for each pen, 9 for each treatment) in the Summax Single-use Containers for Human Venous Blood Specimen Collection (Lithium Heparin Tube), and centrifuged at 200× *g* for 10 min to separate blood cells and plasma. The supernatant was replaced with sterilization PBS, and the sample was mixed uniformly. The mixed sample was added to the same amount of ficol, and centrifuged at 200× *g* for 30 min. The separated PBMC were moved to the new RNase free Eppendorf tube (Gunster Biotech, Co., Ltd., Taipei, Taiwan), and washed by PBS for three times. After this, the sample was centrifuged at 200× *g* for 10 min to remove the suspension PBS. RPMI-1640 (Merck, Darmstadt, Germany) with 10% fetal bovine serum (Merck, Darmstadt, Germany) (for cell test) or PBS (for qPCR) were used to re-suspend PBMCs and dilute it to 10^7 cells/mL.

2.4. *Nitric Oxide Assay*

Isolation diluted PBMC (10^7 cells/mL) were cultured at 37 °C and 5% CO_2 for 2 h. Then 10 µL 1 ng/mL lipopolysaccharides (LPS) and 10 µL sample solution, which had been filtrated by the 0.22 µm filter, were added. Sterilized PBS was used as the control group. After co-incubation for 24 h, 100 µL Griess reagent was added, and the absorbance was detected at 540 nm.

2.5. *3-(4,5-Dimethylthiazol-2-yl)-2,5-Diphenyltetrazolium Bromide (MTT) Assay*

Isolation diluted PBMC (10^7 cells/mL) were cultured at 37 °C and 5% CO_2 for 2 h. Then 10 µL 1 ng/mL lipopolysaccharides (LPS, Merck, Darmstadt, Germany) and 10 µL sample solution, which has been filtrated by 0.22 µm filter, were added. Sterilized PBS was used as the control group. After co-incubating for 48 h, 20 μL 0.5% MTT (Merck, Darmstadt, Germany) was added, and cultured for 4 h. After culturing, 100 μL dimethyl sulfoxide (DMSO, Merck, Darmstadt, Germany) was added, and the absorbance was detected at 570 nm.

2.6. Animal Experimental Design

To clarify the effect of SAFP, AOFP, and SAFP on broilers, the intestinal morphology, blood, and serum characteristics and mRNA expression in broilers' PBMC were measured. This experiment was conducted at National Chung Hsing University, Taiwan. All of the protocols for animal use were followed by the Animal Care and Use Committee (IACUC: 107-013). The methods of animal design, intestinal morphology, blood and serum characteristics, and microbial parameter in intestinal content were slightly modified according to Teng et al. [5]. A total of 240 male one-day-old broiler chickens (Ross 308) were used and divided into four groups: Control, 0.1% SCFP (SCFP), 0.05% SCFP + 0.05% AOFP (SAFP), and 0.1% AOFP (AOFP), respectively. Each pen had 20 broilers, and 3 replicate pens (a total of 60 birds per group). The average body weight (42.0 ± 0.5 g/bird) was similar among every pen initially. All birds were placed in the temperature control house. At 0–3 day-old, the temperature was 33 ± 0.5 °C, and then the temperature decreased as the broiler grew up. Temperature was controlled at 22 ± 1 °C after 30 days. Diets were divided into starter and finisher (Table 1). Both of them met or exceeded the nutrient requirements of broilers (NRC, 1994) with addition of SCFP and AOFP. The proximate composition was analyzed according to the methods of Association of Official Analytical Communities (AOAC) [19]. The starter and finisher diets were offered to the birds from 1–21 day-old and 22–35 day-old respectively. Body weight and feed consumption were recorded at 21 and 35 day-old. Body weight gain and feed conversion ratios (FCR) were calculated. Nine birds of each group were randomly selected to isolate PBMCs for qPCR and cell test.

2.7. Intestinal Morphology

Twenty-four 35-day-old broilers (2 of each pen, 6 for each treatment, fasting for 1 day) were used for the measurement of intestinal morphology. The birds were euthanized, and the middle of the jejunum and ileum were excised, and the content was flushed by PBS and fixed in 10% formalin. Each sample was embedded by paraffin and stained with hematoxylin and eosin. The slice was observed with a light microscope. The villus height, crypt depth, and tunica muscular were measured by Motic Image Plus 2.0 analysis system (Motic Instruments, Richmond, BC, Canada).

2.8. Blood and Serum Characteristics

Thirty six 35-day-old broilers (3 of each pen, 9 for each treatment) blood was collected randomly for the blood and serum characteristic measurement. Blood samples stood for about 4 to 5 h at 4 °C, and then centrifuged at 3000 rpm for 10 min at 4 °C. Blood and serum biochemical parameters analyses were measured by automatic biochemical analyzer (Hitachi, 7150 auto-analyzer, Tokyo, Japan).

2.9. Microbial Parameter in Intestinal Content

Twenty-four 35-day-old broilers (2 of each pen, 6 for each treatment) were randomly selected for microbial analysis. Birds were euthanized, and the contents of ileum and cecum were squeezed out. Next, 1 g chyme was put into 9 mL PBS to sequence a dilution for the microbial parameter. *Lactobacillus* spp. were cultured with DeMan, Rogosa, and Sharpe agar (Difco™ Lactobacilli MRS Agar, BD, Franklin Lakes, NJ); *Clostridium perfringens* were cultured with tryptose sulfite cycloserine agar (GranuCult™ TSCagar, Merck KGaA, 64271, Darmstadt, Germany). Each microbe mentioned above was cultured at 37 °C under anaerobic conditions for 48 h. After culturing, the number of CFU on the agar was counted.

Table 1. Composition and calculated analysis (g/kg as fed) of the basal diet for broilers (1–35 days) [1].

Ingredients	Starter Diet (1–21 Days)	Finisher Diet (22–35 Days)
	g/kg	
Corn, yellow	485.5	535.6
Soybean meal, (CP 44 %)	348	281.4
Full-fat soybean meal	82.9	99.8
Soybean oil	36.5	42.4
Calcium carbonate	16.6	13.3
Monocalcium phosphate	18.2	16.3
DL-Methionine	1.97	1.33
L-Lysine-HCl	3.64	3.21
NaCl	3.88	3.81
Choline-Cl	0.83	0.79
Vitamin premix [2]	0.99	1.03
Mineral premix [3]	0.99	1.03
Total	1000	1000
Calculated nutrient value		
Dry matter, %	88.3	88.2
Crude protein, %DM	23	21
Crude fat, %	6.64	8.95
Calcium, %DM	1.05	0.9
Total Phosphorus, %DM	0.77	0.68
Available Phosphorus, %DM	0.5	0.45
Lysine, %DM	1.43	1.25
Methionine + Cysteine, %DM	1.07	0.96
ME, kcal/kg DM	3050	3175
Chemical analysis value		
Dry matter, %	88.0	88.7
Crude protein, %DM	23.1	20.9
Crude fat, %	6.58	8.78

[1] Maize–soybean base diet for all treatment and add SCFP, SAFP, and AOFP to basal diet for SCFP, SAFP, and AOFP groups, respectively. [2] Vitamin (premix content per kg diet): vit. A, 15,000 IU; vit. D3, 3000 IU; vit. E, 30 mg; vit. K3, 4 mg; thiamine, 3 mg; riboflavin, 8 mg; pyridoxine, 5 mg; vitamin B12, 25 μg; Ca-pantothenate, 19 mg; niacin, 50 mg; folic acid, 1.5 mg; and biotin, 60 μg. [3] Mineral (premix content per kg diet): Co ($CoCO_3$), 0.255 mg; Cu ($CuSO_4{\cdot}5H_2O$), 10.8 mg; Fe ($FeSO_4{\cdot}H_2O$), 90 mg; Mn ($MnSO_4{\cdot}H_2O$), 90 mg; Zn (ZnO), 68.4 mg; Se (Na_2SeO_3), 0.18 mg.

2.10. Total RNA Isolation and qPCR

Total RNA was isolated from PBMCs for determination of the mRNA expression level using SuperScript™ FirstStrand Synthesis System reagent (Thermo Fisher, Waltham, MA, USA) according to the manufacturer's protocol. The method of determination of total RNA purity, cDNA synthesis, and qPCR analysis was slightly modified from Hu et al. [20]. Briefly, 2× SYBR GREEN PCR Master Mix-ROX (Gunster Biotech, Co., Ltd., Taipei, Taiwan), cDNA, deionized water, and each primer were mixed at a ratio of 5:1.2:1.8:1. StepOnePlus™ Real-Time PCR System (Thermo Fisher, Waltham, MA, USA) was used to detect qRT-PCR performance, while the $2^{-\Delta\Delta Ct}$ method was used to calculate the relative mRNA expression level, and ß-actin was used as the housekeeping gene for normalization. Gene-specific primers are according to the genes of Gallus gallus (chicken) given as follows: *beta-actin* (*ß-actin*, 5′-CTGGCACCTAGCACAATGAA-3′ and 5′-ACATCTGCTGGAAGGTGGAC-3′, X00182.1); *interlukin-1 beta* (*IL-1ß*, 5′-GCTCTACATGTCGTGTGTGATGAG-3′ and 5′-TGTCGATGTCCC GCATGA-3′, NM_204524); *inducible nitric oxide synthase* (*iNOS*, 5′-TACTGCGTGTCCTTTCAACG -3′ and 5′-CCCATTCTTCTTCCAACCTC-3′, U46504); *interferon-gamma* (*IFN-γ* (5′-CTCCCGATGAACG

ACTTGAG-3′ and 5′-CTGAGACTGGCTCCTTTTCC-3′, Y07922); *Nuclear factor kappa light chain enhancement of activated B cell p65* (*NFκB p65*, 5′-CCAGGTTGCCATCGTGTTCC-3′ and 5′-GCGTGCGTTTGCGCTTCT-3′, D13719.1).

2.11. Statistical Analysis

Data were analyzed for significance by analysis of variance (ANOVA) using SAS software (SAS® 9.4, 2018, SAS Institue Inc., Cary, NC, USA). Differences between treatment means were separated using Duncan's multiple range test with *p* value less than 0.05.

3. Results

3.1. The General Characteristics of the Probiotics and Fermented Products

After culturing for 24 h, the number of *S. cerevisiae* climbed to over 10^8 CFU/g DM, while *A. oryzae* dry mycelia weight increased to 176.8 mg/100 mL PDB. Both *S. cerevisiae* and *A. oryzae* survived in acid, base solution, and even gastrointestinal fluid, but *A. oryzae* did not survive at 85 °C. Both *S. cerevisiae* and *A. oryzae* adhered to broiler crop epithelial cells. (Figure 1).

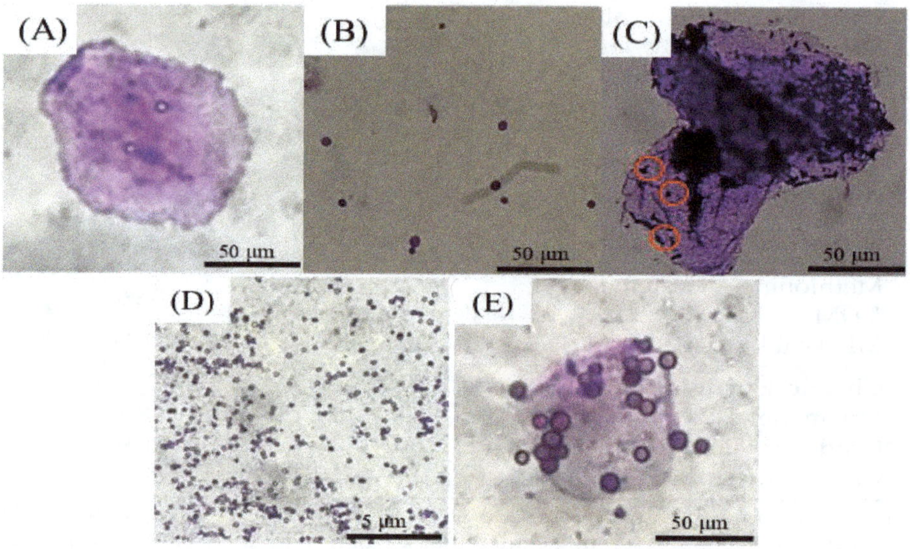

Figure 1. Photo of broiler crop epithelial cell (**A**), *Saccharomyces cerevisiae* (**B**), *Saccharomyces cerevisiae* adhering on the epithelial cell (red cycle) (**C**), spores of *Aspergillus oryzae* (**D**), and spores adhering on the epithelial cell (**E**), screened by an optical microscope.

The results of the enzyme activity of *S. cerevisiae* fermented products (SCFP) and *A. oryzae* fermented products (AOFP) extraction are shown in Table 2. Wheat bran had no detectable enzyme activities; however, after fermentation by *S. cerevisiae* or *A. oryzae*, xylanase, protease, cellulase, and ß-glucanase all exhibited a dramatic increase. Fermentation by *S. cerevisiae* or *A. oryzae*, xylanase, protease, cellulase, and ß-glucanase increased significantly. The enzyme activities of SCFP were 142.26, 777.13, 189.49, and 117.07 U/g DM, respectively, and AOFP are 120, 190, 213.38, and 120.21 U/g DM, respectively. Compared to AOFP, SCFP had a higher protease (777.1 vs. 190.0 U/g DM) and xylanase (142.3 vs. 120.0 U/g DM) production ability. Furthermore, both SCFP and AOFP could produce cellulase and hemicellulase to degrade the fiber content in wheat bran.

The results of nitric oxide (NO) release amounts are shown in Figure 2. After stimulating by LPS, the NO release amounts of PBMCs showed a significant increase. However, adding SCFP or AOFP extraction could reduce the NO release amount stimulated by LPS ($p < 0.05$). Among them, there were not significant differences in the 10 mg/mL LAOFP group, as well as the 50, 100 mg/mL LSCFP group, which means that AOFP has the same effect on reducing NO production of the LPS stimulated PBMC.

Table 2. The xylanase, protease, cellulase, and ß-glucanase activities of SCFP and AOFP extraction (U/g DM).

Enzyme	Products		
	WB [1]	SCFP [2]	AOFP [3]
Xylanase [4]	ND [5]	142.3	120.0
Protease [6]	ND	777.1	190.0
Cellulase [7]	ND	189.5	213.4
ß-glucanase [8]	ND	117.1	120.2

[1] WB: wheat bran. [2] SCFP: *Saccharomyces cerevisiae* fermented products. [3] AOFP: *Aspergillus oryzae* fermented products. [4] One unit of xylanase activity is defined as 1 µmol D-xylose generated from 10 mg/mL xylan in the condition of 37 °C and pH 5.5 in a minute. [5] Not detectable. [6] One unit of protease activity is defined as 1 µg L-tyrosine generated from 10 mg/mL casein in the condition of 40 °C and pH 7.5 in a minute. [7] One unit of cellulase activity is defined as 1 µmol, where reducing sugar is generated from 10 mg/mL CMC in the condition of 37 °C and pH 5.5 in a minute. [8] One unit of ß-glucanase activity is defined as 1 µmol, where reducing sugar is generated from 5 mg/mL ß-glucan in the conditions of 37 °C and pH 5.5 in one minute.

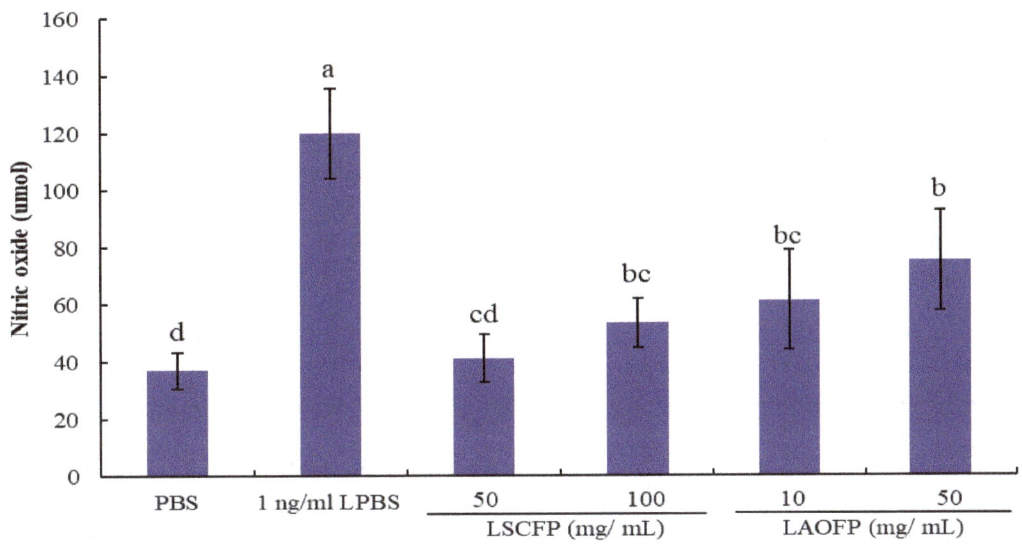

Figure 2. Nitric oxide release amount of phosphate buffer solution (PBS, control), 1 ng/mL lipopolysaccharide (LPBS, negative control), 50 mg/mL *Saccharomyces cerevisiae* fermented products extraction, and 1 ng/mL LPS include (50 mg/mL LSCFP), 100 mg/mL *Saccharomyces cerevisiae* fermented products extraction and 1 ng/mL LPS include (100 mg/mL LSCFP), 10 mg/mL *Aspergillus oryzae* fermented products extraction and 1 ng/mL LPS include (10 mg/mL LAOFP), and 25 mg/mL *Aspergillus oryzae* fermented products extraction and 1 ng/mL LPS include (25 mg/mL LAOFP). [a,b,c,d] Means within the same rows without the same superscript letter are significantly different ($p < 0.05$).

As the results shown in Figure 3, the MTT assay has an opposite result of NO release amount on the data. LPS stimulates PBMCs and causes the survival rate of PBMCs to decrease significantly. However, adding SCFP or AOFP extraction could increase the PBMCs survival rate after stimulation by LPS ($p < 0.05$). Among them, there were not significant differences in the LAOFP group and 100 mg/mL LSCFP group, meaning that AOFP had the same effect on increasing the survival rate in the MTT test with a lower concentration.

3.2. Animals Performance

Growth performances of SCFP, SAFP, and AOFP supplemented are shown in Table 3. All treatments exhibited no significant differences on body weight, feed consumption, weight gain, and feed conversion ratio (FCR) at 1–21 days, 22–35 days, and 1–35 days.

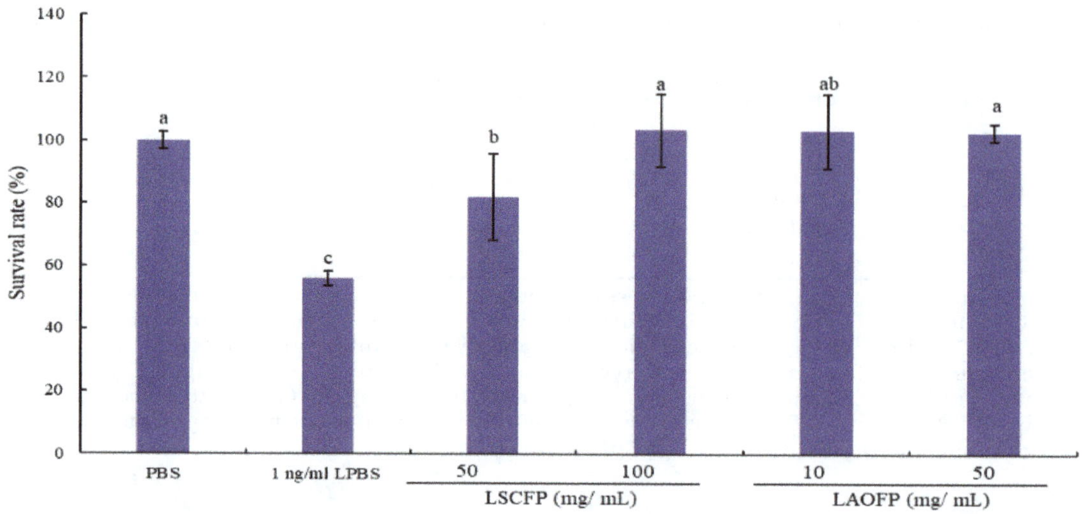

Figure 3. MTT (3-(4,5-dimethylthiazol-2-yl)-2,5-diphenyltetrazolium bromide) assay of phosphate buffer solution (PBS, control), PBS and 1 ng/mL lipopolysaccharide (LPBS, positive control), 50 mg/mL *Saccharomyces cerevisiae* fermented products extraction and 1 ng/mL LPS included (50 mg/mL LSCFP), 100 mg/mL *Saccharomyces cerevisiae* fermented products extraction and 1 ng/mL LPS included (100 mg/mL LSCFP), 10 mg/mL *Aspergillus oryzae* fermented products extraction and 1 ng/mL LPS include (10 mg/mL LAOFP), and 25 mg/mL *Aspergillus oryzae* fermented products extraction and 1 ng/mL LPS include (25 mg/mL LAOFP). ^{a,b,c} Means within the same rows without the same superscript letter are significantly different ($p < 0.05$).

Table 3. Effect of SCFP, SAFP, and AOFP supplemented in diet on growth performance of 1–35 day-old broilers (n = 3).

Item	Treatments					
	Control [1]	SCFP [1]	SAFP [2]	AOFP [3]	SEM [4]	*p* Value
1–21 days						
Body weight, g/bird	837.25	868.12	867.35	844.29	10.31	0.15
Feed consumption, g/bird	964.85	945.83	969.97	907.50	63.68	0.89
Weight gain, g/bird	794.25	825.12	824.35	801.29	10.32	0.15
FCR [5]	1.15	1.09	1.12	1.07	0.08	0.88
22–35 days						
Body weight, g/bird	2163.24	2164.70	2228.59	2168.29	28.90	0.37
Feed consumption, g/bird	2267.14	1915.24	2078.81	1984.48	124.15	0.28
Weight gain, g/bird	1325.99	1296.58	1361.24	1324.00	28.10	0.49
FCR	1.71	1.48	1.53	1.50	0.11	0.48
1–35 days						
Feed consumption, g/bird	3231.99	2861.08	3048.78	2891.98	157.51	0.38
Weight gain, g/bird	2120.24	2121.70	2185.59	2125.29	28.90	0.37
FCR	1.49	1.32	1.37	1.33	0.08	0.45

[1] SCFP: 0.5% *Saccharomyces cerevisiae* fermented products group. [2] SAFP: 0.5% *Saccharomyces cerevisiae* fermented products + 0.5% *Aspergillus oryzae* fermented products group. [3] AOFP: 0.5% *Aspergillus oryzae* fermented products group. [4] SEM: Standard error of the mean. [5] FCR: Feed conversion rate.

Although the results of growth performance showed no significant difference, the results of jejunum villus height and villus:crypt ratio on SAFP and AOFP both had a significant increase (1163.95 vs. 1508.93 and 1505.14 μm, 5.07 vs. 7.67 and 8.13, respectively, $p < 0.0001$). The crypt depth of the SCFP, SAFP, and AOFP groups showed a significant decrease (237.37 vs. 182.37, 203.43 and 188.42 μm, respectively, $p = 0.0002$). At the ileum, only SCFP and AOFP had a significant decrease on crypt depth ($p < 0.0001$), and AOFP had an increase in villus:crypt ratio ($p = 0.0005$) (Table 4). Photomicrography of jejunum and ileum of 35-day-old broilers are shown in Figure 4.

Table 4. Effect of SCFP, SAFP, and AOFP supplemented in diet on intestinal morphology of 35 d-old broilers (n = 6).

Item	Treatments					
	Control	SCFP [1]	SAFP [2]	AOFP [3]	SEM [4]	*p* Value
Jejunum						
Villus height (μm)	1163.95 ^{b}	1124.78 ^{b}	1508.93 ^{a}	1505.14 ^{a}	30.69	<0.0001
Crypt depth (μm)	237.37 ^{a}	182.37 ^{c}	203.43 ^{bc}	188.42 ^{c}	8.59	0.0002
Tunica muscularis (μm)	298.32 ^{a}	285.47 ^{b}	272.12 ^{b}	255.92 ^{b}	14.17	0.004
Villus:crypt	5.07 ^{c}	6.36 ^{b}	7.67 ^{a}	8.13 ^{a}	0.3	<0.0001
Ileum						
Villus height (μm)	1109.77	1055.33	1074	1050.13	19.04	0.1185
Crypt depth (μm)	209.25 ^{a}	173.25 ^{b}	201.29 ^{a}	157.32 ^{b}	6.82	<0.0001
Tunica muscularis (μm)	650.8	439.38	366.38	318.68	108.05	0.1453
Villus:crypt	5.52 ^{b}	6.25 ^{ab}	5.44 ^{b}	7.08 ^{a}	0.28	0.0005

[1] SCFP: 0.5% *Saccharomyces cerevisiae* fermented products group. [2] SAFP: 0.5% *Saccharomyces cerevisiae* fermented products + 0.5% *Aspergillus oryzae* fermented products group. [3] AOFP: 0.5% *Aspergillus oryzae* fermented products group. [4] SEM: Standard error of the mean. ^{a,b} Means within the same rows without the same superscript letter are significantly different ($p < 0.05$).

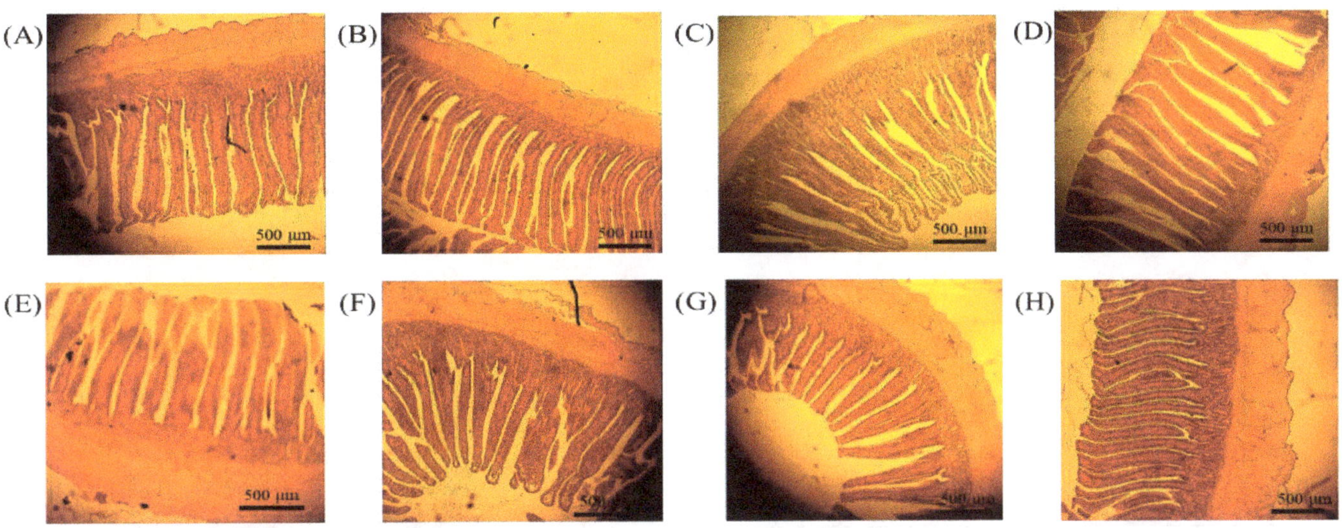

Figure 4. Photomicrography of jejunum and ileum of 35-day-old broiler feed with control, *Saccharomyces cerevisiae* fermented products (SCFP) or *Aspergillus oryzae* fermented products (AOFP). (**A–D**) jejunum, (**E–H**) ileum. Photos are respectively Control (**A**,**E**), 0.1% SCFP (**B**,**F**), 0.05% SCFP + 0.05% AOFP (**C**,**G**), and 1% AOFP (**D**,**H**). Hematoxylin and eosin stain 40×.

The white blood cells (WBC) increased significantly on SAFP ($p < 0.0001$) compared to other group. The concentration of uric acid (UA) significantly decreased following the treatment of SCFP and SAFP ($p = 0.001$). Furthermore, the SCFP group had a better effect on decreasing the UA content in serum compared to the AOFP group (3.36 vs. 4.69 mg/dL). However, there were no significant effects on triglyceride (TG), high-density lipoprotein (HDL), high-density lipoprotein (LDL), Ca, and P (Table 5).

The results of mRNA expression level of immunomodulatory genes in chicken PBMCs are shown in Figure 5. As seen, the *IL-1ß*, *iNOS*, *IFN-γ*, and *NFκB* mRNA expression significantly decreased in SCFP, SAFP, and AOFP ($p < 0.05$). Furthermore, it is worth noting that AOFP can reduce the inflammation-related mRNA content most effectively.

Table 5. Effect of SCFP, SAFP and AOFP supplemented in diet on serum characteristics of broilers (35 d) (n = 9).

Treatments	Items [1]							
	WBC	UA	CHOL	TG	HDL-C	LDL-C	Ca	P
Units	10^3/uL	mg/dL	mg/dL	mg/dL	mg/dL	mg/dL	mg/dL	mg/dL
Control	189.69 ^{b}	5.57 ^{a}	119.22	77	73.44	38.33	9.86	9.13
SCFP [2]	193.69 ^{b}	3.36 ^{c}	117.44	61.33	74.56	38.11	9.51	9
SAFP [3]	234.46 ^{a}	4.16 ^{bc}	133.22	89.11	81.33	41.78	9.99	8.94
AOFP [4]	185.40 ^{b}	4.69 ^{ab}	125.33	66.56	77.11	42.89	9.77	8.72
SEM [5]	4.35	0.35	5.84	8.19	3.67	2.95	0.12	0.34
p Value	<0.0001	0.001	0.2359	0.1017	0.4479	0.576	0.0641	0.8596

[1] WBC: white blood cell; UA: uric acid; CHOL: cholesterol; TG: triglycerides; HDL: cholesterol-high-density lipoprotein; LDL-C: cholesterol-low-density lipoprotein; Ca: calcium; P: phosphorus. [2] SCFP: 0.5% *Saccharomyces cerevisiae* fermented products group. [3] SAFP: 0.5% *Saccharomyces cerevisiae* fermented products + 0.5% *Aspergillus oryzae* fermented products group. [4] AOFP: 0.5% *Aspergillus oryzae* fermented products group. [5] SEM: standard error of the mean. ^{a,b,c} means within the same rows without the same superscript letter are significantly different (p < 0.05).

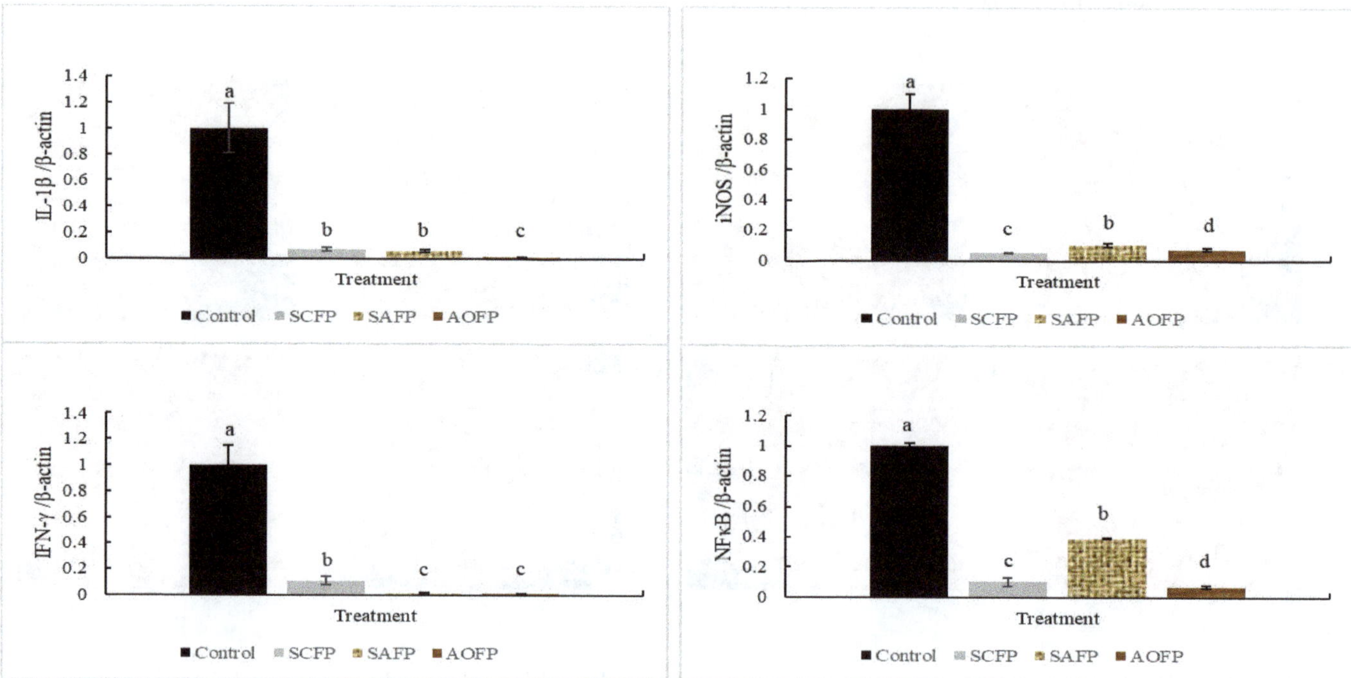

Figure 5. The mRNA expression level of immunomodulatory genes in chicken peripheral blood mononuclear cells at 35d. *IL-1ß* (top left), *iNOS* (top right), *IFN-γ* (bottom left), and *NFκB* (bottom right). The treatments are control, 0.5% *Saccharomyces cerevisiae* fermented products group (SCFP), 0.5% *Saccharomyces cerevisiae* fermented products + 0.5% *Aspergillus oryzae* fermented products group (SAFP), 0.5% *Aspergillus oryzae* fermented products group (AOFP), respectively. Data is presented in mean ± SE (n = 9). ^{a,b,c,d} Means within the same rows without the same superscript letter are significantly different (p < 0.05).

3.3. Microbial Parameter in Intestinal Content

The treatment of SCFP, SAFP, and AOFP significantly decrease the number of *Clostridium perfringens* (p = 0.0014), but did not increase the number of *Lactobacillus* spp. (Table 6). In the caecum, although the number of *C. perfringens* did not significantly decrease (p = 0.6770), the number of *C. perfringens* still decreased in the data (7.39, 7.00, 7.01, and 7.22 log CFU/g in the Control, SCFP, SAFP, and AOFP groups, respectively), especially for the SCFP group.

Table 6. Effect of SCFP, SAFP, and AOFP supplemented in diet on microbial parameter in the intestinal content of 35 d-old broilers (n = 3).

Microbial Parameter	Treatments					
	Control	SCFP [1]	SAFP [2]	AOFP [3]	SEM [4]	*p* Value
	log CFU/g					
Ileum						
Clostridium perfringens	7.84 ^{a}	7.02 ^{b}	7.05 ^{b}	7.22 ^{b}	0.12	0.0014
Lactobacillus spp.	8.35	8.84	8.25	8.91	0.39	0.1765
Caecum						
Clostridium perfringens	7.39	7.00	7.01	7.22	0.24	0.6770
Lactobacillus spp.	8.95	9.13	8.92	8.91	0.23	0.7039

[1] SCFP: 0.5% *Saccharomyces cerevisiae* fermented products group. [2] SAFP: 0.5% *Saccharomyces cerevisiae* fermented products + 0.5% *Aspergillus oryzae* fermented products group. [3] AOFP: 0.5% *Aspergillus oryzae* fermented products group. [4] SEM: Standard error of the mean. ^{a,b} Means within the same rows without the same superscript letter are significantly different ($p < 0.05$).

4. Discussion

The high content of insoluble fiber and phytate could reduce the utilization of nutrition in animals. SCFP and AOFP contain quite a large amount of enzymes, such as cellulase, ß-glucanase, xylanase, and protease, thereby improving the nutrient utilization of broilers [21]. Nevertheless, adding phytase to the fermentation process of SCFP and AOFP may also help to reduce the phytate content [12]. From the results, the addition of SCFP and AOFP did not improve body weight, feed consumption, weight gain, and FCR. However, Santos et al. [22] reported that phytase addition (500 FTU/kg) improved the body weight gain of 1–21 day-old broilers, but did not improve the body weight gain and FCR in 22–35 day-old broilers. Mountzouris et al. [23] pointed out that *S. cerevisiae* powder (50 mg (10^9 CFU)/kg diet) addition did not enhance the growth performance. Otherwise, there are few studies investigating the effects of *A. oryzae* addition on broilers growth performance. As a previous study shows, replacing soybean meal by *A. oryzae* fermented soybean meal (29.5% replacement in whole diet) can improve average daily gain and feed intake of broilers [24]. As the replacement is far higher than in the current study, it is speculated that the addition of *A. oryzae* should be sufficient to prove effective in growth performance.

The improvement of white blood cells (WBC) may be attributed to the joint effects of SCFP and AOFP because only SAFP increase WBC count in broilers blood. The SCFP and SAFP groups can significantly reduce the uric acid (UA) content in the blood, and the data number of SAFP is between SCFP and AOFP. This is because the protease yielded by SCFP was higher than AOFP and protease improved the protein utilization of the host, thereby decreasing the UA content in the blood. UA is a metabolite of protein that has an antioxidant function but is converted to a pro-oxidant in the cell or cytoplasm, and may be associated with cardiovascular disease when the concentration increases [25]. In addition, UA also promotes inflammation and is associated with insulin resistance and metabolic disorders [25]. Therefore, the addition of SCFP or AOFP to feed may prevent the development of cardiovascular disease, and reduce the oxidative stress of animal cells by the pathway of reduced UA.

In this study, villus height and the villus:crypt ratio in the jejunum were significantly increased and crypt depth significantly decreased on SAFP and AOFP, indicating that SAFP and AOFP may have an effect on improving nutrition absorption. One possible reason is that the enzyme yielded by SCFP and AOFP may degrade the anti-nutritional factor and make the energy utilization focus on villi growth and nutrient absorption [26]. Kalantar et al. [21] also reported that adding 0.1% commercial enzyme COMBO® (including 1000 FTU/g phytase, 200 U/g xylanase, 200 U/g β-glucanase, and 80 U/g hemicellulase) in broilers diet can decrease the crypt depth and increase the villus length/width ratio and villus length/crypt depth ratio. Gao et al. [27] showed that yeast supplement (2.5, 5.0, and 7.5 g/kg diet) can increase the villus height of the small intestine. Feng et al. [24] found that replacing soybean meal by *A. oryzae* fermented soybean meal can increase villus height on both duodenum and jejunum and decrease the crypt depth in jejunum. In addition, a decrease in the crypt depth indicates that the animal can spend less energy on the formation of the crypt, thereby reducing the energy loss [26]. Furthermore, the results showed that Tunica muscularis decreased significantly on jejunum after adding SCFP, SAFP, and AOFP. Tunica muscularis consists of smooth muscle and is related to digestion and absorption of chyme. Cil et al. [28] reported that inflammation causes the intestine to swell and increases Tunica muscularis weight as well as thickness. Therefore, by the anti-inflammatory effect of SCFP, SAFP, and AOFP confirmed by cell test and broilers PBMCs mRNA expression, the decreased thickness of Tunica muscularis is expected.

Previous studies pointed out that probiotic can reduce the number of harmful microbes in the gut [5]. Mountzouris et al. [23] indicated that *S. cerevisiae* powder (50 mg (1×10^9 CFU)/kg diet) can protect broilers from *Salmonella Enteritidis* (2×10^6 CFU/birds) challenge at day 15 and increase the growth performance of broilers. In this study, adding SCFP, SAFP, and AOFP can reduce *C. perfringens* in the ileum. Among them, the SCFP is better than the AOFP based on the data, but with no significant difference, while the SAFP effect was in the middle. The antibacterial effect may be due to the growth of probiotics that crowd out the growth resources of *C. perfringens*, or the glucan and mannan produced by the probiotics encapsulate the toxins yield from *C. perfringens* and reduce the competitiveness of *C. perfringens*. Furthermore, the number of *C. perfringens* is positively correlated with the histochemistry score and necrotizing enterocolitis [29], thereby decreasing the number of *C. perfringens* may decrease the amount of pro-inflammatory cytokine.

LPS is an extracellular structure of gram-negative bacteria that can cause inflammation and stimulate the release amount of nitric oxide (NO) from PBMCs [11]. However, SCFP and AOFP can protect PBMCs from LPS stimulus and decrease the NO production. The reason may be that the glucan and mannan in *S. cerevisiae* and *A. oryzae* cell wall acted as antitoxin [3]. In addition, it is well-known that low-grade inflammatory response can protect the host from pathogens; however, high-grade inflammatory causes cell damage and apoptosis, so it is important to decrease the level of inflammation to prevent unexpected cell death. It is worth noting that AOFP has a similar effect as SCFP on lowering the concentration on NO production. Therefore, AOFP has similar effects as SCFP on lowering the concentration, which may represent that AOFP has a better anti-inflammatory effect than SCFP does.

For broilers, inflammation is a double-edged sword depending of its level [11]. Lee et al. [30] indicated that LPS challenge on broilers (1 mg/kg body weight) causes significant weight loss because of the immune response and decrease in feed intake. In broilers, there are two ways to decrease the inflammation level: one is to decrease the stimulants, such as LPS or bacterial toxin, and the other way is to block the pro-inflammatory cytokine pathway. From the results, SCFP and AOFP reduced the inflammation-related stimulant, including UA, *C. perfringens*, and NO production from LPS stimulated PBMCs.

As a pro-inflammatory cytokine, *IL-1ß* mediates many pathways involved in apoptosis or inflammation [31]. It is well-known that *S. cerevisiae* and *A. oryzae* cell wall are rich in β-glucans [3], which can inhibit *IL-1ß* production [32] and thereby reduce *NF-κB*-mediated inflammatory responses [33]. Therefore, *IL-1ß* is positively correlated with *NF-κB*. *NF-κB* is a main inflammatory factor, which promotes reactive nitrogen species (RNS) content [11]. In normal conditions, *NF-κB* binds with the inhibitory protein named inhibitor of kappa B (IκB) and has no effect [33]. With stimulants such as heat stress or infection, *IκB* phosphorylates and releases the *NF-κB* as a nuclear transcription factor to induce an inflammatory response [33]. SCFP, SAFP, and AOFP addition can decrease the *IL-1ß* and *NF-κB* mRNA expression. Hegazy and Bedewy [9] had similar results, indicating that 1×10^{10} CFU of *Lactobacillus delbruekii* and *L. fermentum* can decrease the interleukin-6 (IL-6), tumor necrosis factor-α (TNF-α), and *NF-κB* expression. The results show that probiotics reduced inflammation by suppressing the *NF-κB* and *IL-1ß* expression. Furthermore, the expression of *iNOS* can increase NO production, and is associated with infection [34], while SCFP and AOFP reduced the *iNOS* mRNA expression. *NF-κB* can induce *iNOS* expression, and increase the RNS production [35]; therefore, decreasing the *NF-κB* content can also suppress *iNOS* expression. *IFN-γ* is associated with infection and high concentration of *IFN-γ* may contribute to autoimmune disease [36]. When animals suffer from infection, *IFN-γ* expression increases and induces *NF-κB* expression to produce reactive oxygen species (ROS) and RNS to resist pathogens [37]. The data shown above indicate that SCFP and AOFP can decrease the damage caused by LPS and the number of *C. perfringens* in ileum, which are both related to the infection. Therefore, *IFN-γ* suppressed by SCFP, SAFP, and AOFP in broilers also decreases the *NF-κB* expression and reduces the inflammatory response [37]. Decreasing the stimulant and blocking the inflammation-related mRNA expression mentioned above confirm that both SCFP and AOFP addition can reduce the inflammation, especially AOFP.

5. Conclusions

Based on the above results, this study confirmed that SCFP, SAFP, and AOFP have positive effects on a decrease in *C. perfringens* number in ileum and suppress the mRNA relation of *IL-1ß*, *NF-κB*, *iNOS*, and *IFN-γ* on broilers' PBMCs. Among them, SCFP, SAFP, and AOFP have different effects. Results showed that the addition of AOFP is much more effective than the addition of SCFP, while the effect of SAFP is somewhere between them, and there is no multiplication effect. Therefore, compared to the SCFP and SAFP, the AOFP is suggested to be a functional feed additive.

Author Contributions: Conceptualization, W.Y.C. and T.-T.L.; materials analysis, W.Y.C. and C.M.H.; methodology, W.Y.C. and Y.C.H.; supervision, T.-T.L.; investigation, W.Y.C. and T.-T.L.; writing—original draft preparation, W.Y.C.; W.C.L., and T.-T.L.; writing—review and editing, W.Y.C.; W.C.L.; S.C.C., and T.-T.L.

Acknowledgments: The authors thank the Ministry of Science and Technology (MOST 107-2313-B-005-037-MY2) and the iEGG and Animal Biotechnology Center from The Feature Areas Research Center Program within the framework of the Higher Education Sprout Project by the Ministry of Education (MOE) in Taiwan for supporting this study.

Abbreviations

SCFP	*Saccharomyces cerevisiae* fermented products
AOFP	*Aspergillus oryzae* fermented products
PBMC	Peripheral blood mononuclear cells
LPS	Lipopolysaccharides
SAFP	The mix of SCFP and AOFP at a ratio of 1:1;
WB	Wheat bran
GRAS	Generally recognized as safe
Lys	Lysine
Met	Methionine
Cys	Cysteine
Thr	Threonine
YM	Yeast-Mald
CFU	Colony-forming unit
PDB	Potato dextrose broth
PBS	Phosphate buffered saline
FTU	Phytase units
RPMI	Roswell park memorial institute
FBS	Fetal bovine serum
MTT	3-(4,5-Dimethylthiazol-2-yl)-2,5-Diphenyltetrazolium Bromide
DMSO	Dimethyl sulfoxide
MRS	Man, Rogosa, and Sharpe agar
TSC	Tryptose sulfite cycloserine
DM	Dry matter
NO	Nitric oxide

FCR	Feed conversion ratio
WBC	White blood cells
UA	Uric acid
TG	Triglyceride
HDL	High-density lipoprotein
LDL	High-density lipoprotein
IL-1ß	Interleukin-1 beta
iNOS	Inducible nitric oxide synthase
IFN-γ	Interferon-gamma
NFκB	Nuclear factor kappa light chain enhancement of activated B cell
RNS	Reactive nitrogen species
IκB	Inhibitor of kappa B
IL-6	Interleukin-6
TNF-α	Tumor necrosis factor-α
ROS	Reactive oxygen species

References

1. Hemery, Y.; Rouau, X.; Lullien, V.; Barron, C.; Abécassis, J. Dry processes to develop wheat fractions and products with enhanced nutritional quality. *J. Cereal Sci.* **2007**, *46*, 327–347. [CrossRef]
2. Shenoy, A.; Prakash, J. Wheat bran (*Triticum aestivum*): Composition, functionality and incorporation in unleavened bread. *J. Food Qual.* **2007**, *25*, 197–211.
3. Eeckhout, W.; Paepe, M.D. Total phosphorus, phytate phosphorus and phytase activity in plant feedstuffs. *Anim. Feed Sci. Technol.* **1994**, *47*, 19–29. [CrossRef]
4. Awaad, M.; Atta, A.M.; El-Ghany, W.A.; Elmenawey, M.; Ahmed, A.; Hassan, A.A.; Nada, A.; Abdelaleem, G.A.; Kawkab, A.A. Effect of a specific combination of mannan-oligosaccharides and β-glucans extracted from yeast cell wall on the health status and growth performance of ochratoxicated broiler chickens. *J. Am. Sci.* **2011**, *7*, 82–96.
5. Teng, P.Y.; Chung, C.H.; Chao, Y.P.; Chiang, C.J.; Chang, S.C.; Yu, B.; Lee, T.T. Administration of *Bacillus amyloliquefaciens* and *Saccharomyces cerevisiae* as direct-fed microbials improves intestinal microflora and morphology in broiler chickens. *J. Poult. Sci.* **2017**, *54*, 134–141. [CrossRef]
6. Shirazi, S.H.; Rahman, S.R.; Rahman, M.M. Production of extracellular lipases by saccharomyces cerevisiae. *World J. Microbiol. Biotechnol.* **1998**, *14*, 595–597. [CrossRef]
7. Khempaka, S.; Thongkratok, R.; Okrathok, S.; Molee, W. An evaluation of cassava pulp feedstuff fermented with *A. oryzae*, on growth performance, nutrient digestibility and carcass quality of broilers. *J. Poult. Sci.* **2013**, *51*, 71–79. [CrossRef]
8. Wang, C.C.; Lin, L.J.; Chao, Y.P.; Chiang, C.J.; Lee, M.T.; Chang, S.C.; Yu, B.; Lee, T.T. Antioxidant molecular targets of wheat bran fermented by white rot fungi and its potential modulation of antioxidative status in broiler chickens. *Br. Poult. Sci.* **2017**, *58*, 262–271. [CrossRef]
9. Hegazy, S.K.; El-Bedewy, M.M. Effect of probiotics on pro-inflammatory cytokines and *NF-κB* activation in ulcerative colitis. *World J. Gastroenterol.* **2010**, *16*, 4145–4151. [CrossRef]
10. Sears, B. Anti-inflammatory Diets. *J. Am. Coll. Nutr.* **2015**, *34*, 14–21. [CrossRef]
11. Lee, Y.; Lee, S.; Gadde, U.D.; Oh, S.; Lillehoj, H.S. Relievable effect of dietary Allium hookeri on LPS induced intestinal inflammation response in young broiler chickens. *J. Immunol.* **2017**, *198* (Suppl. S1), 226.3.
12. Cowieson, A.J.; Ruckebusch, J.P.; Sorbara, J.O.B.; Wilson, J.W.; Guggenbuhl, P.; Roos, F.F. A systematic view on the effect of phytase on ileal amino acid digestibility in broilers. *Anim. Feed Sci. Technol.* **2017**, *225*, 182–194. [CrossRef]
13. Hamdi, M.; Perez, J.F.; L'etourneau-Montminy, M.P.; Franco-Rossell'o, R.; Aligue, R.; Sol'a-Oriol, D. The effects of microbisal phytase and dietary calcium and phosphorus levels on the productive performance and bone mineralization of broilers. *Anim. Feed Sci. Technol.* **2018**, *243*, 41–51. [CrossRef]
14. Annika, M.M.; Manninen, M.; Gylienberg, H. The adherence of lactic acid bacteria to the columnar epithelial cells of pigs and calves. *J. Appl. Bacteriol.* **1983**, *55*, 241–245.

15. Miller, G.L. Use of dinitrosalicylic acid reagent for determination of reducing sugar. *Anal. Chem.* **1959**, *31*, 426–428. [CrossRef]
16. Sandhya, C.; Alagarsamy, S.; Szakacs, G.; Pandey, A. Comparative evaluation of neutral protease production by *Aspergillus oryzae* in submerged and solid-state fermentation. *Process. Biochem.* **2005**, *40*, 2689–2694. [CrossRef]
17. Tian, S.; Wang, Z.; Fan, Z.; Zuo, L.; Wang, J. Determination methods of cellulase activity. *Afr. J. Biotechnol.* **2011**, *10*, 7122–7125.
18. Kaiser, M.G.; Cheeseman, J.H.; Kaiser, P.; Lamont, S.J. Cytokine expression in chicken peripheral blood mononuclear cells after in vitro exposure to *Salmonella enterica* serovar enteritidis. *Poult. Sci.* **2006**, *85*, 1907–1911. [CrossRef]
19. AOAC. *Official Methods of Analysis of AOAC International*, 19th ed.; AOAC International: Gaithersburg, MD, USA, 2012.
20. Hu, X.; Chi, Q.; Wang, D.; Chi, X.; Teng, X.; Li, S. Hydrogen sulfide inhalation-induced immune damage is involved in oxidative stress, inflammation, apoptosis and the Th1/Th2 imbalance in broiler bursa of Fabricius. *Ecotox Environ. Safe.* **2018**, *164*, 201–209. [CrossRef]
21. Kalantar, M.; Schreurs, N.M.; Raza, S.H.A.; Khan, R.; Ahmed, J.Z.; Yaghobfar, A.; Shah, M.A.; Kalantar, M.H.; Hosseini, S.M.; Rahman, S.U. Effect of different cereal-based diets supplemented with multi-enzyme blend on growth performance villus structure and gene expression (SGLT1, GLUT2, PepT1 and MUC2) in the small intestine of broiler chickens. *Gene Rep.* **2019**, *15*, 100367. [CrossRef]
22. Santos, T.T.; Srinongkote, S.; Bedford, M.R.; Walk, C.L. Effect of high phytase inclusion rates on performance of broilers fed diets not severely limited in available phosphorus. *Asian-Australas J. Anim. Sci.* **2013**, *26*, 227–232. [CrossRef] [PubMed]
23. Mountzouris, K.C.; Dalaka, E.; Palamidi, I.; Paraskeuas, V.; Demey, V.; Theodoropoulos, G.; Fegeros, K. Evaluation of yeast dietary supplementation in broilers challenged or not with Salmonella on growth performance, cecal microbiota composition and Salmonella in ceca, cloacae and carcass skin. *Poult. Sci.* **2015**, *94*, 2445–2455. [CrossRef] [PubMed]
24. Feng, J.; Liu, X.; Xu, Z.R.; Wang, Y.Z.; Liu, J.X. Effects of fermented soybean meal on digestive enzyme activities and intestinal morphology in broilers. *Poult. Sci.* **2007**, *86*, 1149–1154. [CrossRef] [PubMed]
25. Ndrepepa, G. Uric acid and cardiovascular disease. *Eur. Caediol.* **2018**, *484*, 150–163. [CrossRef]
26. Hong, K.J.; Lee, C.H.; Kim, S.W. *Aspergillus oryzae* GB-107 fermentation improves nutritional quality of food soybeans and feed soybean meals. *J. Med. Food.* **2004**, *7*, 430–434. [CrossRef]
27. Gao, J.H.; Zhang, J.; Yu, S.H.; Wu, S.G.; Yoon, I.; Quigley, J.; Gao, Y.P.; Qi, G.H. Effects of yeast culture in broiler diets on performance and immunomodulatory functions. *Poult. Sci.* **2008**, *87*, 1377–1384. [CrossRef]
28. Cil, O.; Phuan, P.W.; Gillespie, A.M.; Lee, S.; Tradtrantip, L.; Yin, J.; Tse, M.; Zachos, N.C.; Lin, R.; Donowitz, M.; et al. Benzopyrimido-pyrrolo-oxazine-dione CFTR inhibitor (R)-BPO-27 for anti-secretory therapy of diarrheas caused by bacterial enterotoxins. *FASEB J.* **2016**, *31*, 751–760. [CrossRef]
29. Lacey, J.A.; Stanley, D.; Keyburn, A.L.; Ford, M.; Chen, H.; Johanesen, P.; Lyras, D.; Moore, R.J. *Clostridium perfringens*-mediated necrotic enteritis is not influenced by the pre-existing microbiota but is promoted by large changes in the post-challenge microbiota. *Vet. Microbiol.* **2018**, *227*, 119–126. [CrossRef]
30. Lee, M.T.; Lin, W.C.; Lee, T.T. Potential crosstalk of oxidative stress and immune response in poultry through phytochemicals—A review. *Asian-Australas J. Anim. Sci.* **2019**, *32*, 309–319. [CrossRef]
31. Gabay, C.; Lamacchia, C.; Palmer, G. IL-1 pathways in inflammation and human diseases. *Nat. Rev. Rheumatol.* **2010**, *6*, 232–241. [CrossRef]
32. Municio, C.; Alvarez, Y.; Montero, O.; Hugo, E.; Rodrı'guez, M.; Domingo, E.; Alonso, S.; Ferna'ndez, N.; Crespo, M.S. The response of human macrophages to ß-glucans depends on the inflammatory milieu. *PLoS ONE* **2013**, *8*, e62016. [CrossRef] [PubMed]
33. Huang, C.M.; Lee, T.T. Immunomodulatory effects of phytogenics in chickens and pigs—A review. *Asian-Australas. J. Anim. Sci.* **2017**, *31*, 617–627. [CrossRef] [PubMed]
34. Suschek, C.V.; Schnorr, O.; Bachofen, V.K. The role of *iNOS* in chronic inflammatory processes in vivo: Is it damage-promoting, protective, or active at all? *Curr. Mol. Med.* **2004**, *4*, 763–775. [CrossRef] [PubMed]
35. Arias-Salvatierra, D.; Silbergeld, E.K.; Acosta-Saavedra, L.C.; Calderon-Aranda, E.S. Role of nitric oxide produced by *iNOS* through *NF-κB* pathway in migration of cerebellar granule neurons induced by lipopolysaccharide. *Cell Signal.* **2011**, *23*, 425–435. [CrossRef] [PubMed]

36. Ivashkiv, B.; Donlin, L.T. Regulation of type I interferon responses. *Nat. Rev. Immunol.* **2014**, *14*, 36–49. [CrossRef] [PubMed]
37. Harada, K.; Isse, K.; Nakanuma, Y. Interferon gamma accelerates *NF-κB* activation of biliary epithelial cells induced by Toll-like receptor and ligand interaction. *J. Clin. Pathol.* **2006**, *59*, 184–190. [CrossRef]

14

Iodine Supplemented Diet Positively Affect Immune Response and Dairy Product Quality in Fresian Cow

Marco Iannaccone [1,†], Andrea Ianni [2,†], Ramy Elgendy [3], Camillo Martino [4], Mery Giantin [5], Lorenzo Cerretani [6], Mauro Dacasto [5] and Giuseppe Martino [1,*]

1 Faculty of Bioscience and Technology for Food, Agriculture, and Environment, University of Teramo, Via R. Balzarini 1, 64100 Teramo, Italy; m.iannaccone@unina.it
2 Department of Medical, Oral and Biotechnological Sciences, "G. d'Annunzio" University Chieti-Pescara, Via dei Vestini 31, 66100 Chieti, Italy; andreaianni@hotmail.it
3 Department of Immunology, Genetics and Pathology, Uppsala University, Uppsala 75185, Sweden; ramy.elgendy@igp.uu.se
4 Istituto Zooprofilattico Sperimentale dell'Abruzzo e del Molise "G. Caporale" Via Campo Boario, 64100 Teramo, Italy; c.martino@izs.it
5 Department of Comparative Biomedicine and Food Science, University of Padua, Viale dell'Università 16, 35020 Legnaro (PD), Italy; mery.giantin@unipd.it (M.G.); mauro.dacasto@unipd.it (M.D.)
6 Pizzoli SPA, Via Zenzalino Nord, 40054 1 Budrio (BO), Italy; l.cerretani@pizzoli.it
* Correspondence: gmartino@unite.it.
† These authors contributed equally to this work.

Simple Summary: Iodine represents an important micronutrient and plays a fundamental role in animal biology. This trace element is currently supplied to animal diet to investigate its potential effects on productive and reproductive performances. However, little is known about its role in the regulation of gene expression in ruminants. In this study, the dietary iodine supplementation in dairy cows showed effective modification of the expression of several molecular targets, with an improvement of the pathways involved in immune response and oxidative stress and undoubted positive repercussions on animal health.

Abstract: The effects of iodine supplementation on the whole-transcriptome of dairy cow using RNA sequencing has been investigated in this study. Iodine did not influence the milk composition, while an improvement was observed in the immune response as well as in the quality of dairy product. Indeed, the iodine intake specifically influenced the expression of 525 genes and the pathway analysis demonstrated that the most affected among them were related to immune response and oxidative stress. As a consequence, we indirectly showed a better response to bacterial infection because of the reduction of somatic cell counts; furthermore, an improvement of dairy product quality was observed since lipid oxidation reduced in fresh cheese. Such findings, together with the higher milk iodine content, clearly demonstrated that iodine supplementation in dairy cow could represent a beneficial practice to preserve animal health and to improve the nutraceutical properties of milk and its derived products.

Keywords: Iodine; immune response; oxidative stress; somatic cell count; fresh cheese; transcriptomics

---

1. Introduction

Micronutrients are essential to orchestrate all physiological functions. Among them, iodine (I) plays a unique role because it is the main component of the thyroid hormones, i.e., thyroxine (T4) and triiodothyronine (T3) [1]. Since the thyroid gland regulates many metabolic processes, the extent of I requirement strongly depends on the age and stage of development of the individual [2]. Thus, when the I demand is not satisfied, reduced functionality of the thyroid gland could occur (hypothyroidism) with negative consequences for proper mental development, body growth, and fertility. For these reason, diet is often integrated with I supplements, generally provided through iodized salt [3]. For infants, milk constitutes the only I source; therefore, the dietary calibration of this micronutrients in dairy animals assumes relevant importance in order to obtain I-rich milk without inducing variations in animal performances [4].

In animal husbandry, I supplementation as calcium iodate, sodium iodide, and other iodine compounds is needed since the native iodine content of plant straight feed-stuffs is low; moreover, the increasing use of rapeseed meal (RSM) in livestock diets is associated with the intake of glucosinolates, which are known to be iodine antagonists inhibiting the activity of sodium iodide symporter [5]. For this reason, the European community has recently brought to 5 mg/kg the maximum level of I supplementation for milk-producing ruminants (milk intended for human consumption), while it remained at 10 mg/kg of complete feed for other ruminant categories [6]. Consequently, several studies have been carried out to evaluate the productive performances in animals fed the I supplementation. Weiss et al. [7] showed that I concentration increased in serum but not in milk after supplementation of this element in diets of dairy cows. In contrast, studies performed in small ruminants showed that I supplementation doubled the milk iodine content when compared with the control group, even though no evident effect was observed in the milk gross composition [8]. I supplementation showed also beneficial effects on healthy status. Indeed, early study on feedlot cattle demonstrated an increased resistance to foot rot [9], which is due to improved phagocytic cell function [10]. Moreover, it was shown in lambs that high-dose potassium iodide supplementation can be effective to decrease the severity of airway viral infections, supposedly through the augmentation of mucosal oxidative defenses [11,12].

Recently, the RNA-sequencing approach was shown to be useful to elucidate which molecular pathways are affected by I supplementation in sheeps [13] and showed the positive effects of olive pomace-supplemented diet on inflammation and cholesterol in laying hens [14]. However, to date, little is known about the effects of I supplementation on transcriptomic profiles in dairy cows. Moreover, we tried to correlate the information concerning the signaling pathways influenced by I supplementation, with the qualitative parameters of milk and derivatives, taking into account those studies in which it was shown that dietary iodine supplementation in ruminants contributes to an improvement in the quality of dairy products [15].

2. Materials and Methods

2.1. Animal and Study Design

The study design was approved by the Teramo University Institutional Animal Care and Use Committee. Animals were managed according to Directive 2010/63/EU of the European Parliament regarding the protection of animals used for experimentation or other scientific purposes [16].

Twenty-two Friesian cattle, homogenous for age (range between 39 and 42 months), number of births (2 calves), and lactation length (70 ± 5 days), have been enrolled in this study. Animals, belonging to the same farm and bred in the same way were randomly divided in 2 groups of 11 cows each. During the 3-week acclimatization period (21 days), both the control (CTR) and the experimental iodine groups (IG) received a basal diet that mainly consisted of alfalfa hay plus a custom-formulated concentrate supplemented with 20 mg/day/animal I in order to guarantee the daily micronutrient requirement for each animal; then, the IG animals were fed for 56 days (during April and May) with a custom-formulated concentrate supplemented with additional 65 mg/day/animal of I in order to obtain a total intake of about 85 mg; this amount has been set not to exceed the maximum level allowed by law [17]. Ingredients and composition of total mixed ration (TMR) administered to the animals during the experimental period are reported in Table 1.

Table 1. Ingredients and composition of total mixed ration (TMR) administered to each animal of both group of study.

Ingredients of TMR	
Corn silage, %	23.7
First cut, alfalfa hay, %	5.3
Corn meal, %	3.4
Soybean, meal, %	3.2
Fine bran, %	3.0
Barley, meal, %	1.9
$CaCO_3$, %	0.2
Vitamins and minerals, %	0.4
Kg of dry matter/head per day	22.41
Chemical Composition of TMR	
Dry Matter, %	56.76
Crude protein [1], %	15.34
Ether extract [1], %	2.97
Ash [1], %	5.31
Neutral detergent fiber [1], %	32.51
Acid detergent fiber [1], %	20.03
Starch [1], %	27.02
Iodine (mg/head/day) [1]	20 (+65) *

[1] On a dry matter (DM) basis; * In brackets is the amount of iodine added to the diet for the experimental group.

2.2. *Blood and Milk Sampling*

Individual whole blood (WB) samples were collected at the beginning (T0) and at the end of the dietary supplementation for the evaluation of the hematochemical parameters. In the case of the RNA-Seq analysis, 2.5 mL of jugular venous blood was collected in duplicate from each animal only at the end of the experimental period. In this case, samples were collected in PAXgene™ tubes (Qiagen SpA, Milan, MI, Italy), stored overnight at room temperature, and then placed at −20 °C until RNA isolation, following the manufacturer's instructions.

Regarding milk, at the beginning (T0) and after 8 weeks (T8) of dietary supplementation, 50 mL of individual samples was collected from each group in triplicate in 80 mL polypropylene tubes containing a bronopol/sodium azide preservative solution, useful to extend the storage time before the analysis up to 48–72 h. Milking was performed through mechanical support (DeLaval, Milan, Italy), and teat disinfection occurred both before and after milking by using iodine-free solutions (DeLaval, Milan, Italy).

2.3. *Blood Analysis*

Complete blood cell count with leukocyte formula (total white blood cells, monocyte, lymphocyte, basophils, neutrophils, and eosinophils) for both the CTR and IG groups at T0 and T8 were performed at the Veterinary and Public Health Institute (Teramo, Italy) using a laser-based hematology analyzer with software applications for animal species (ADVIA 120 hematology system, Siemens, Munich, Germany). Plasma samples were analyzed for thyroid hormones (Thyroid-stimulating hormone (TSH), T3, and T4) with an automatic biochemistry analyzer (ILAB 650, Instrumentation Laboratory-Werfen, Milan, Italy) and following the routine procedure of the institute (Veterinary and Public Health Institute "G. Caporale", Teramo, Italy).

2.4. *Chemical Analysis of Milk*

Chemical composition of milk (fat, protein, casein, lactose, and urea) was determined by MilkoScan FT 6000 (Foss Integrator IMT; Foss, Hillerød, Denmark), whereas the somatic cells count (SCC) was performed using the Fossomatic FC (Foss).

2.5. *Iodine Determination in Milk*

The amount of I in the milk at T0 and T8 was determined according to Fecher et al. (1998) with some modifications [18]. Briefly, for each sample, 0.5 g of milk were homogenized with tetramethylammonium hydroxide (0.25 M) and 2 mL of deionized water (30%). Then, samples were heated in a microwave (800 W) at 170 °C for 30 min. After cooling, samples were transferred into a sterile tube and diluted with distilled water to a final volume of 15 mL. After centrifugation at 12,000 rpm × min^{-1} at room temperature for 10 min, samples were filtered through polytetra-fluoroethylene (PTFE) syringe filters (0.45 μm) and, finally, stored at 4 °C until analysis. A standard calibration curve was created using six calibration points equals to concentrations of 0, 5, 10, 25, 50, and 100 mg/L of I in tetramethylammonium hydroxide. For the carrier and gas formation, argon gas was used at flow rates of 1.05 and 0.2 L/min, respectively. The iodine content was determined by an inductively coupled plasma mass spectrometer Agilent 7500ce ICP-MS (Agilent Technologies, Palo Alto, CA, USA) at m/z = 127 and a total acquisition time of 21 s. Before the sequence analysis, the ICP-MS was auto-tuned by a solution containing 1 ppb of different metals (Li, Y, Ce, Tl, and Co).

2.6. *Library Preparation and RNA-Seq Analysis*

Total RNA was isolated from blood using the PAXgene blood RNA kit (Qiagen, Milan, Italy) as per the manufacturer's instructions. Total RNA concentration was determined by the NanoDrop ND-1000 spectrophotometer (NanoDrop Technologies Inc., Wilmington, DE, USA), and its quality was measured by the 2100 Bioanalyzer and RNA 6000 Nano kit (Agilent Technologies, Santa Clara, CA, USA). Strand-specific RNA-Seq libraries were prepared using the SureSelect strand-specific mRNA library preparation kit (Agilent Technologies, Santa Clara, CA, USA) as per the manufacturer's protocol. In brief, poly(A) RNA was purified from 1 μg of total RNA using two serial rounds of binding to oligo(dT) magnetic particles and, then, fragmented and reverse transcribed to generate cDNA. Illumina-specific adaptor was sequentially ligated to the 3′ end of cDNA fragments and purified using the AMPure XP beads (Beckman Coulter, Brea, CA, USA), and finally PCR-amplified (13 cycles) using an appropriate indexing primer to allow further samples multiplexing. The PCR-amplified libraries were purified with the AMPure XP beads (Beckman Coulter, Brea, CA, USA) and then assessed for their quality and fragments distribution using the 2100 Bioanalyzer DNA 1000 assay (Agilent Technologies, Santa Clara, CA, USA). In the presence of adaptor-dimers (Electropherogram's peak at 100 to 150-bp), another round of magnetic beads purification was performed. Libraries were quantified by both the Qubit® Fluorometer (Life Technologies, Carlsbad, CA, USA) and the qPCR-based NEBNext library quantification kit (New England BioLabs, Hitchin, UK). Finally, libraries were pooled and then sequenced by an Illumina HiSeq 2500 for 50 sequencing cycles.

The raw 50-bp single-end sequences (Sanger/Illumina 1.9 encoding) were quality controlled using FastQC (v.0.11.4; Babraham Institute, Cambridge, UK), and the low-quality bases (quality scores < 30) and adaptor contamination (if present) were removed by Trimmomatic v.0.36. The high-quality reads were mapped by HISAT v.2.0.5 against the Bos taurus reference genome (Ensembl Bos_taurus_UMD_3.1.1). The uniquely mapped reads aligned to exons were counted with HTSeq v.0.6.1 [19] and then tested by the DESeq2 R package v.1.14.1 [20] for the presence of differentially expressed genes (DEGs) in the IG group compared with the CTR one (e.g., T8 I vs T8 CTR). Genes with a false discovery rate (FDR) less than 0.05 were considered as DEGs. All analyses were performed using the software Artificial Intelligence RNA-seq (A.I.R; developed by Sequentia Biotech, Barcelona, Spain) and the sequencing data (FASTQ files) associated with this project are deposited in the GenBank's Sequence Read Archive (SRA) under the accession number PRJNA516565.

2.7. *Enriched Pathway Analysis*

The STRING software (Version 11.0, http://string-db.org/) was used to identify canonical pathways using the dataset of 525 DEGs identified between the IG group and the CTR one with a FDR < 0.05.

We set the interaction score as 0.9, the highest value permitted by the software to avoid false positives. The significance of the canonical pathway was measured with the *p*-value and the ratio of DEG/number of genes in the pathway.

2.8. Ricotta Cheese-Making Procedure and Lipid Oxidation by Thiobarbituric Acid Reactive Substance Test

At the end of the supplementation period, an aliquot of about 150 L of bulk milk from each experimental group was separately collected and manipulated according to common cheese-making procedures. The whey separately collected from each experimental group during the cheese-making was used to produce "ricotta" cheese. This was obtained by acid-thermal coagulation of whey, through heating up to 80 °C and adding 50 g of lactic acid. The whey flocs were collected in holed baskets with a total capacity equal to 0.5 L and left to drain for 30 min. Then, baskets still containing the ricotta were transferred in a refrigerator and stored under chilling conditions (4 °C). From each experimental group, 9 ricotta forms (single weight = about 400 g) were obtained. In order to evaluate lipid peroxidation, thiobarbituric acid reactive substances (TBARS) were measured at day 0 and day 7. The analysis was performed according to the procedure reported by Ianni et al. with slight modifications [21]: 4.5 g of frozen ricotta cheese were mixed within 2 min of sample withdrawal from the freezer, with 450 µL of 0.1% of butylated hydroxytoluene in methanol with the aim to stop the oxidation process. The mixture was homogenized with UltraTurrax T-25 high speed homogenizer (IKA, Staufen, Germany) in 40 mL of an aqueous solution of 7% trichloroacetic acid and, then, distilled; 1.5 mL of each distillate was mixed with an equal volume of a 0.02 M thiobarbituric acid (TBA) in 90% acetic acid, and the solution was kept for one hour in a thermostatic bath at 80 °C. Only after cooling, the absorbance at 534 nm was evaluated with a spectrophotometer (JENWAY 6305 UV/vis, Jenway, Essex, UK). The amount of malondialdehyde (MDA) of each sample was calculated by using a calibration curve ranging from 0 to 100 ppm ($R^2 = 0.986$), and results were expressed in µg of MDA per g of cheese.

2.9. Statistics

GraphPad Prism version 6.0 (GraphPad Software, La Jolla, CA, USA) was used for statistical analysis. Differences in milk parameters, as well as the amount of thyroid hormone in blood sera and malondialdehyde (MDA) concentrations in ricotta fresh cheese were statistically evaluated by using ordinary two-way ANOVA.

3. Results

3.1. Serum Thyroid Hormone and Iodine Concentrations

All animals maintained a good state of health for the entire duration of the trial, and no significant variations in milk yield have been observed between the two groups at the end of the experimental period.

Because I affects production and secretion of thyroid hormones, we quantified the levels of thyrotropin (TSH), triiodothyronine (T3), and thyroxine (T4) at the beginning (T0) and at the end (T8) of the supplementation in the serum blood samples. As shown in Figure 1, no differences in hormone levels were ever noticed between groups, suggesting that the thyroid functionality was not affected by the iodine supplementation and that the CTR group was not in iodine-deficiency condition.

As expected (Figure 2), we appreciate a higher amount of iodine in the I group, indicating that consumption of milk from dairy cows fed with high iodine intake is helpful to integrate diets where physiological stages like infancy and/or pregnancy require higher iodine intake [22].

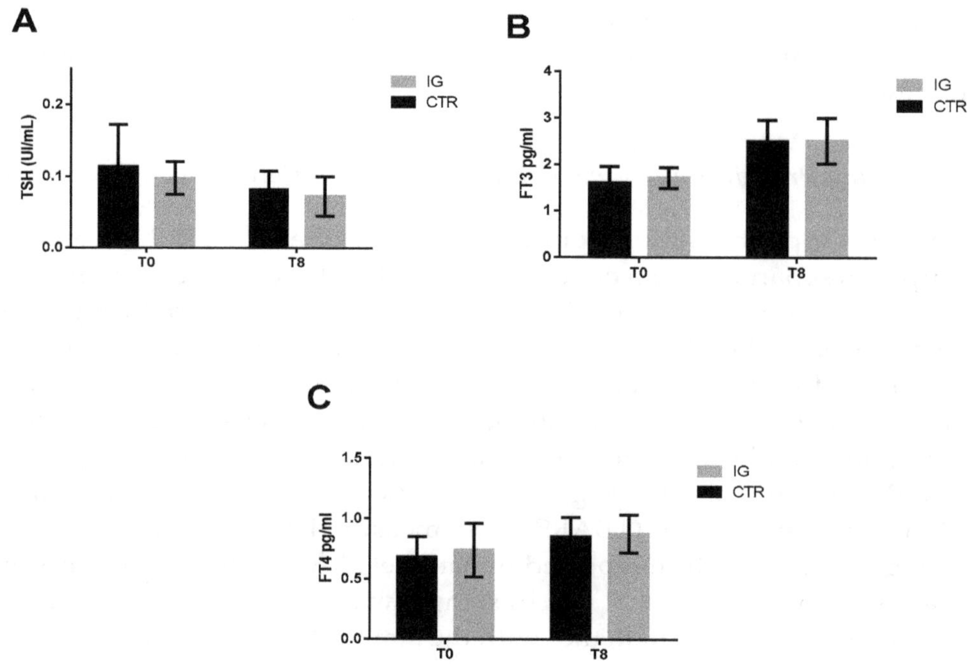

Figure 1. Thyroid hormone concentrations: Thyroid-stimulating hormone (**A**), free thyroxine (**B**), and free triiodothyronine (**C**) were measured in serum samples from control (CTR) (n = 11) and iodine group (IG) (n = 11) at the beginning and at the end of iodine supplementation. Data are expressed as mean ± SD, and differences were assessed using 2-way ANOVA.

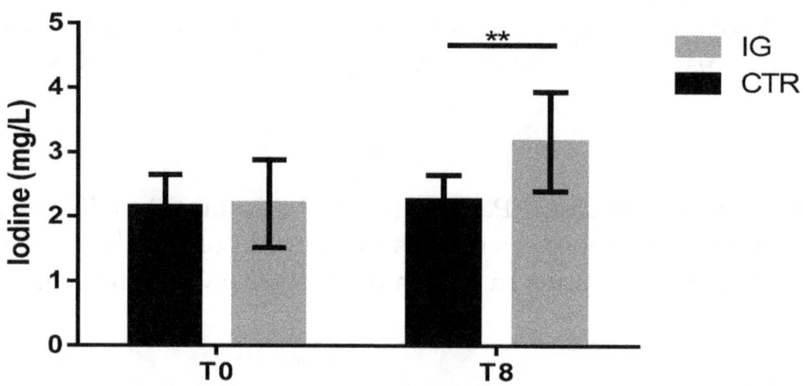

Figure 2. Iodine quantification in milk samples: Iodine was quantified in milk samples of both CTR (n = 11) and IG (n = 11) at the beginning and at the end of iodine supplementation. Data are shown as mean ± SD, and differences were assessed using 2-way ANOVA. ** p-value < 0.01.

However, it is worth mentioning that milk parameters were not influenced by iodine supplementation (Table S1).

3.2. Influence of I-supplemented Diet on Blood Transcriptome

To identify the molecular networks associated to I supplementation, we collected peripheral blood from both groups at the end of I supplementation period and performed a transcriptomic analysis by RNA-Seq. After trimming and quality control, on average, 96% of reads resulted of high-quality reads (Table S2) and were mapped against Bos taurus reference genome (Bos_taurus_UMD_3.1.1).

Filtering our data by using a FDR < 0.05, we identified 525 DEGs (Table S3); in particular, 274 and 248 genes were respectively down- and upregulated in the I group compared to CTR; moreover, they enabled us to discriminate the two groups not only on a heat map scale but also on a hierarchical clustering analysis, thereby indicating the robustness of analysis (Figures 3 and 4).

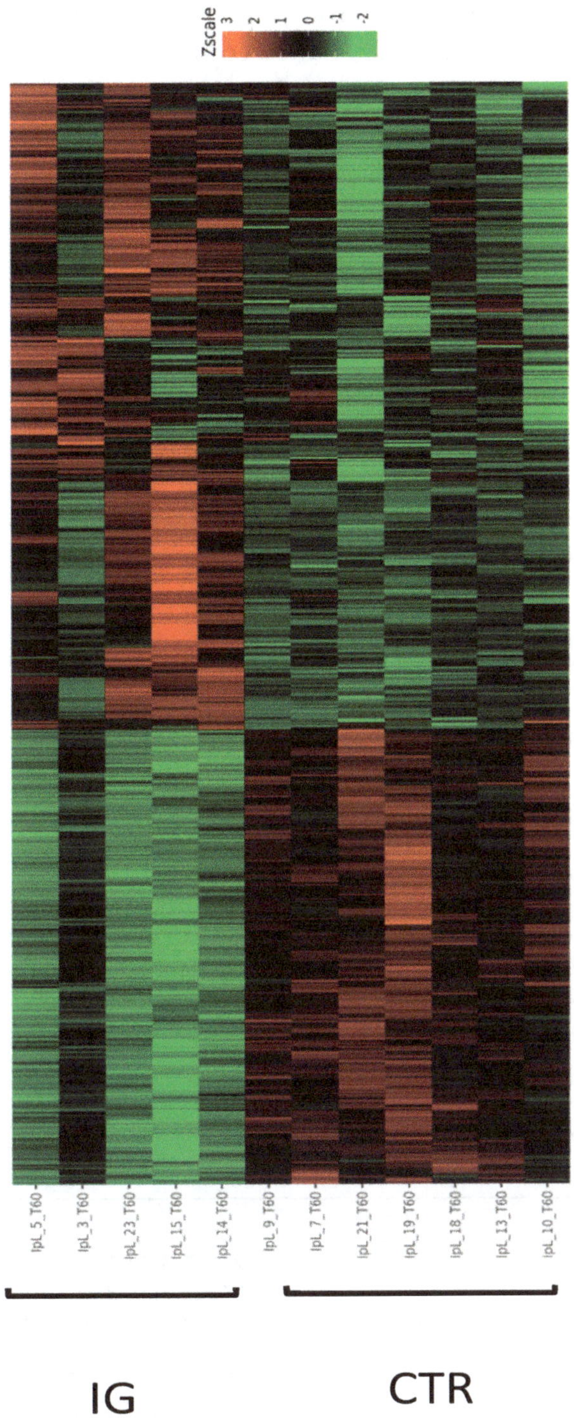

Figure 3. Heat map analysis of the 525 differentially expressed genes identified at the end of iodine supplementation period.

In order to identify the enriched pathways and putative interactions among DEGs, we interrogated the STRING software by using the up- and downregulated genes. To increase stringency and confidence in our data set, we applied the highest available interaction score (0.9), and in Table 2 (downregulated) and Table 3 (upregulated) are listed the most significant pathways (FDR < 0.05). The highest enriched pathway was "Fc gamma R-mediated phagocytosis" (FDR: 2.63×10^{-6}), which was associated with clearance of opsonized particles, suggesting that iodine supplementation could help the immune system against invading bacteria [23]. Interestingly, also, the "oxidative phosphorylation" pathway was dysregulated by iodine supplementation, indicating a positive effect of iodine on mitochondrial activity with consequent reduction of oxidative stress, which could potentially act on quality dairy products.

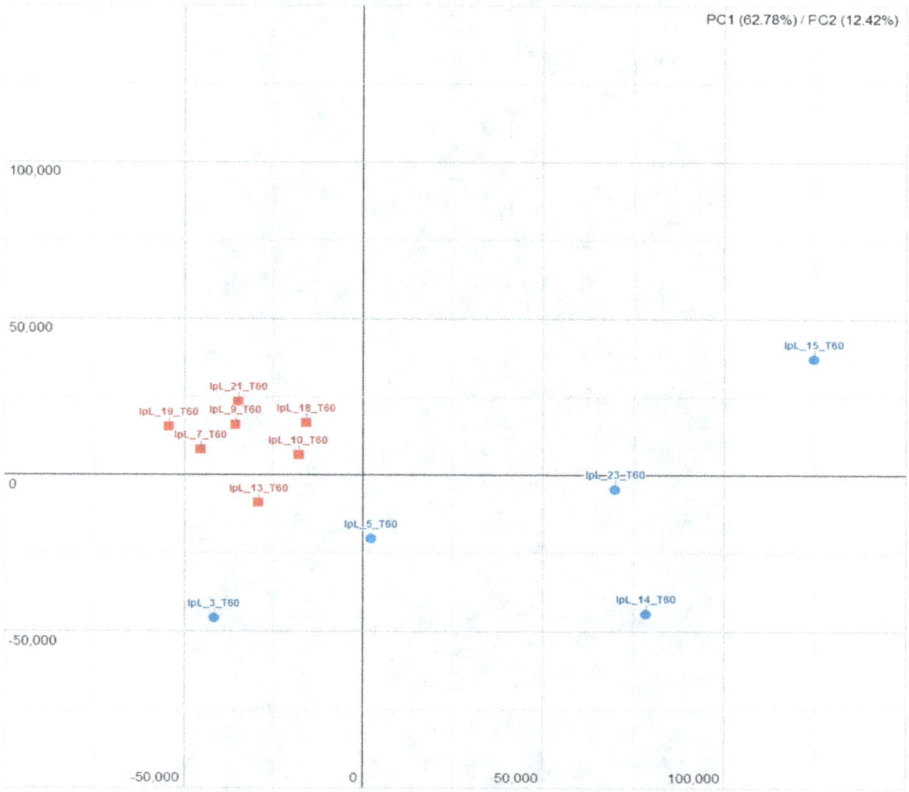

Figure 4. Principal components analysis using the 525 differentially expressed gene identified at the end of the iodine supplementation period: The blue circles identify the iodine samples while the red squares identify CTR samples.

Table 2. List of significantly enriched pathways obtained using the downregulated genes following iodine supplementation.

Pathway	FDR	Genes
regulation of transcription, DNA-templated	1.30×10^{-5}	ZNF93, SBNO1, ZBTB6, SNAPC3, ZFX, ZNF12, LOC530973, ZNF655, LOC104968476, ZNF175, MAP3K7, ZNF184, JADE1, ZNF182, ZNF148, RBAK, ZNF248, ZNF286A, DNTTIP2, MYNN, ZNF605, ZNF572, ZNF436
cellular response to DNA damage stimulus	7.45×10^{-4}	SHPRH, HELB, ZMAT3, FMR1, RNF168, SPRTN, USP16, NEK4
transcription, DNA-templated	0.0057	CCNT2, ZKSCAN8, ZBTB6, SNAPC3, ZNF131, ZFX, SCAI, ZNF518A, MAP3K7, ZNF184, JADE1, ZNF148, PSIP1, DNTTIP2, USP16, MYNN, ZNF572
RNA processing	0.0153	DHX29, U2SURP, YTHDC2, DHX36
negative regulation of transforming growth factor beta receptor signaling pathway	0.0186	ZNF451, LEMD3, SMURF2, SIRT1
cell division	0.0473	CDC40, USP37, USP16, CENPJ, SPICE1, SMC3

FDR: false discovery rate.

Table 3. List of significantly enriched pathways obtained using the upregulated genes following iodine supplementation.

Pathway	FDR	Genes
Fc gamma R-mediated phagocytosis	2.63×10^{-6}	AKT1, AKT2, ARF6, ARPC1B, CFL1, CRKL, GSN, LIMK1, RAC2, SCIN, VASP
Non-alcoholic fatty liver disease (NAFLD)	1.33×10^{-5}	AKT1, AKT2, COX8A, GSK3A, MAP3K11, NDUFA7, NDUFB7, NDUFS6, NDUFS7, NDUFS8, SDHA, TGFB1, UQCR11
Oxidative phosphorylation	0.0005	ATP5D, ATP5G2, COX8A, NDUFA7, NDUFB7, NDUFS6, NDUFS7, NDUFS8, SDHA, UQCR11
Rap1 signaling pathway	0.0008	ACTB, ACTG1, AKT1, AKT2, CRKL, MAP2K3, RAC2, RAPGEF1, RASSF5, SIPA1, TLN1, VASP
Bacterial invasion of epithelial cells	0.0016	ACTB, ACTG1, ARPC1B, CRKL, PXN, RHOG, SEPT9
Proteasome	0.0078	PSMB10, PSMB4, PSMB8, PSMB9, PSMD4
Carbon metabolism	0.0079	ENO1, GPI, HK1, IDH2, PFKL, SDHA, TPI1
Focal adhesion	0.0079	ACTB, ACTG1, AKT1, AKT2, CRKL, PXN, RAC2, RAPGEF1, TLN1, VASP
Regulation of actin cytoskeleton	0.0079	ACTB, ACTG1, ARPC1B, CFL1, CRKL, GSN, LIMK1, PXN, RAC2, SCIN
B cell receptor signaling pathway	0.0024	AKT1, AKT2, CD81, NFKBIB, RAC2

FDR: false discovery rate.

3.3. *Effects of I-Supplementation on Quality of Dairy Products*

To test the hypothesis if I supplementation could positively influence quality of dairy products, we evaluated the milk somatic cell count (SCC) at the beginning T0 and the end T8 of the supplementation. Clearly, in Figure 5, we demonstrated a significant reduction of SCC, indicating that the I treatment ameliorated immune response against bacteria responsible for infectious diseases like mastitis [24].

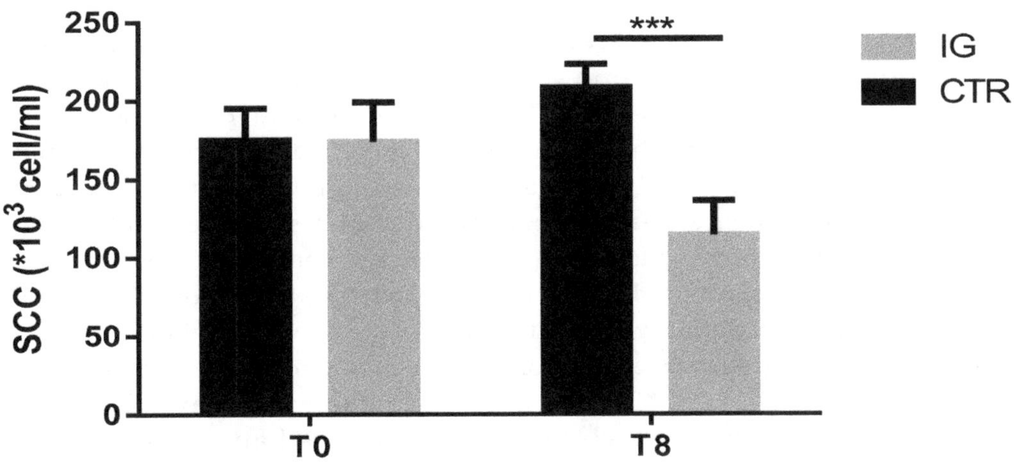

Figure 5. Somatic cell count in individual milk samples from CTR (n = 11) and IG (n = 11): Data are shown as mean ± SD, and differences were assessed using 2-way ANOVA. *** p-value < 0.001.

Then, we evaluated also the quality of fresh dairy cheese (ricotta) obtained using milk derived by the CTR and I groups. Lipid oxidation is strongly influenced by the redox homeostasis, and in our data set, iodine influences mitochondrial activity, which is highly involved in controlling oxidative stress. Thus, we measured the time-dependent variation (at day 0 and day 7) in malondialdehyde

(MDA) levels in fresh ricotta cheese kept at 4 °C. As showed in Figure 6, ricotta from the I group showed statistically lower amounts of MDA, which confirms the favorable effect of I supplementation on cheese quality.

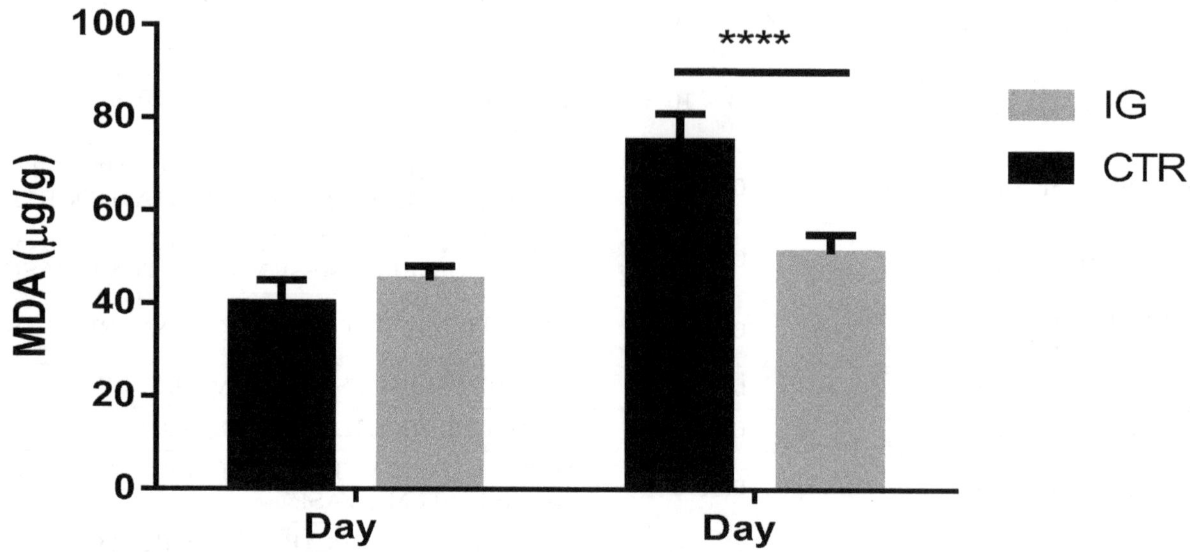

Figure 6. Malondialdehyde (MDA) levels at 0 (40 ± 5 vs. 45 ± 3) and 7 (75 ± 6 vs. 51 ± 4) days from fresh cheese ricotta stored at 4 °C: Data represent mean ± SD, **** $p < 0.0001$, two-way ANOVA (5 samples/group).

4. Discussion

In this study, we provide evidence that dairy cows fed an I-supplemented diet for a limited time showed transcriptional changes related to immune response and oxidative stress. In our experience, whole blood is a good starting point to understand in ruminants the effects of different diet supplements such as agro-industrial by-products and microelements [25–27]. To avoid that differences identified in gene expression in this study that could be influenced by different composition in white blood cell, we measured the complete blood cell count both in CTR and IG at the beginning T0 and at the end of supplementation, and we did not find any difference (Table S4). Then, because I is the major component of thyroid hormones, we measured the free hormone levels in the sera of both groups. In agreement with previous studies, the thyroid hormone levels did not differ between the two groups, clearly indicating that I supplement used in this study does not affect thyroid functionality [28].

In our study, RNA-sequencing analysis confirmed that I supplementation deregulated the expression of numerous genes, showing a significant biological connection confirmed by a very small p-value for protein–protein interaction, indicating no random nodes within our data set (PPI enrichment p-value $< 1.0 \times 10^{-16}$). More in detail, we identified many pathways related to immune response (Table 2), and the most significant one was that of "Fc gamma R-mediated phagocytosis" (FDR: 2.36 $\times 10^{-6}$) and, as a consequence, also of "B cell receptor signaling pathway" (FDR: 0.0024), which is in agreement with previous study in which I exposure produced an increase in immunoglobulin synthesis by lymphocytes [29]. Moreover, phagocytes are the lymphocyte subsets which express the higher level of sodium iodide symporter [28]. Thus, iodide supplementation reinforces immune response via strengthened antibody production and phagocytosis. Moreover, iodide also interacts with myeloperoxidase and H2O2, reinforcing the killing of bacterial activity [30]. Altogether, these positive effects are potentially associated with pathogen clearance and we have demonstrated that milk collected from animals belonging to the experimental group showed a reduced somatic cell count, which has directly associated with reduced bacterial load [24].

Also, oxidative phosphorylation is one of the activated pathways in our study (4.75×10^{-4}). The effects of I and thyroid hormones have been previously fully reviewed [31]; thus, it is not surprising to find the activation of genes belonging to the multimeric protein ATP synthase (ATP5G2 and ATP5D), which is responsible for ATP production from ADP. Indeed, it is been demonstrated that increased production of ATP was inversely associated with production of cellular damage enzyme associated biomarkers (Serum lactate dehydrogenase, LHD; creatine phosphokinase, CPK; Alkaline phosphatase, ALP) following I supplementation. Also, several genes encoding for different subunits of the NADH: ubiquinone oxidoreductase were upregulated in our data set. These results further confirm the effect of I-supplementation on oxidative phosphorylation with a potential favourable energy balance.

In addition, I has been considered, in an evolutionary sense, one of the most ancient antioxidant because it can easily interact with reactive oxygen species (ROS) [32]. In our study, milk samples collected from the experimental groups have shown increased I content in agreement with previous studies on I supplementation in ruminants [13,33]. Thus, when we analyzed ricotta cheese produced by using milk from the CTR and I groups, we appreciated a reduced production of MDA in IG & CTR, indirectly confirming a reduction in lipid oxidation.

In conclusion, in our study, we have demonstrated that iodine supplementation has multiple positive effects on animal health and dairy products quality. Indeed, on one side, I supplementation ameliorates immune response and resistance to infectious disease, improving phagocytosis, and on the other side, it improves the quality of fresh cheese, reducing lipid oxidation with beneficial effects also on consumer's health. Moreover, because of the higher I content in milk, make potentially this product suitable also for subjects that, for particular physiological conditions (i.e., pregnancy) or life stage (i.e., infant), need more I intake.

Author Contributions: Conceptualization, G.M., M.I., and A.I.; methodology, R.E., C.M., and M.G.; validation, M.I., A.I., and R.E.; formal analysis, M.I. and A.I.; investigation, M.I., A.I., and R.E.; resources, G.M., L.C., and M.D.; data curation, M.I. and A.I.; writing—original draft preparation, M.I. and A.I.; writing—review and editing, R.E., M.D., and G.M.; supervision, G.M. and M.D.; project administration, G.M.; funding acquisition, G.M.

References

1. Zimmermann, M.B. Iodine deficiency. *Endocr. Rev.* **2009**, *30*, 376–408. [CrossRef] [PubMed]
2. Gunnarsdottir, I.; Dahl, L. Iodine intake in human nutrition: A systematic literature review. *Food Nutr. Res.* **2012**, *56*. [CrossRef] [PubMed]
3. Tetens, I. EFSA Panel on Dietetic Products Nutrition and Allergies. Scientific Opinion on Dietary Reference Values for iodine. *EFSA J.* **2014**, *12*, 3660.
4. Ghirri, P.; Lunardi, S.; Boldrini, A. Iodine Supplementation in the Newborn. *Nutrients* **2014**, *6*, 382–390. [CrossRef] [PubMed]
5. Flachowsky, G.; Franke, K.; Meyer, U.; Leiterer, M.; Schöne, F. Influencing factors on iodine content of cow milk. *Eur. J. Nutr.* **2014**, *53*, 351–365. [CrossRef] [PubMed]
6. EU. Commission Implementing Regulation 2015/861 of 3 June 2015 concerning the authorisation of potassium iodide, calcium iodate anhydrous and coated granulated calcium iodate anhydrous as feed additives for all animal species. *Off. J. Eur. Union.* **2015**, *L137*, 1–7.
7. Weiss, W.P.; Wyatt, D.J.; Kleinschmit, D.H.; Socha, M.T. Effect of including canola meal and supplemental iodine in diets of dairy cows on short-term changes in iodine concentrations in milk. *J. Dairy Sci.* **2015**, *98*, 4841–4849. [CrossRef] [PubMed]
8. Nudda, A.; Battacone, G.; Decandia, M.; Acciaro, M.; Aghini-Lombardi, F.; Frigeri, M.; Pulina, G. The effect of dietary iodine supplementation in dairy goats on milk production traits and milk iodine content. *J. Dairy Sci.* **2009**, *92*, 5133–5138. [CrossRef]
9. Maas, J.; Davis, L.E.; Hempstead, C.; Berg, J.N.; Hoffman, K.A. Efficacy of ethylenediamine dihydriodide in the prevention of naturally occurring foot rot in cattle. *Am. J. Vet. Res.* **1984**, *45*, 2347–2350.

10. Kulow, M.; Merkatoris, P.; Anklam, K.S.; Rieman, J.; Larson, C.; Branine, M.; Döpfer, D. Evaluation of the prevalence of digital dermatitis and the effects on performance in beef feedlot cattle under organic trace mineral supplementation. *J. Anim. Sci.* **2017**, *95*, 3435–3444. [CrossRef]
11. Antaya, N.T.; Soder, K.J.; Kraft, J.; Whitehouse, N.L.; Guindon, N.E.; Erickson, P.S.; Conroy, A.B.; Brito, A.F. Incremental amounts of *Ascophyllum nodosum* meal do not improve animal performance but do increase milk iodine output in early lactation dairy cows fed high-forage diets. *J. Dairy Sci.* **2015**, *98*, 1991–2004. [CrossRef] [PubMed]
12. Derscheid, R.J.; van Geelen, A.; Berkebile, A.R.; Gallup, J.M.; Hostetter, S.J.; Banfi, B.; McCray, P.B., Jr.; Ackermann, M.R. Increased Concentration of Iodide in Airway Secretions Is Associated with Reduced Respiratory Syncytial Virus Disease Severity. *Am. J. Respir. Cell Mol. Biol.* **2014**, *50*, 389–397. [CrossRef] [PubMed]
13. Iannaccone, M.; Elgendy, R.; Ianni, A.; Martino, C.; Palazzo, F.; Giantin, M.; Grotta, L.; Dacasto, M.; Martino, G. Whole-transcriptome profiling of sheep fed with a high iodine-supplemented diet. *Animal* **2019**. [CrossRef] [PubMed]
14. Iannaccone, M.; Ianni, A.; Ramazzotti, S.; Grotta, L.; Marone, E.; Cichelli, A.; Martino, G. Whole Blood Transcriptome Analysis Reveals Positive Effects of Dried Olive Pomace-Supplemented Diet on Inflammation and Cholesterol in Laying Hens. *Animals* **2019**, *9*, 427. [CrossRef] [PubMed]
15. Lopez, C.C.; Serio, A.; Rossi, C.; Mazzarrino, G.; Marchetti, S.; Castellani, F.; Grotta, L.; Fiorentino, F.P.; Paparella, A.; Martino, G. Effect of diet supplementation with *Ascophyllum nodosum* on cow milk composition and microbiota. *J. Dairy Sci.* **2016**, *99*, 6285–6297. [CrossRef] [PubMed]
16. EU. Directive 2010/63/EU of the European Parliament and of the Council of 22 September 2010 on the protection of animals used for scientific purposes. *Off. J. Eur. Union.* **2010**, *L276*, 33–79.
17. EC. No 1459/2005 of 8 September 2005 amending the conditions for authorisation of a number of feed additives belonging to the group of trace elements. *Off. J. Eur. Union.* **2005**, *L233*, 8–10.
18. Fecher, P.A.; Goldmann, I.; Nagengast, A. Determination of iodine in food samples by inductively coupled plasma mass spectrometry after alkaline extraction. *J. Anal. At. Spectrom.* **1998**, *13*, 977–982. [CrossRef]
19. Anders, S.; Pyl, P.T.; Huber, W. HTSeq-A Python framework to work with high-throughput sequencing data. *Bioinformatics* **2015**, *31*, 166–169. [CrossRef]
20. Love, M.I.; Huber, W.; Anders, S. Moderated estimation of fold change and dispersion for RNA-seq data with DESeq2. *Genome Biol.* **2014**, *15*, 550. [CrossRef]
21. Ianni, A.; Innosa, D.; Martino, C.; Grotta, L.; Bennato, F.; Martino, G. Zinc supplementation of Friesian cows: Effect on chemical-nutritional composition and aromatic profile of dairy products. *J. Dairy Sci.* **2019**, *102*, 2918–2927. [CrossRef] [PubMed]
22. Zimmermann, M.B. The Effects of Iodine Deficiency in Pregnancy and Infancy. *Paediatr. Perinat. Epidemiol.* **2012**, *26*, 108–117. [CrossRef] [PubMed]
23. De Vito, P.; Incerpi, S.; Pedersen, J.Z.; Luly, P.; Davis, F.B.; Davis, P.J. Thyroid Hormones as Modulators of Immune Activities at the Cellular Level. *Thyroid* **2011**, *21*, 879–890. [CrossRef] [PubMed]
24. Sharma, N.; Singh, N.K.; Bhadwal, M.S. Relationship of Somatic Cell Count and Mastitis: An Overview. *Asian-Australas. J. Anim. Sci.* **2011**, *24*, 429–438. [CrossRef]
25. Iannaccone, M.; Elgendy, R.; Giantin, M.; Martino, C.; Giansante, D.; Ianni, A.; Dacasto, M.; Martino, G. RNA Sequencing-Based Whole-Transcriptome Analysis of Friesian Cattle Fed with Grape Pomace-Supplemented Diet. *Animals* **2018**, *8*, 188. [CrossRef]
26. Elgendy, R.; Giantin, M.; Castellani, F.; Grotta, L.; Palazzo, F.; Dacasto, M.; Martino, G. Transcriptomic signature of high dietary organic selenium supplementation in sheep: A nutrigenomic insight using a custom microarray platform and gene set enrichment analysis. *J. Anim. Sci.* **2016**, *94*, 3169–3184. [CrossRef]
27. Elgendy, R.; Palazzo, F.; Castellani, F.; Giantin, M.; Grotta, L.; Cerretani, L.; Dacasto, M.; Martino, G. Transcriptome profiling and functional analysis of sheep fed with high zinc-supplemented diet: A nutrigenomic approach. *Anim. Feed Sci. Tech.* **2017**, *234*, 195–2042. [CrossRef]
28. Bilal, M.Y.; Dambaeva, S.; Kwak-Kim, J.; Gilman-Sachs, A.; Beaman, K. A Role for Iodide and Thyroglobulin in Modulating the Function of Human Immune Cells. *Front. Immunol.* **2017**, *8*, 1–13. [CrossRef]
29. Weetman, A.P.; Mcgregor, A.M.; Campbell, H.; Lazarus, J.H.; Ibbertson, H.K.; Hall, R. Iodide enhances IgG synthesis by human peripheral blood lymphocytes in vitro. *Eur. J. Endocrinol.* **1983**, *103*, 210–215. [CrossRef]

30. Klebanoff, S.J.; Kettle, A.J.; Rosen, H.; Winterbourn, C.C.; Nauseef, W.M. Myeloperoxidase: A front-line defender against phagocytosed microorganisms. *J. Leukoc. Biol.* **2013**, *93*, 185–198. [CrossRef]
31. Soboll, S. Thyroid hormone action on mitochondrial energy transfer. *Biochim. Biophis. Acta Bioenerg.* **1993**, *1144*, 1–16. [CrossRef]
32. Venturi, S. Evolutionary Significance of Iodine. *Curr. Chem. Biol.* **2011**, *5*, 155–162. [CrossRef]
33. Grace, N.D.; Waghorn, G.C. Impact of iodine supplementation of dairy cows on milk production and iodine concentrations in milk. *N. Z. Vet. J.* **2005**, *53*, 10–13. [CrossRef] [PubMed]

15

Microbial and Fungal Phytases can Affect Growth Performance, Nutrient Digestibility and Blood Profile of Broilers Fed Different Levels of Non-Phytic Phosphorous

Youssef A. Attia [1,*], Fulvia Bovera [2,*], Francesco Iannaccone [2], Mohammed A. Al-Harthi [1], Abdulaziz A. Alaqil [3], Hassan S. Zeweil [4] and Ali E. Mansour [5]

1 Arid Land Agriculture Department, Faculty of Meteorology, Environment and Arid Land Agriculture, King Abdulaziz University, P.O. Box 80208, Jeddah 21589, Saudi Arabia; malharthi@kau.edu.sa
2 Department of Veterinary Medicine and Animal Production, University of Napoli Federico II, via F. Delpino 1, 80137 Napoli, Italy; francesco.iannaccone@unina.it
3 Department of Animal and Fish Production, King Faisal University, Al-Hufof, Al-Hassa 31982, Saudi Arabia; aalaqil@yahoo.com.
4 Department of Animal and Fish Production, Faculty of Agriculture, Saba Basha, Alexandria University, Alexandria 21527, Egypt; hszeweil@yahoo.com
5 Ministry of agriculture, Animal Production section, Behiri Governorate 22951, Egypt; alierfatmansour@gmail.com
* Correspondence: yaattia@kau.edu.sa (Y.A.A.); bovera@unina.it (F.B.);

Simple Summary: To reduce the environmental pollution is a must to preserve the health of the world. The environmental impact of poultry farming is receiving an increasing attention due to several emissions among these is phosphorus. This element is in general present in the commercial diets of broilers or laying hens in an amount exceeding the real needing of the animals, and, therefore, a great amount of phosphorus ends in the excreta. Thus, optimizing the amount of phosphorous in the diets of poultry could partially alleviate the environmental impact of these farms.

Abstract: A total of 420 day old chicks were divided into seven groups (5 replicates of 12 chicks/group) fed isoproteic and isoenergetic diets. The control group was fed diets containing 0.50%, 0.45% and 0.40% of non-phytic phosphorous (nPP) in starter (1–35), grower (37–56) and finisher (57–64 d) periods, respectively. The three intermediate nPP (IntnPP) groups were fed diets with 0.40%, 0.35% and 0.30% nPP according to the growth period and were submitted to three dietary treatments: unsupplemented; supplemented with 500 FTU/kg diet of an *Aspergillus niger* phytase (IntnPP_fp) and supplemented with 500 FTU/kg diet of an *Escherichia coli* phytase (IntnPP_bp). The three low nPP groups fed diets contained 0.30%, 0.25% and 0.20% nPP and were submitted to the same dietary treatments than IntnPP to obtain LnPP, LnPP_fp and LnPP_bp groups. IntnPP and LnPP groups had lower body weight gain and feed, crude protein (CP) and metabolizable energy (ME) intake ($p < 0.05$) than the control. Feed conversion ratio of IntnPP was more favorable ($p < 0.01$) than the LnPP group. CP and ME conversion ratios worsened ($p < 0.01$) in IntnPP and LnPP groups in comparison to the control. The nPP conversion ratio improved ($p < 0.01$) from the control to the LnPP group. Fungal phytase reduced ($p < 0.05$) feed, CP, ME and nPP intake than the bacterial one. IntnPP and LnPP diets had a lower digestibility of CP ($p < 0.01$) and CF ($p = 0.01$) than the control. IntnPP and LnPP groups showed a higher ($p < 0.05$) economic efficiency than the control. Blood total protein was the lowest ($p < 0.05$) in the LnPP group, the control group showed the lowest ($p < 0.05$) level of albumin and IntnPP group had the lowest ($p < 0.01$) globulin level. The use of bacterial phytase increased ($p < 0.01$) total protein and globulin and decreased ($p < 0.05$) the plasma cholesterol in comparison to fungal phytase. Decreasing nPP levels in colored slow-growing broilers diet negatively affects growth performance and the use of phytase can partly alleviate these negative effects, but the efficiency of different enzyme sources (bacterial or fungal) was tied to the dietary nPP levels.

Keywords: phytase; Sasso broiler; non-phytic phosphorous; productive performance; blood profiles

1. Introduction

Phosphorus (P) is an essential nutrient for plants and animals and is critically important in the production of poultry. Phosphorus is well represented in the vegetables ingredients commonly used in poultry diet preparation (as soybean or corn), however 50%–80% of total phosphate in plant seeds is stored as phytate P [1]. Phosphorous phytate is poorly available to intestine absorption of monogastric animals and can also reduce the digestibility of other nutrients as well as the performance of animals owing to its antinutritional effect [2,3]. It is possible to add a further amount of non-phytic P to the diet, but this allows an increase of the amount of P released, especially from the waste of intensive poultry farms. It is well known that excess of P in the water led to its eutrophication and the subsequent enrichment of surface water by plant nutrients is a form of pollution [4]. This problem has led to numerous studies to reduce the P amount in the waste from the poultry industry. Several studies indicated great differences in the non-phytate phosphorus (nPP, inorganic) requirement of broilers compared to the data presented in the 1994 by the National Research Council (NRC). Waldroup et al. [5] showed that the nPP requirement for the starter phase of broilers ranges from 0.37% to 0.39%. Angel et al. [6,7] fixed the nPP requirement of the broiler between 0.32% and 0.28% in the growing period (18–32 d), 0.24% and 0.19% in the finisher period (32–42 d) and 0.16% and 0.11% in the withdrawal period (42–49 d). Dhandu and Angel [8] reported a requirement of 0.20% nPP for the finisher period and 0.16% nPP in the withdrawal period of broiler. All these nPP requirements are substantially lower than that reported by the NRC [9]. These data suggested that it is possible to reduce the P concentration in the diet and this can reduce the amount of P in the manure without negative effects on poultry performance. The addition of phytase to the diet can help to release P from phytate and thus reduce the amount of non-phytic P supplemented to the diets. Phytase is the most common enzyme used in the feed for monogastric animals: it can reduce the antinutritional effect of phytate and improve the digestibility of phosphorous, calcium, amino acids and energy, as well as reduce the negative impact of inorganic P excretion to the environment [10]. Phytase are phosphatases hydrolyzing one or more phosphate group. Depending on the position of the phosphate group on the myo-inositol ring, which they firstly hydrolyze, they belong to one of two subclasses, 3-phytase and 6-phytase [11]. Ghazalah et al. [12] and Selim [13] indicated that phytase supplementation to the diet of fast-growing chickens can replace around 0.1% of dietary P; however, the data are inconsistent due to type of diet, dietary composition, type and age of chicks, type and dose of phytase. So, there are many areas of research that need to be addressed such as the effect of type of phytase and the efficacy of bacterial (phytase-6) vs. fungal phytase (phytase-3) on phosphorus utilization, due to the exist contradictory results [6,14–18].

The aim of our research was to study the effect of microbial or fungal phytase on growth performance, nutrient digestibility, carcass and meat quality and blood profile of colored slow-growing Sasso broilers fed diets with different supplementation of non-phytic phosphorous.

2. Materials and Methods

2.1. Chickens, Experimental Design and Husbandry

The department committee of Animal and Poultry Production accepts all the procedures done in the current study. These procedures suggest minimal stress to the animal to ensure rights and welfare by eliminating harm or suffering to animals according to the official decrees of the Ministry of Agriculture in Egypt regarding animal welfare (Decree No. 27 (1967) that enforces the humane treatment of animals generally).

A total of 420 unsexed day-old colored broilers (Sasso strain) were homogeneously divided into 7 groups (5 replicates of 12 chicks each). Chicks were housed in 35 floor pens (1.0 m × 1.0 m/pen) furnished with rice hulls as a litter. Along the experimental period, chicks were fed three different diets, according to the animal age: starter (1–35 d of age); grower (36–56) and finisher diet (57–64). The experimental period was listed from 1 to 64 d of age. The control group fed diets supplemented with 0.50%, 0.45% and 0.40% of non-phytic phosphorus (nPP), respectively for each period. The three groups fed intermediate nPP (IntnPP) diets supplemented with 0.40%, 0.35% and 0.30% nPP in the starter, grower and finisher periods, respectively were submitted to three dietary treatments: unsupplemented, supplemented with 500 FTU of a fungal phytase (IntnPP_fp, BASF Germany, n-3 Aspergillus niger phytase)/kg diet and supplemented with 500 FTU of a bacterial phytase (IntPP_bp, Phyzyme® XP, n-6 *Escherichia coli* phytase, Danisco Animal Nutrition)/kg diet. The three groups fed low nPP (LnPP) diets were supplemented with 0.30%, 0.25% and 0.20% nPP in the starter, grower and finisher periods, respectively and were submitted to the same dietary treatments than IntnPP groups to obtain LnPP, Lnpp_fp and LnPP_bp fed groups.

The experimental diets (Table 1) were formulated according to NRC [9]. Washed building sand was used at small amount (0.14%–0.50%) to keep nutrients profiles of different diets similar as far as possible as, otherwise the change in feedstuffs contents may affect nutrients profiles.

Table 1. Ingredients profiles and nutrient compositions (as the fed basis) of the experimental diets fed during the starter, grower and finisher diets.

Items	Starter			Grower			Finisher		
	0.50 nPP	**0.45 nPP**	**0.35 nPP**	**0.45 nPP**	**0.35 nPP**	**0.25 nPP**	**0.40 nPP**	**0.30 nPP**	**0.20 nPP**
Ingredients									
Yellow corn	58.30	58.35	58.35	62.65	62.65	62.70	63.00	63.00	63.55
Soybean meal	32.00	32.00	32.00	27.50	27.60	27.50	28.25	28.25	28.20
Fish meal 65 % CP	3.00	3.00	3.00	3.00	3.00	3.00	-	-	-
Limestone	1.00	1.35	1.73	0.90	1.25	1.60	0.85	1.20	1.55
Dicalcium Phosphate	1.85	1.25	0.65	1.60	1.00	0.35	1.75	1.10	0.50
Vit + Min premix [1]	0.30	0.30	0.30	0.30	0.30	0.30	0.30	0.30	0.30
NaCl	0.30	0.30	0.30	0.30	0.30	0.30	0.30	0.30	0.30
Methionine	0.15	0.15	0.15	0.15	0.15	0.15	0.15	0.15	0.15
Lysine	0.10	0.10	0.10	0.10	0.10	0.10	0.10	0.10	0.10
Vegetable oils	3.00	3.00	3.00	3.50	3.50	3.50	5.30	5.30	5.30
Washed building sand	0.00	0.20	0.42	0.00	0.15	0.50	0.00	0.30	0.14
Chemical-nutritional characteristics									
Dry matter, % [2]	89.61	89.83	89.67	89.61	89.83	89.67	89.57	89.72	89.66
CP, % [2]	21.03	21.04	21.04	19.49	19.42	19.40	17.80	17.83	17.78
ME, MJ/kg [3]	12.33	12.33	12.33	12.65	12.66	12.65	12.93	12.93	12.97
SAA, % [3]	0.85	0.85	0.85	0.80	0.80	0.80	0.74	0.74	0.74
Lysine, % [3]	1.24	1.24	1.24	1.13	1.13	1.13	1.00	1.00	1.00
Calcium, % [3]	1.00	1.00	1.00	0.90	0.90	0.90	0.80	0.80	0.80
Av. P, %[2]	0.47	0.38	0.26	0.41	0.33	0.23	0.37	0.28	0.17
Ether extract, % [2]	5.47	5.51	5.45	6.21	6.18	6.22	7.63	7.68	7.59
Crude fiber, % [2]	3.47	3.47	3.49	3.31	3.36	3.28	3.33	3.36	3.29
Ash, % [2]	9.24	9.24	9.70	9.30	9.41	9.88	9.18	9.50	9.37
NFE, % [2]	50.4	50.57	49.99	51.3	51.46	50.89	51.63	51.35	51.63

[1] Vit+Min mixture provides per kilogram of the diet: vitamin A (retinyl acetate) 24 mg, vitamin E (dl-α-tocopheryl acetate) 20 mg, menadione 2.3 mg, Vitamin D3 (cholecalciferol) 0.05mg, riboflavin 5.5 mg, calcium pantothenate 12 mg, nicotinic acid 50 mg, choline chloride 600 mg, vitamin B12 10 *g, vitamin B6 3 mg, thiamine 3 mg, folic acid 1 mg, d biotin 0.50 mg. Trace mineral (milligrams per kilogram of diet): Mn 80 Zn 60, Fe 35, Cu 8, Se 0.60. [2] Analyzed values. [3] Calculated values. nPP: non phytic phosphorous; CP: crude protein; ME: metabolizable energy; SAA: sulphate amino-acids; Av. P: available phosphorous; NFE: nitrogen free extract

Feed samples of the starter, grower and finisher diets were chemically analyzed for dry matter, crude protein, ether extract, crude fiber and ash according to the Association of Official Agricultural Chemists (AOAC) [19] official methods. The nitrogen free extract was determined by subtracting the sum of all fractions mentioned above from the dry matter (DM).

Mash form diets and clean water were offered ad libitum. Birds were illuminated with 24 h light cycle during the first 3 days of age and then with 23:1 light-dark cycle, according to Attia et al. [20]. The vaccinations and medical care were conducted according to veterinary indications. Chicks were farmed under similar environmental, managerial, and hygienic condition and breed according to Hendrix Genetic indication for the Sasso strain (available at: https://www.hendrix-genetics.com/en/news/new-brand-promise-sasso/).

2.2. Data Collection

Birds were weighed (g) in the morning before offering feeds at the beginning (day 1) and at the end of the trial (day 64) to calculate body weight gain for each replicate. Each group was provided daily with enough pre-weighed feed of its corresponding diet. The remainder and scattered feed as well as the consumed feed were weekly calculated for each replicate to calculate feed intake in the periods 1–64 days of age. The consumption of CP, ME and nPP was calculated by multiplying the CP, ME and nPP contents of the experimental diets by its corresponding feed consumption during the entire experimental period.

Conversion indexes of feed (feed conversion ratio (FCR), g feed/g gain), crude protein (crude protein conversion ratio, CPCR, g protein/g gain) and ME (metabolizable energy conversion ratio, MECR, MJ/g gain) of the diets were calculated during the experimental period. Utilization of P as the nPP conversion ratio was calculated as g of nPP consumption required to produce 1000 g body weight gain. Mortality rate was calculated as the percentage of birds dead in each treatment during the entire experimental period.

At 64 days of age, 6 broilers per treatment as three replicates of two males each were used to measure the nutrient digestibility of the diets according to the total collection method described by Attia et al. [20]. The broilers were fasted for 24 h, thus fed on their experimental diets for 120 h; feed intake and excreta were measured during 72 h. The excreta, collected for each replicate, sprayed with 4% boric acid to capture the ammonia in the form of ammonium borate. The excreta were cleaned from feathers and feed, weighed and dried in a forced air oven at 70 °C until constant weight. Samples were finely ground and placed in screw-top glass jars until analyses.

Fecal nitrogen was separated from urine in the excreta samples according to Jakobsen et al. [21]. Total nitrogen, fecal nitrogen, fat, crude fiber and ash contents were determined in the excrement and feeds according to AOAC [19] procedures (dry matter, method number 934.01; crude protein, method number 954.01; ether extract, method number 920.39; crude fiber, method number 954.18 and ash, method number 942.05) and expressed on a dry matter basis. Phosphorus concentrations were determined utilizing flame spectrophotometric techniques using the method of Haraguchi and Fuwa [22].

The apparent digestibility of the nutrients (dry matter, crude protein, fat and fiber) and the apparent retention of ash were calculated by dividing the daily amount retained (g/d) by the amount intake (g/d). The daily amount retained is equal to the amount intake (% nutrient in feed × g of feed consumed) minus that lost in the excreta (% nutrient in excreta × g of excreta voided). Metabolic nitrogen was equal to excreta nitrogen − fecal nitrogen.

At 64 d of age, six birds (3 males and 3 females) were randomly collected from each treatment, weighed after fasting overnight, slaughtered, their feathers were plucked and the inedible parts (head, legs and inedible viscera) were taken aside. Thus, the remaining carcass (dressed carcasses) was weighed and expressed as a percentage of the body weight. The internal organs (liver, pancreas and spleen) were weighed and expressed as percentage of live body weight. Abdominal fat (located in the abdominal cavity, surrounding the intestines and around the heart) was separated, weighed and expressed as a percentage of the live body weight.

A sample of 50% breast muscle + 50% thigh muscle was weighed and kept in an electric drying oven at 70 °C until constant weight. The dried meat was ground to pass through a sieve (1 mm^2) and then carefully mixed. The air-dried samples were kept into tight glass container for the subsequent chemical analysis [19]. Protein content of meat was calculated as 100 − moisture − ash − fat. The ability of meat to hold water (WHC), meat tenderness, pH, color intensity and drip loss were measured according to Bovera et al. [23].

The right tibia was removed from the carcass of 6 birds per group, cleaned from tissues, set in hexane for 48 h to remove fat and dry in an oven until constant weight. Their length (mm), width (mm), weight (g) as well as ash, Ca and P contents were determined. Length and width of tibia were determined using a Vernier caliber. The width of tibia was measured at three points in the middle and at the end of both sides. Ca and P contents were determined after ashing at 600 °C, according to the method of AOAC [24].

Blood samples were collected at 64 d of age from 6 broilers per group in heparinized tubes. Plasma was separated by centrifugation at 3000 rpm for 15 minutes and stored at −20 °C until analysis. Concentrations of plasma Ca, inorganic P and alkaline phosphatase were determined according to Attia et al. [25]. All biochemical traits of blood plasma (total protein, albumin, alaninine aminotransferase (ALT), aspartate aminotransferase (AST), total lipids and cholesterol) were determined using commercial diagnosing kits (Diamond Diagnostics Company, Egypt) as reported by Attia et al. [26–28]. Globulin concentration was calculated as the difference total protein − albumin.

Economic evaluation for all experimental diets was made. Economic efficiency was calculated as described by Attia et al. [29].

2.3. Statistical Analysis

Data were analyzed using the GLM procedure of Statistical Analysis Software (SAS) version 6.11 (SAS Institute Inc.: Cary, NC, USA) [30] by one-way design as the phytases was added to two nPP levels only resulted in an unbalanced experimental deign as the recommended phosphorus levels usually not supplemented with phytases. In addition, the use of factorial analyses in this case can generate main effects that were confounded by the interactions effect. The effect of nPP was compared between the unsupplemented levels and phytases within different nPP levels. Mean difference at $p \leq 0.05$ was tested using the Student Newman–Keuls test. The replicate was the experimental unit. Data in percentage were transformed to log 10 before running the analyses of variance.

3. Results

No deaths were observed in the groups along the trial and all the animals were healthy.

Data on the broilers' growth performance and feed efficiency in the entire experimental period (1–64 d) are reported in Table 2. The control and IntnPP_bp groups showed higher ($p < 0.01$) body weight gain than the other groups, except for LnPP_fp. The LnPP group showed also lower ($p < 0.01$) body weight gain (BWG) than LnPP_fp group. Feed, crude protein and energy intakes were lower ($p < 0.01$) in LnPP and LnPP_fp than the other groups. nPP intake was higher ($p < 0.01$) in the control, followed by all the IntnPP and then all the LnPP groups. The feed conversion ratio of LnPP group was higher ($p < 0.01$) than the control, IntnPP_bp and LnPP_fp groups. Crude protein and metabolizable energy conversion ratios of LnPP were higher ($p < 0.01$) than the control, IntnPP_bp and _fp and LnPP_fp. IntnPP_bp and LnPP_fp groups showed lower crude protein and metabolizable energy conversion ratios ($p < 0.01$) than the control and LnPP groups.

Table 2. Growth performance, feed, crude protein, energy intakes and their conversion during days 1-64 of age of Sasso chickens fed diets containing different non-phytate phosphorus levels with or without two types of commercial phytases.

Groups	BWG (g)	FI (g)	CP Intake (g)	ME intake (MJ)	nPP intake (g)	FCR	CPCR	MECR
Control	1769 ^{a}	4025 ^{a}	794 ^{a}	50.78 ^{a}	17.17 ^{a}	2.36 ^{b}	0.449 ^{b}	6.86 ^{b}
IntnPP	1552 ^{bc}	3704 ^{a}	732 ^{a}	46.76 ^{a}	13.16 ^{b}	2.39 ^{ab}	0.472 ^{abc}	7.20 ^{abc}
IntnPP_bp	1668 ^{a}	3706 ^{a}	733 ^{a}	46.83 ^{a}	13.19 ^{b}	2.22 ^{b}	0.439 ^{c}	6.72 ^{c}
IntnPP_fp	1598 ^{bc}	3709 ^{a}	731 ^{a}	46.69 ^{a}	13.10 ^{b}	2.32 ^{ab}	0.457 ^{bc}	6.98 ^{bc}
LnPP	1478 ^{c}	3675 ^{b}	726 ^{b}	46.37 ^{b}	9.37 ^{c}	2.49 ^{a}	0.491 ^{a}	7.50 ^{a}
LnPP_bp	1586 ^{bc}	3738 ^{a}	742 ^{a}	47.35 ^{a}	9.59 ^{c}	2.36 ^{ab}	0.468 ^{abc}	7.14 ^{abc}
LnPP_fp	1646 ^{ab}	3689 ^{b}	725 ^{b}	46.34 ^{b}	9.37 ^{c}	2.24 ^{b}	0.441 ^{c}	6.73 ^{c}
RMSE	204.2	42.7	8.56	2.15	0.376	0.088	0.0173	0.267
p value	0.01	0.01	0.05	0.04	0.0001	0.001	0.001	0.001

^{abc} means within a column under the same treatment with different superscripts are significantly different, NS= not significant, RMSE= root mean square error. BWG = body weight gain; FI= feed intake, CP= crude protein, ME= metabolizable energy, nPP= Non-phytate phosphorus, FCR= feed conversion ratio, CPPR= crude protein conversion ratio, MECR= metabolizable energy conversion ratio; IntnPP: intermediate nPP group; LnPP: low nPP group; bp: bacterial phytase; fp: fungal phytase.

Table 3 shows the effect of dietary treatments on nutrient digestibility, ash retention and fate of nitrogen in broilers. Both IntnPP and LnPP diets had a lower digestibility of CP ($p < 0.01$) and CF ($p = 0.01$) and a lower ($p < 0.01$) ash retention. Excreta and metabolic nitrogen were higher in IntnPP and LnPP diets in comparison to the control. The effect of the treatments revealed that the use of both phytases improved CP digestibility only when supplemented to IntnPP diets. CF digestibility improved due to the addition of both phytases to IntnPP diets, while fungal phytase supplementation only improved CF digestibility in the LnPP groups. The excreta nitrogen in IntnPP groups decreased ($p < 0.01$) due to use bacterial phytase in comparison to the unsupplemented group while within the LnPP groups the bacterial phytase increased the percentage of excreta nitrogen in comparison to fungal phytase. Fecal nitrogen in the IntnPP and LnPP groups was similar to that of the control and lowered ($p < 0.01$) due to the use of both types of phytase.

Table 3. Apparent digestibility, ash retention and excreta and fecal nitrogen of 64-day-old Sasso chickens fed diets containing different non-phytate phosphorus levels with or without two types of commercial phytases.

Groups	Apparent Digestibility, %				Ash Retention, %	Excreta Nitrogen, %	Fecal Nitrogen, %
	DM	CP	CF	EE			
Control	80.6	77.6 ^{a}	30.7 ^{a}	78.6	31.4 ^{a}	4.99 ^{b}	2.53 ^{a}
IntnPP	81.0	75.3 ^{b}	28.8 ^{b}	77.4	30.4 ^{b}	5.32 ^{a}	2.54 ^{a}
IntnPP_bp	80.7	77.6 ^{a}	31.7 ^{a}	77.9	33.5 ^{a}	4.05 ^{b}	2.33 ^{b}
IntnPP_fp	81.2	77.5 ^{a}	31.4 ^{a}	78.7	33.1 ^{a}	5.00 ^{ab}	2.30 ^{b}
LnPP	80.4	75.3 ^{b}	28.6 ^{b}	77.7	30.9 ^{b}	5.29 ^{ab}	2.50 ^{a}
LnPP_bp	82.2	75.5 ^{b}	29.0 ^{ab}	79.6	34.1 ^{a}	5.42 ^{a}	2.33 ^{b}
LnPP_fp	81.8	76.7 ^{ab}	31.6 ^{a}	78.0	33.7 ^{a}	5.04 ^{b}	2.27 ^{b}
RMSE	2.05	1.27	1.69	2.19	1.24	0.854	0.883
p value	NS	0.007	0.01	NS	0.0001	0.001	0.007

^{ab} means within a column under the same treatment with different superscripts are significantly different, NS= not significant, RMSE= root mean square error; IntnPP: intermediate nPP group; LnPP: low nPP group; bp: bacterial phytase; fp: fungal phytase; DM = dry matter; CP = crude protein; CF = crude fiber; EE = ether extract.

Table 4 shows carcass traits and economic indexes of broilers as affected by dietary treatments. LnPP diets had a lower ($p = 0.01$) liver percentage than the other groups. Liver percentage, reduced due to use of fungal phytase in IntnPP groups only.

Table 4. Carcass characteristics of 64-day-old Sasso chickens fed diets containing different non-phytate phosphorus levels with or without two types of commercial phytases.

Groups	Dressing %	Liver %	Pancreas %	Spleen %	Abdominal Fat %	Economic Efficiency %
Control	70.8	2.40 [a]	0.200	0.165	1.93	41.2
IntnPP	69.7	2.47 [a]	0.186	0.144	2.52	38.7
IntnPP_bp	69.9	2.43 [a]	0.240	0.183	2.36	45.5
IntnPP_fp	70.6	2.20 [b]	0.193	0.181	2.41	47.8
LnPP	70.4	2.22 [b]	0.217	0.162	1.66	42.4
LnPP_bp	71.4	2.30 [ab]	0.199	0.132	1.54	47.8
LnPP_fp	70.5	2.14 [b]	0.222	0.154	3.00	46.1
RMSE	14.7	0.43	0.045	0.056	1.17	0.553
p value	NS	0.01	NS	NS	NS	NS

[ab] means within a column under the same treatment with different superscripts are significantly different, NS= not significant, RMSE= root mean square error; IntnPP: intermediate nPP group; LnPP: low nPP group; bp: bacterial phytase; fp: fungal phytase.

Tibia characteristics were presented in Table 5. Reducing nPP level in the diets progressively reduced ($p < 0.05$) tibia weight and phosphorus. LnPP diets had a lower ($p < 0.05$) tibia diameter and ash percentage than the other groups, while both IntnPP and LnPP diets reduced ($p = 0.01$) Ca percentage in tibia in comparison to the control. An effect of the treatments was observed for almost all the parameter in the Table except for tibia length. No differences were observed within IntnPP groups for tibia weight, but within LnnPP groups there is a progressive increase of tibia weight from unsupplemented to bacterial and fungal phytase groups with fungal phytase showed stronger effect. Tibia diameter was decreased due to supplementation of fungal phytase in comparison to the other groups in IntnPP diets while both phytases increased tibia diameter in LnPP diets in comparison to the unsupplemented group. Tibia ash was unaffected by dietary treatments in IntnPP groups while both phytases increased ($p < 0.05$) this criterion in comparison to the unsupplemented group when broilers fed LnPP diets. Regarding tibia calcium, no differences were observed among LnPP groups while in IntnPP groups both phytases increased tibia calcium percentage in comparison to the unsupplemented group. Tibia P was unaffected by dietary treatments in IntnPP diets, while in LnPP diets the addition of bacterial phytase increased P percentage in comparison to the unsupplemented group.

Table 5. Tibia characteristics of 64-day-old Sasso chickens fed diets containing different non-phytate phosphorus levels with or without two types of commercial phytases.

Groups	Tibia Length (mm)	Tibia Weight (g)	Tibia Diameter (mm)	Tibia Ash (%)	Tibia Calcium (%)	Tibia Phosphorus (%)
Control	11.0	7.59 [a]	3.59 [a]	45.1 [a]	37.3 [a]	14.40 [a]
IntnPP	10.8	7.21 [ab]	3.40 [a]	44.8 [a]	34.7 [b]	13.59 [ab]
IntnPP_bp	10.7	7.22 [ab]	3.50 [a]	45.1 [a]	37.1 [a]	15.42 [a]
IntnPP_fp	10.6	7.03 [b]	3.12 [b]	45.0 [a]	37.4 [a]	14.43 [ab]
LnPP	10.6	6.20 [c]	3.18 [b]	43.8 [b]	35.3 [b]	13.15 [b]
LnPP_bp	10.9	7.04 [b]	3.51 [a]	44.8 [a]	36.5 [ab]	14.42 [a]
LnPP_fp	11.0	7.58 [a]	3.52 [a]	45.0 [a]	36.6 [ab]	14.36 [ab]
RMSE	0.739	1.57	0.407	1.01	0.888	1.08
p value	NS	0.04	0.04	0.03	0.01	0.02

[abc] means within a row under the column treatment with different superscripts are significantly different, NS= not significant, RMSE= root mean square error, IntnPP: intermediate nPP group; LnPP: low nPP group; bp: bacterial phytase; fp: fungal phytase.

No effects of dietary treatments were observed for chemical and physical characteristics of meat (Table 6).

Table 6. Chemical composition and physical characteristics of meat of 64-day-old Sasso chickens fed diets containing different non-phytate phosphorus levels with or without two types of commercial phytases.

Groups	Dry Matter %	Protein %	Lipid %	Ash %	pH	Color Intensity	Tenderness $cm^2/0.3$ g	WHC $cm^2/0.3$ g
Control	27.0	75.1	18.9	4.90	6.71	0.270	2.98	5.82
IntnPP	27.3	74.4	19.5	4.80	6.74	0.288	2.91	5.74
IntnPP_bp	26.4	74.4	19.6	4.60	6.75	0.295	2.89	5.70
IntnPP_fp	26.8	75.5	18.6	4.60	6.70	0.257	2.93	5.94
LnPP	26.6	75.2	18.3	4.80	6.69	0.270	2.90	5.61
LnPP_bp	26.5	74.7	18.6	5.30	6.70	0.288	2.98	5.74
LnPP_fp	26.2	74.9	18.9	4.80	6.63	0.275	2.91	5.65
RMSE	0.916	1.14	6.02	0.537	0.132	0.034	0.148	0.323
p value	NS	NS	NS	NS	NS	NS	NS	NS

NS= not significant, RMSE= root mean square error, pH=hydrogen power, WHC= water holding capacity; IntnPP: intermediate nPP group; LnPP: low nPP group; bp: bacterial phytase; fp: fungal phytase.

Table 7 shows the effect of dietary treatments on blood criteria. Total protein was the lowest ($p < 0.05$) in LnnP group, the control group showed the lowest ($p < 0.05$) level of albumin and IntnPP group had the lowest ($p < 0.01$) globulin level in blood. Cholesterol was reduced ($p < 0.05$) in IntnPP than the other two groups; the control group showed the highest ($p < 0.05$) level of phosphorus.

Table 7. Metabolic profiles of 64-day-old Sasso chickens fed diets containing different non-phytate phosphorus levels with or without two types of commercial phytases.

Groups	TP, g/dL	Alb, g/dL	Glo, g/dL	TL, mg/dL	Chol, mg/dL	Calcium mg/dL	Phosphorus mg/dL	ALP U/L	AST U/L	ALT U/L
Control	4.15 ^{a}	1.11 ^{b}	3.00 ^{a}	631	111.8 ^{a}	10.9	6.70 ^{a}	52.6	10.5	5.24
IntnPP	4.05 ^{a}	1.55 ^{a}	1.88 ^{c}	692	83.7 ^{b}	9.8 ^{b}	6.04 ^{b}	49.2	10.5	5.33
IntnPP_bp	4.36 ^{a}	1.43 ^{b}	3.93 ^{a}	694	97.8 ^{b}	12.3 ^{a}	6.68 ^{a}	54.1	10.4	5.57
IntnPP_fp	4.29 ^{a}	1.57 ^{a}	2.73 ^{b}	634	105.0 ^{ab}	12.4 ^{a}	6.63 ^{a}	50.5	10.7	5.37
LnPP	3.31 ^{b}	1.67 ^{a}	2.38 ^{b}	691	117.2 ^{a}	10.0 ^{b}	6.10 ^{b}	50.9	10.7	5.34
LnPP_bp	4.50 ^{a}	1.69 ^{a}	2.84 ^{b}	689	87.2 ^{b}	12.2 ^{a}	6.30 ^{ab}	51.5	11.0	5.57
LnPP_fp	3.86 ^{ab}	1.48 ^{b}	2.36 ^{b}	687	119.1 ^{a}	12.0 ^{a}	6.64 ^{a}	50.0	10.3	5.04
RMSE	0.771	0.133	0.775	53.2	17.84	4.76	0.167	3.74	3.99	0.675
p value	0.04	0.01	0.04	NS	0.01	0.03	0.01	NS	NS	NS

^{ab} means within a column under the same treatment with different superscripts are significantly different, NS= not significant, RMSE= root mean square error, TP: total protein, Alb: albumin; Glob: globulin; TL: total lipids; Chol: cholesterol; ALP=alkaline phosphatase; AST= asparate amino transferase; ALT= alainine amino transferase; IntnPP: intermediate nPP group; LnPP: low nPP group; bp: bacterial phytase; fp: fungal phytase.

The use of bacterial phytase increased the total protein in comparison to the unsupplemented group when LnPP albumin was detected due to the use of bacterial phytase in IntnPP diets and of fungal phytase in LnPP diets. Bacterial phytase was able to decrease plasma cholesterol in comparison to the other groups when a broiler was fed LnPP diets while no differences were observed among IntnPP groups. Both phytases increased ($p < 0.05$) the phosphorus level in comparison to the unsupplemented group when broilers were fed IntnPP diets while with LnPP diets the addition of fungal phytase increased phosphorus level in comparison to the unsupplemented group.

4. Discussion

The decrease of broiler performance due to the reduction of nPP levels in the diets without enzyme supplementation, confirmed the inability of birds to utilize the phytic sources of P, and also confirmed that 0.50, 0.45 and 0.40 nPP levels in the starter, grower and finisher diet (without phytase), respectively were adequate to maintain growth performance.

The worsening of broiler growth performance as weight gain, feed and nutrient intakes are in line with findings by other authors [31–33]. According to Karimi et al. [34] the FCR worsened when nPP levels were under 30 mg/kg considering the average value of the three experimental diets. However, the yield of protein and energy was decreased already at the intermediate level of nPP. This result can be ascribed to the combination of low protein and energy intake and a low digestibility of crude protein and fiber found at the intermediate and low nPP levels. Our results disagree with the findings of Karimi et al. [34] who concluded that feeding 0.40% followed by 0.35 % nPP diets during starter and grower periods respectively results in sufficient feed intake of broilers, with such levels actually being in excess of that required for maintaining FCR or mortality rate comparable to the controls. In fact, in our study the reduction of feed intake at the intermediate level of nnP is important and tied to a severe reduction of body weight gain of broilers (−12.27% and −16.50% than the control, respectively for IntnPP and LnPP groups). This discrepancy is probably tied to the age of animals used in our trial. In fact, as a slow-growth genetic type (colored broilers), the slaughter ages of our broilers was 64 days more than double than the slaughter age of broilers used by Karimi et al. [34] in their trial. It is well known that phosphorus requirements are tied to animal age and probably the amounts of nnP at the intermediate and low levels are not sufficient to allow a sufficient storage of the phosphorus in the broilers body, specifically in the bones, with negative effects on animal growth. This is confirmed by a reduction of circulating phosphorus, tibia weight and tibia Ca and P contents observed in our trial with both IntnPP and LnPP diets. In this regard, we must consider that bone density and strength were positively correlated to calcium and phosphorus content in the tibia [35] and that bone characteristics are important in broilers also as indicators of animal health and welfare. The balance between Ca and P in poultry diets is of the utmost importance, particularly in starter diets; Ca: P should be approximately 2:1 [36,37]. In our trial the Ca:P ratio based on calculated and analyzed values of phosphorus in the starter diets was 2:1 vs. 2: 0.94 in the control and 2:0.8 vs. 2:0.76 and 2:0.6 vs. 2:0.54, respectively for IntnPP and LnPP groups.

The plasma protein showed differences between the different fraction among nPP levels but, very interesting, there is also a difference in albumin to globulin ratio, which is the lowest (0.37) for the control group, while IntnPP and LnPP groups showed similar values (0.82 and 0.70, respectively): high globulin levels and low albumin/globulin ratios indicates better disease resistance and immune response of birds [38].

The results indicate that the effect of type of phytase was correlated with nPP concentrations showing that the differences between the two types of phytase were maximized when the amount of nPP increase in the broiler diet. It is well established that bacterial phytase have a higher efficacy than fungal phytase due to its resistance to low pH an digestive protease [10], so when the amount of nPP in the diet decreases, the bacteria phytase would show a higher efficiency in broiler performance improvement. However, the opportunity to observe improvements in phytase activity is influenced by the level of nPP also in another way: as the nPP level in the diet decreases the efficiency of its utilization increases. In our trial, the phosphorus retention reaches high values and thus the potential to show an improvement in efficiency activity of phytase is greatly reduced. This agrees to other studies in broilers [39,40] and in turkeys [41], which showed that digested P per unit of phytase decreases as the ratio available or non-phytate P/phytase decreases.

This double mechanism of action could be responsible of the results obtained in our trial. As an example, feed, protein and metabolizable energy intakes were not different among unsupplemented, bacterial or fungal phytase groups when nPP level was intermediate, while at the low nPP level bacterial phytase increase all the three criteria when compared to unsupplemented or fungal phytase groups.

The decrease of CP digestibility due to lower nPP availability is in line with the findings of Xue et al. [42], who observed that, like to pigs, also in poultry there is a linear interrelationship between N and P digestion, quantified by the authors as 10:1. This relationship between protein and P could be linked in various aspects. For example, both muscle and bone developments occur concurrently in growing animals, and both require an adequate amount of AA and P; but also a protein deficiency may affect gut development and therefore the absorption and utilization of P in the small intestine [42]. In general, the researches on this topic focused on the effect of diet protein level modification on P absorption in the developing animals, but as there is a linear correlation, it is also possible that deficiencies of P can affect protein digestion, as shown in our research. A further confirmation is that the use of phytase has a positive impact on CP digestion. As known, phytase has a positive effect on protein digestibility as phytate can bind an amount of free amino acids, such as lysine [43]. In our trial, fungal phytase at the low nPP level was more effective than bacterial phytase on protein availability for metabolic purposes. The increase of feed and nutrient intake in broiler fed bacterial phytase at low nPP levels could be tied to low bioavailability. In this regard, the differences in CP digestibility among low nPP groups were not significant but the group supplemented with fungal phytase had a CP digestibility not different from the groups supplemented with both phytases in broiler fed intermediate nPP levels. This trend to decrease nutrient digestibility as nPP decreases, with beneficial effects due to the use of phytase was observed also for crude fiber and could be ascribed to a higher efficiency of fungal than bacterial phytase. This point needs further study to correlate the phytase activity to different metabolic condition of broilers. On the other hand, our results found a further confirmation in the improvement of nutrient conversion indexes in the fungal phytase low nPP group when compared to the unsupplemented one. However, when low-nPP diets were fed to the broiler, the addition of both phytase had no effects on nPP intake or the conversion ratio.

It is not easy to understand the decrease of the blood cholesterol level in the IntnPP unsupplemented group, but it is very interesting to observe that in broilers that were fed an LnPP diet supplemented with bacterial phytase, the levels of blood cholesterol decreased in comparison to the control and the other LnPP group, suggesting that bacterial phytase had a greater potential than fungal phytase in lowering cholesterol. Several studies [44–46] reported that dietary phytate decreased the concentrations of serum total and low-density lipoprotein (LDL) cholesterol in mice or rats. Jariwalla et al. [47] suggested that phytate decreased serum cholesterol by affecting the zinc: copper ratio, as phytate is a strong chelator of divalent cations. Perhaps more likely, the ratio of these cations to each other may affect cholesterol absorption or metabolism. Thus, the addition of phytase can act increasing the cholesterol levels in blood as shown by Liu et al. [48].

The fungal phytase also showed a higher activity than bacterial when low-nPP diets were fed to the broiler as the levels of P in plasma were not different between the two type of phytases at both intermediate and low-nPP diets but, with low-nPP levels fungal phytase was able to increase plasma P levels in comparison to the unsupplemented group. However, at the same nPP levels in the diet, the percentage of P in tibia was not different between phytase levels but bacterial phytase had a higher percentage than the control.

5. Conclusions

Decreasing non phytic phosphorous levels in the colored slow-growing broilers diet negatively affects the growth performance and the digestibility of some nutrients (crude protein and fiber). The use of phytase can partly alleviate these negative effects but the efficiency of different enzyme sources (bacterial or fungal) was tied to the nPP levels in the diet. When the amount of non-phytic phosphorous was decreased, the efficiency of P utilization increased and thus in some cases was difficult to appreciate the differences reported in the literature on enzyme efficiency due to the different source and, in other cases, the fungal phytase seemed to show a higher efficiency in comparison to the bacterial phytase. Further investigations need to understand well the interaction between non phytic phosphorus in the broiler diet and efficiency of bacterial or fungal phytase.

Author Contributions: Conceptualization, Y.A.A. and F.B.; methodology, A.E.M.; data collection, and formal analysis, M.A.A.-H. and A.A.A.; ; statistical anaylses, A.A.A.; investigation, H.S.Z.; resources, F.I.; writing—original draft preparation, Y.A.A., M.A.A.-H., and F.B.; writing—review and editing, F.I.; visualization, Y.A.A.; supervision All authors have read and agreed to the published version of the manuscript.

Acknowledgments: In this section you can acknowledge any support given which is not covered by the author contribution or funding sections. This may include administrative and technical support, or donations in kind (e.g., materials used for experiments).

References

1. Kumar, V.; Sinha, A.H.; Makkar, H.P.S.; De Boeck, G.; Becker, K. Phytate and phytase in fish nutrition. *J. Anim. Physiol. Anim. Nutr.* **2012**, *96*, 335–364. [CrossRef]
2. Woyengo, T.A.; Nyachoti, C.M. Review: Anti-nutritional effects of phytic acid in diets for pigs and poultry: Current knowledge and directions for future research. *Can. J. Anim. Sci.* **2013**, *93*, 9–21. [CrossRef]
3. Morgan, N.K.; Walk, C.L.; Bedford, M.R.; Scholey, D.V.; Burton, E.J. Effect of feeding broilers diets differing in susceptible phytate content. *Anim. Nutr.* **2016**, *2*, 33–39. [CrossRef]
4. Withers, P.J.; Davidson, I.A.; Foy, R.H. The contribution of agricultural phosphorus to eutrophication. In *The Fertilizer Society*; Proc. No. 365; Greenhill House: Thorpe Wood, Petersborough, UK, 1995.
5. Waldroup, P.W.; Kersey, J.H.; Saleh, E.A.; Fritts, C.A.; Yan, F.; Stilborn, H.L.; Crum, R.C., Jr.; Raboy, V. Nonphytate phosphorus requirement and phosphorus excretion of broiler chicks fed diets composed of normal or high available phosphate corn with and without microbial phytase. *Poult. Sci.* **2000**, *79*, 1451–1459. [CrossRef]
6. Angel, R.; Applegate, T.J.; Christman, M. Effect of dietary non-phytate phosphorous (nPP) on performance and bone measurements in broilers fed on a four-phase feeding system. *Poult. Sci.* **2000**, *79*, 21–22.
7. Angel, R.; Applegate, T.J.; Christman, M.; Mitchell, A.D. Effect of dietary non-phytate phosphorous (nPP) on broiler performance and bone measurements in the starter and grower phase. *Poult. Sci.* **2000**, *79* (Suppl. 1), 22.
8. Dhandu, A.S.; Angel, R. Broiler non-phytinphosphorus requirement in the finisher and withdrawal phases of four phases feeding program. *Poult. Sci.* **2003**, *82*, 1257–1265. [CrossRef]
9. NRC National Research Council. *Nutrient Requirements of Poultry*, 9th ed.; National Academy Press: Washington, DC, USA, 1994.
10. Dersjant-Li, Y.; Awati, A.; Schulze, H.; Partridge, G. Phytase in non-ruminant animal nutrition: A critical review on phytase activities in the gastrointestinal tract and influencing factors. *J. Sci. Food Agric.* **2015**, *30*, 878–896. [CrossRef]
11. Żyła, K.; Mika, M.; Stodolak, B.; Wikiera, A.; Koreleski, J.; Świątkiewicz, S. Towards complete dephosphorylation and total conversion of phytases in poultry feeds. *Poult. Sci.* **2004**, *83*, 1175–1186. [CrossRef]
12. Ghazalah, A.A.; Abd EI-Gawad, A.H.; Soliman, M.S.; Youssef, A.W. Effect of enzyme preparation on performance of broilers fed corn-soybean meal based diets. *Egypt. Poult. Sci. J.* **2005**, *25*, 295–316.
13. Selim, W. Effect of Two Sources of Microbial Phytase on Productive Performance Meat Quality and Dietary Protein and Energy Utilization by Sasso Chickens. Master's Thesis, Faculty of Agriculture-Damanhour, Alexandria University, Alexandria, Egypt, 2007.
14. Valaja, J.; Perttila, S.; Partanen, K.; Kiiskinen, T. Effect of two microbial phytase preparations on phosphorus utilisation in broilers fed maize-soybean meal based diets. *Agric. Food Sci.* **2001**, *10*, 19–26. [CrossRef]
15. Camden, B.J.; Morel, P.C.H.; Thomas, D.V.; Ravindran, V.; Bedford, M.R. Effectiveness of exogenous microbial phytase in improving the bioavailabilities of phosphorus and other nutrients in maize-soybean diets for broilers. *Anim. Sci.* **2001**, *73*, 289–297. [CrossRef]
16. Augspurger, N.R.; Webel, D.M.; Lei, X.G.; Baker, D.H. Efficacy of an E. coli phytase expressed in yeast for releasing phytate-bound phosphorus in young chicks and pigs. *J. Anim. Sci.* **2003**, *81*, 474–483. [CrossRef]
17. Roberson, K.D.; Kalbfleisch, J.L.; Pan, W.; Charbeneau, R.A. Effect of corn distiller's dried grains with solubles at various levels on performance of laying hens and egg yolk color. *Int. J. Poult. Sci.* **2005**, *4*, 44–51.
18. Attia, Y.A.; Addeo, N.F.; Al-Hamid, A.A.; Bovera, F. Effects of phytase supplementation to diets with or without zinc addition on growth performance and zinc utilization of white pekin ducks. *Animals* **2019**, *9*, 280. [CrossRef]
19. Association of Official Analytical Chemists, AOAC. *Official Methods of Analysis of the Association of Official Analytical Chemists*, 18th ed.; AOAC: Washington, DC, USA, 2004.

20. Attia, Y.A.; El-Tahawy, W.S.; Abd El-Hamid, A.E.; Hassan, S.S.; Nizza, A.; El-Kelaway, M.I. Effect of phytase with or without multienzyme supplementation on performance and nutrient digestibility of young broiler chicks fed mash or crumble diets. *Ital. J. Anim. Sci.* **2012**, *11*, 303–308. [CrossRef]
21. Jakobsen, P.E.; Gertov, K.; Nilsen, S.H. Frdjelighed frogmed fierbrae. "Digestibility trails with poultry". Bereting fra for sogslabortoriet. *Kabenhaven* **1960**, *56*, 1–34.
22. Haraguchi, H.; Fuwa, K. Determination of phosphorus by molecular absorption flame spectrometry using the phosphorus monoxide band. *Anal. Chem.* **1976**, *48*, 784–786. [CrossRef]
23. Bovera, F.; Lestingi, A.; Iannaccone, F.; Tateo, A.; Nizza, A. Use of dietary mannanoligosaccharides during rabbit fattening period: Effects on growth performance, feed nutrient digestibility, carcass traits, and meat quality. *J. Anim. Sci.* **2012**, *90*, 3858–3866. [CrossRef]
24. Association of Analytical Communities (AOAC) International. *Official Methods of Analysis*, 15th ed.; AOAC International: Washington, DC, USA, 1990.
25. Attia, Y.A.; Qota, M.A.; Bovera, F.; Tag El-Din, A.E.; Mansour, S.A. Effect of amount and source of manganese and/or phytase supplementation on productive and reproductive performance and some physiological traits of dual-purpose cross-bred hens in the tropics. *Br. Poult. Sci.* **2010**, *51*, 235–245. [CrossRef]
26. Attia, Y.A.; Abd El-Hamid, A.E.; Bovera, F.; El-Sayed, M.I.M. Reproductive performance of rabbit does submitted to an oral glucose supplementation. *Animal* **2009**, *3*, 1401–1407. [CrossRef] [PubMed]
27. Attia, Y.A.; Al-Hanoun, A.; Bovera, F. Effect of different levels of bee pollen on performance and blood profile of New Zealand White bucks and growth performance of their offspring during summer and winter months. *J. Anim. Physiol. Anim. Nutr.* **2011**, *95*, 17–26. [CrossRef]
28. Attia, Y.A.; Al-Hanoun, A.; Tag El-Din, A.E.; Bovera, F.; Shewika, E. Effect of bee pollen levels on productive, reproductive and blood traits of NZW rabbits. *J. Anim. Physiol. Anim. Nutr.* **2011**, *95*, 294–303. [CrossRef]
29. Attia, Y.A.; Hamed, R.S.; Abd El-Hamid, A.E.; Al-Harthi, M.A.; Shahba, H.A.; Bovera, F. Performance, blood profile, carcass and meat traits and tissue morphology in growing rabbits fed mannanoligosaccharides and zinc-bacitracin continuously or intermittently. *Anim. Sci. Pap. Rep.* **2015**, *33*, 85–101.
30. SAS Institute. *SAS/STAT User's Guide Statistics*; SAS Institute Inc.: Cary, NC, USA, 2002.
31. Singh, P.K.; Khatta, V.K.; Thakur, R.S.; Dey, S.; Sangwan, M.L. Effects of Phytase Supplementation on the Performance of Broiler Chickens Fed Maize and Wheat Based Diets with Different Levels of Non-phytate Phosphorus. *Asian-Aust. J. Anim. Sci.* **2003**, *16*, 1642–1649. [CrossRef]
32. Karimi, A. Responses of broiler chicks to non-phytate phosphorus levels and phytase supplementation. *Int. J. Poult. Sci.* **2006**, *5*, 251–254.
33. Wilkinson, S.J.; Bedford, M.R.; Cowieson, A.J. Effect of dietary nonphytate phosphorus and calcium concentration on calcium appetite of broiler chicks. *Poult. Sci.* **2014**, *93*, 1695–1703. [CrossRef]
34. Karimi, A.; Bedford, M.R.; Sadeghi, G.; Ghobadi, Z. Influence of Dietary Non-phytate Phosphorous Levels and Phytase Supplementation on the Performance and Bone Characteristics of Broilers. *Braz. J. Poult. Sci.* **2011**, *1*, 43–51. [CrossRef]
35. Barreiro, F.R.; Sagula, A.L.; Junqueira, O.M.; Pereira, G.T.; Baraldi-Artoni, S.M. Densitometric and biochemical values of broiler tibias at different ages. *Poult. Sci.* **2009**, *88*, 2644–2648. [CrossRef]
36. Whitehead, C.C. Nutrition and poultry welfare. *World's Poult. Sci. J.* **2001**, *58*, 349–356. [CrossRef]
37. Fleming, R.H. Nutritional factors affecting poultry bone health. *Proc. Nutr. Soc.* **2008**, *67*, 177–183. [CrossRef]
38. Bovera, F.; Loponte, R.; Pero, M.E.; Cutrignelli, M.I.; Calabrò, S.; Musco, N.; Vassalotti, G.; Panettieri, V.; Lombardi, P.; Piccolo, G.; et al. Laying performance, blood profiles, nutrient digestibility and inner organs traits of hens fed an insect meal from Hermetia illucens larvae. *Res. Vet. Sci.* **2018**, *120*, 86–93. [CrossRef]
39. Dembow, D.M.; Ravindran, V.; Kornegay, T. Improving phosphorus availability in soybean meal for broilers by supplemental phytase. *Poult. Sci.* **1995**, *74*, 1831–1842. [CrossRef]
40. Kornegay, E.; Denbow, D.; Ravindran, V. Responses of broilers to graded levels of Natuphos phytase added to corn soybean meal based diets containing three levels of non-phytate phosphorus. *Br. J. Nutr.* **1996**, *75*, 839–852. [CrossRef]
41. Ravindran, V.; Bryden, W.; Kornegay, E. Phytates: Occurrence, bioavailability and implications in poultry nutrition. *Poult. Avian Biol. Rev.* **1995**, *6*, 125–143.
42. Xue, P.C.; Ajuwon, K.M.; Adeola, O. Phosphorus and nitrogen utilization responses of broiler chickens to dietary crude protein and phosphorus levels. *Poult. Sci.* **2016**, *95*, 2615–2623. [CrossRef]

43. Ravindran, V.; Selle, P.H.; Bryden, W.L. Effect of phytase supplementation individually and in combination with glycanase on the nutritive value of wheat and barley. *Poult. Sci.* **1999**, *78*, 1588–1595. [CrossRef]
44. Lee, D.H.; Choi, S.U.; Hwang, Y.H. Culture Conditions and Characterisations of a new phytase-producing fungal isolate, Aspergillus sp. L117. *Mycobiology* **2005**, *33*, 223–229. [CrossRef]
45. Lee, J.; Yunjaie, C.; Peter-Changwhan, L.; Seungha, K.; Jinduck, B.; Jaiesoon, C. Recombinant production of Penicillium oxalicum PJ3 phytase in Pichia pastoris. *World J. Microbiol. Biotechnol.* **2007**, *23*, 443–446. [CrossRef]
46. Szkudelski, T. Phytic acid-induced metabolic changes in the rat. *J. Anim. Physiol. Anim. Nutr.* **2005**, *89*, 397–402. [CrossRef]
47. Jarivalla, R.J.; Sabin, R.; Lawson, S.; Herman, Z.S. Lowering of serum cholesterol and tryglicerides and modulation of divalent cations by dietary phytase. *J. Appl. Nutr.* **1990**, *42*, 18–28.
48. Liu, N.; Ru, Y.J.; Li, F.D. Effect of dietary phytate and phytase on metabolic change of blood and intestinal mucosa in chickens. *J. Anim. Physiol. Anim. Nutr.* **2010**, *94*, 368–374. [CrossRef]

16

Dietary Supplementation of Lysophospholipids Affects Feed Digestion in Lambs

Qin Huo [1], Bo Li [1], Long Cheng [2], Tingting Wu [1], Peihua You [3], Shuanghua Shen [4], Yiyong Li [3], Yuhua He [1], Wannian Tian [1], Rongquan Li [1], Changsheng Li [1], Jianping Li [1], Baijun Song [1], Chunqing Wang [1] and Xuezhao Sun [1,*]

1 The Innovation Centre of Ruminant Precision Nutrition and Smart and Ecological Farming, Jilin Agricultural Science and Technology University, Jilin 132109, China; huoqin0605@163.com (Q.H.); ma_lee@163.com (B.L.); ttw528@126.com (T.W.); yhyttt0235@sina.com (Y.H.); wannian2000@hotmail.com (W.T.); 16643429184@sohu.com (R.L.); csli-64@163.com (C.L.); sxljp2004@126.com (J.L.); jlsbj2005@126.com (B.S.); chunqingw@sina.com (C.W.)

2 Faculty of Veterinary and Agricultural Sciences, Dookie Campus, The University of Melbourne, Victoria 3647, Australia; long.cheng@unimelb.edu.au

3 Portal Agri-Industries Co., Ltd., Nanjing 211803, China; ph-u@163.com (P.Y.); liyiyongfeed@163.com (Y.L.)

4 Floor 15, Building B, 650 Xinzhuang Avenue, Xinqiao Town, Songjiang District, Shanghai 201612, China; sshqz@sina.com

* Correspondence: xuezhaos@hotmail.com.

Simple Summary: Previous works showed that supplementation of lysophospholipid as a feed additive improves animal growth and milk yield in beef and dairy cattle production. However, its effects on fattening lambs have not been reported before. In this study, we fed fattening lambs a diet with no or 0.5 g lysophospholipid in a kilogram of diet. We found that lysophospholipid did not or slightly improved the growth of fattening lambs. Feed digestibility, ruminal fermentation parameters and rumen bacterial community were altered, which may be associated with decreased fiber digestion. However, lipase concentration in serum was decreased, which might enhance fat deposition in muscle and thus may increase meat quality. Effects of lysophospholipid on sheep observed in this study are different from those on cattle, which warrants further study.

Abstract: Five experiments were conducted to examine effects of lysophospholipids (LPL) on live weight gain, nutrient digestibility, ruminal fermentation parameters, serum biochemical parameters and rumen bacterial community profile in fattening lambs. Two dietary treatments (pelleted complete feed supplemented without (control diet; CON) or with 0.05% LPL on dry matter basis) were tested in these experiments. Feed and water were provided *ad libitum* to lambs. The results showed that average daily gain (ADG) tended to increase or was not affected by LPL supplementation. Compared with CON, the supplementation of LPL resulted in an increase in dry matter, crude protein and organic matter digestibilities, and a decrease in neutral detergent fiber and acid detergent fiber digestibilities. Ruminal pH values did not change with LPL supplementation, but the concentrations of ammonia and total short chain fatty acids (SCFAs) were increased. The molar proportion of major individual SCFAs and the ratio of acetate to propionate were not affected by LPL supplementation. While the activity of lipase was decreased with LPL supplementation, all other serum biochemical parameters did not change. Rumen bacterial community was altered by LPL supplementation with the relative abundance of fibrolytic bacteria in the total bacterial population, such as *Prevotella*, decreased. In conclusion, LPL supplementation can alter feed digestion, but may not result in consistent positive responses in animal growth performance.

Keywords: lysophospholipid; growth performance; digestibility; rumen function; blood metabolite; sheep; pelleted complete feed

1. Introduction

Pelleted complete feed has been increasingly applied to sheep farming in recent years [1,2]. The indoor sheep production system based on pelleted complete feed provides us with more opportunities than the grazing system, as it allows better manipulation of rumen fermentation, as well as increased feed efficiency and animal growth performance. A wide range of feed additives has been studied and applied for rumen fermentation manipulation in sheep production [3–6]. In addition to directly providing nutrients, such as rumen-protected lysine, the supplementation of microbial additives and antibiotics has been the major nutritional approach to altering rumen function and animal performance [7,8]. However, microbial additives are normally unstable in their quality and consequently in the response of animal performance [7]. Antibiotics, such as monesin, have been banned in some countries due to concerns over the formation of antibiotic-resistant microbial mutants [8,9]. This leads animal scientists to keep seeking new feed additives to increase animal production.

Lysophospholipids (LPL) are produced from the enzymatic hydrolysis of phospholipid with a chain of fatty acids removed, mainly containing lysophosphatidylcholine, lysophosphatidic acid, lysophosphatidylethanolamine, and lysophosphatidylinositol [10]. The LPL can selectively inhibit the growth of Gram-positive but not Gram-negative bacteria [11]. They also could emulsify dietary fat [12] and upregulate genes in the intestinal epithelium [13] to increase fat absorption. This group of compounds has been used as feed additives in pig and poultry for improved feed efficiency and increased growth rate [14–17].

To our best knowledge, limited study has been conducted with LPL supplementation to sheep, though a few trials were performed on dairy cows [18–20] and beef cattle [21]. Lee et al. [19] found that adding 0.05−0.075% LPL (dry matter (DM) basis) to the diet of lactating dairy cows can increase milk production. Our preliminary trials showed that adding 0.5 mg LPL/kg DM to the diet for lactating cows increased milk yield from 28.0 to 29.5 L/day/cow. The positive animal performance might result from the emulsifying effect of LPL as emulsifying agents could increase enzymatic activities in the rumen [22] and also from the selective inhibition of Gram-positive bacteria, such as *Roseburia* spp. in the rumen [11].

The hybrid of thin-tailed sheep and Northeast fine-wool sheep was the most commonly farmed breed in Northeast China [23]. This breed has a relatively high growth performance and is well accommodated to the local environment.

We hypothesized that the addition of LPL to pelleted formulated complete feed for fattening lambs may result in more efficient feed digestion and rumen fermentation towards more propionate and less acetate formation, which consequently may lead to more nutrients available for growth. This study aimed to investigate the effect of the addition of LPL to the pelleted diet of fattening lambs on animal growth performance, blood parameters, feed digestibility, ruminal fermentation parameters and rumen bacterial community.

2. Materials and Methods

The use of animals and all animal handling and management in this study were approved in advance by the Committee of Animal Ethics and Welfare of Jilin Agricultural Science and Technology University (Approval number 2018001).

2.1. Experimental Design and Animals

The study included two dietary treatments: a control diet (CON) and a control diet supplemented with 0.5 g/kg DM of LPL. It was conducted in five experiments at three experimental sites. Experiment

1 (Exp 1) was for the determination of growth performance, digestibility, blood parameters, ruminal fermentation parameters and rumen bacterial community, while the other experiments (Exps 2–5) were for the determination of growth performance only.

Healthy hybrids of thin-tailed sheep and Northeast fine-wool sheep with similar liveweight were used for the study. All sheep were negative in the rose bengal plate test for brucellosis (*Bruceila* spp.) [24].The slow-growing sheep with daily liveweight gain less than 100 g/day during the adaptation period were excluded from the study. Sheep were stratified by liveweights measured at the end of the adaptation period and randomly allocated to one of the two dietary treatments. During the formal experimental period, sick lambs or pregnant female lambs were excluded from the study.

Exp 1 was conducted at the Animal Experimental Station of Jilin Agricultural Science and Technology University, Jilin city, Jilin province, China. Eighteen three-month-old ram lambs with a liveweight of 24.7 ± 1.9 (mean ± SD) kg were adapted for 10 days and then 14 of them with the best growth performance were randomly allocated to one of two groups with eight lambs each. After three more days for further adaptation, the formal experimental period started and lasted for 68 days (from 19 August to 26 October 2018). Sheep were fed in a half-opened feedlot and transferred to individual metabolic cages on day 15 (3 September 2018) for digestibility measurements. They were adapted to the cage housing conditions for 4 days and then the digestibility sample collection period lasted for 7 days. After the digestibility trial, animals were transferred back to the feedlot. Rumen samples were taken on days 29 and 68 (17 September and 26 October 2018) and blood samples were taken on day 42 (30 September 2018).

Exp 2 had the same experimental design as the Exp 1 and was conducted at the same site as the Exp 1, but the ram lambs were heavier and older than those in the Exp 1. The initial liveweight was at 43.8 ± 4.0 kg and aged at five months. After two weeks of isolation for the detection of brucellosis, 16 lambs were allocated to one of the two dietary treatments with seven each, housed in a pen for each treatment, and adapted to their assigned diets. The formal experimental period lasted for 28 d (from 27 September to 25 October 2018).

Exp 3 was conducted at Jiaogeermiao sheep farm in Tongyu county, Jilin province, China. Thirty six ewe lambs with a liveweight of 25.0 ± 3.1 kg were adapted to pelleted complete feed for 10 days from 11 to 21 August 2018 and then assigned to one of the two treatment groups. The formal experimental period lasted for 29 days and ended on 19 September 2018.

Exp 4 and Exp 5 were conducted at the Portal Chifeng Experimental Station, Inner Mongolia, China. Exp 4 had ram lambs with a liveweight of 29.9 ± 5.4 kg and in Exp 5 ewe lambs had a liveweight of 27.0 ± 3.5 kg. Lambs were adapted to pelleted complete feed for 14 days and then were weighed and grouped by weight. Lambs from each treatment were fed in two pens with the equal number of animals. The formal experimental period lasted for 62 days from 7 November 2018 to 8 January 2019. Animals were weighed on days 0, 24 and 62. At the end of the experiment, all lambs were slaughtered and hot carcasses were weighed.

2.2. Diets and Feeding Management

The experimental diets were formulated according to the Chinese Feeding Standard for Lamb Finishing (NY/T 816-2004). The ingredients and nutrient contents of the diets are listed in Table 1. Sorghum husk was ground using a 2 mm sieve and maize, maize germ meal, sunflower seed meal, peanut shell, rice hull, cottonseed meal, and barley malt root using a 4/6 mm sieve (half sieve with holes having a diameter of 4 mm and another half 6 mm).

Table 1. Ingredients and nutrient contents of experimental diets supplemented without (CON) or with (LPL) lysophospholipids.

Item	Diet	
	CON	LPL
Ingredient (kg/t of fresh weight)		
Maize	350	350
Maize germ meal	120	120
Sunflower seed meal	120	120
Peanut shell	113	113
Rice hull	70	70
Cottonseed meal	30	30
Bentonite	20	20
Barley malt root	100	100
Limestone	14	14
Sorghum husk	10	10
Calcium hydrogen phosphate	7	7
Soybean powder	20	20
Sodium chloride	6	6
Premix (Trace mineral salt and vitamins) [1]	20	20
Lysophospholipids		0.5
Nutrient contents [2] (g/kg of DM)		
Dry matter (DM) (g/kg of fresh weight)	880	874
Organic matter (OM)	903	905
Crude protein (CP)	158	162
Neutral detergent fiber (NDF)	427	435
Acid detergent fiber (ADF)	174	178
Ether extract (EE)	16	16
Metabolizable energy (MJ/kg of DM) [3]	11.7	11.8

[1] Premix per kg contained 24,000 IU vitamin A, 4800 IU vitamin E, 120 mg Fe, and 24 mg Cu; [2] The nutrient contents were measured values. [3] Metabolizable energy was estimated from NRC (2007) [25].

All ingredients were thoroughly mixed, conditioned at 90 °C for 45 s and pelletized at 85 °C using a pelleting machine with the compression ratio of ring to die at 1:7 (model YPM508E; Jiangsu Yongli Machinery Co., Ltd., Liyang City, Jiangsu Province, China). The pellets were air-cooled. Feed pellets were 5 mm in diameter and 8–10 mm in length. All feed used in the whole period of the experiments was pelleted in one batch at the Chifeng Subsidiary Company of Jiangsu Portal Agri-Industries Co., Ltd.

During the adaptation period, feed provided to lambs was gradually changed from hay to pellets by an increase of 100 g of pellets a day. Pellets provided in the first week contained an antiparasitic agent (triclabendazole, 250 mg/kg feed) and then experimental pellets were provided. Feed was provided twice a day with equal portions at 8:00 and 17:00. Once animals were completely on pelleted feed, feed allowance was set to allow leftover at ca. 10% of total feed provided on the previous day to achieve feeding *ad libitum*. Leftovers were cleaned up and weighed before morning feeding for intake calculation. Drinking water was available all the time. Weather, temperature and humidity were recorded daily, and animal behaviors were observed.

2.3. Liveweight Measurement and Sampling

Liveweight was measured before morning feeding after 16 h fasting with an accuracy of 0.05 kg using an electronic scale for living bodies (YongHuang, Jinhua, Zhejiang, China) at the beginning and the end of the formal experimental period, and in the middle of the experiment if the experimental period was longer than one month. Average daily gain (ADG) was calculated as the slope of liveweight against date.

Rumen sample (50 mL) was taken from each lamb in the Exp 1 via mouth using stomach tubing (1500 mm in length, 8 mm in inner diameter, 6 openings with a diameter of 2 mm in the end to the rumen, made with polyurethane material) at 0 and 3 h after morning feeding. Ruminal pH value was measured immediately after sampling using a pH meter with the accuracy of 0.01 (LICHEN pH-100A, Shanghai Lichen Scientific Laboratory Instrument Ltd., Shanghai, China). The rumen sample was placed in a 2-mL cryogenic vial with the circular bottom (Corning Inc., New York, USA) and then stored at −20 °C for the analysis of short chain fatty acid (SCFA) and ammonia concentrations. The remaining samples (3 h after morning feeding samples only) were stored at −80 °C for the characterization of rumen microbial profile.

Blood sample (5 mL) was collected into a coagulation promoting tube with separating gel (Sanli Industrial Co., Ltd., Huizhou, China) from the jugular vein of each lamb in the Exp 1 before morning feeding.

2.4. Digestibility Trial

Digestibility measurement was conducted in metabolic cages using the total feces collection technique. The digestibility trial lasted for 10 days, including a 3-day adaptation and 7-day collection periods. Feed allowance was adjusted to allow 5−10% refused and feed provided at 8:00 and 15:00 with equal portions. Feed sample (50 g) was collected daily. Refusals, feces and urine were collected and quantified daily. Refusals were all kept and feces and urine subsampled. One fifth of collected feces was stored at −20 °C, with half acidified with 10% H_2SO_4 at a ratio of 1:10 for crude protein (CP) determination and another half for the determination of other nutrients. Total urine was collected in a tank with 100 mL of 10% H_2SO_4 added and diluted to 3000 mL, from which 30 mL was taken and kept at −20 °C. At the end of the experiment, samples were pooled over individual animals. Feed, refusal and feces samples were dried at 65 °C for 48 h and stored in cold for later analysis. Data from one sheep in the LPL group were discarded due to health problem of the sheep.

2.5. Laboratory Analysis

Feed samples were analyzed for DM, organic matter (OM), CP, neutral detergent fiber (NDF), acid detergent fiber (ADF), ether extract (EE), Ca and P, refusal and feces samples for DM, OM, CP, NDF and ADF, and urine samples for CP. Crude protein was determined in the modified method of Kjeldahl (method GB/T 6432-2018) using an automatic nitrogen determination apparatus (model K9860; Haineng Instruments Ltd., Jinan, Shandong, China). Dry matter was determined by the loss of water during drying at 105 °C for 3 h (method GB/T 6435-2014). Ash was determined by samples heated until no smoking and then in a muffle at 550 °C for 4 h (GB/T 6438-2007). Ether extract was determined as dry matter loss after extraction in petroleum ether for 24 h at 30–60 °C (GB/T 6433-2006). NDF and ADF were consecutively determined after boiling in 3% neutral detergent for 1 h (GB/T 20806-2006) and 2% acid detergent for 1 h (NY/T 1459-2007), respectively.

A cryogenic vial with a rumen sample was thawed in tap water and centrifuged at 2000× *g* for 10 min. The supernatant was transferred to a 5 mL centrifugation tube, mixed with 25% metaphosphoric acid at a 4:1 ratio, vortexed and centrifuged at 4 °C in 8000× *g* for 10 min. The supernatant from this centrifugation was filtered by a 0.22 μm filter for water solution (Shanghai Rebus Biotechnology Co., Ltd., Shanghai, China) and transferred to a new 5 mL centrifugal tube for the determination of SCFA and ammonia concentrations.

Short chain fatty acids were identified and quantified using a gas chromatograph system (GC9790, Fuli Instruments Ltd., Wenling, Zhenjiang, China) fitted with a flame ionisation detector (FID). The GC was fitted with a FFAP 30 m × 3 mm × 0.25 μm capillary polar column. The column temperature was set at 140 °C, the FID set at 250 °C and vaporization chamber temperature at 250 °C with N_2 as a carrier gas and shunt ratio at 1:50. The injection volume was 1 μL.

The above filtered rumen sample was diluted with double evaporated water in a ratio of 1:7 and vortexed in a 5 mL centrifugal tube. Ammonia concentration was determined using the Indigo phenol blue-spectrophotometry method [26] modified by Feng and Gao [27]. Briefly, the sample solution was mixed with a color developer 37 °C for 20 min, cooled down with tap water to room temperature, and measured for absorbance at 637 nm.

The collected blood sample was centrifuged at 1000× *g* for 5 min (Model TDL-80-2B; Anting Scientific Instrument Factory, Shanghai, China) and the serum analyzed for blood biochemical parameters using an automatic biochemical analyzer (Model 7160; Hitachi Ltd., Tokyo, Japan). The reagents for the analysis were purchased from Mairui Biomedical Electronics Co., Ltd. (Shenzhen, China).

Rumen bacterial populations were analyzed at Shanghai Biozeron Biotechnology Co., Ltd. (Shanghai, China). Total genome DNA was extracted using the CTAB/SDS method, monitored for DNA concentration and purity on 1% agarose gels and then diluted to 1 ng/μL. The 16S rRNA genes were amplified using hypervariable V3-V4 region PCR primers (341F: 5′- CCTAYGGGRBGCASCAG-3′; 806R: 5′- GGACTACNNGGGTATCTAAT-3′) with the barcode. All PCR reactions were carried out in 30 μL reactions with 15 μL of Phusion®High-Fidelity PCR Master Mix (New England Biolabs, Ipswich, MA, USA), 0.2 μM of forward and reverse primers and about 10 ng template DNA. Thermal cycling consisted of initial denaturation at 98 °C for 1 min, followed by 30 cycles of denaturation at 98 °C for 10 s, annealing at 50 °C for 30 s, elongation at 72 °C for 60 s and finalization at 72 °C for 5 min. After amplification, PCR products were mixed with the same volume of SYBR green (QIAGEN Inc., Alameda, CA, USA) containing buffer and then operated electrophoresis on 2% agarose gel. Samples with bright main strip between 400–450 bp were purified with GeneJET Gel Extraction Kit (Thermo Fisher Scientific (China) Ltd., Shanghai, China). Sequencing libraries were generated using NEB Next®Ultra™DNA Library Prep Kit for Illumina (New England Biolabs, Ipswich, MA, USA) following the manufacturer's recommendations. The library quality was assessed on the Qubit@ 2.0 Fluorometer (Thermo Fisher Scientific (China) Ltd., Shanghai, China) and Agilent Bioanalyzer 2100 system (Agilent Technologies, Inc., Santa Clara, CA, USA). At last, the library was sequenced on an Illumina MiSeq platform (Illumina, Inc., San Diego, CA, USA) and 250 bp/300 bp paired-end reads were generated. Paired-end reads from the original DNA fragments were merged using FLASH [28], which was designed to merge paired-end reads when at least some of the reads overlap the read generated from the opposite end of the same DNA fragment. Sequences analysis were performed by UPARSE software package using the UPARSE-OTU and UPARSE-OTU ref algorithms [29]. In-house Perl scripts were used to analyze alpha (within samples) and beta (among samples) diversity. Sequences with ≥97% similarity were assigned to the same OTUs and the RDP classifier was used to annotate taxonomic information for the representative sequence in each OTU according to the database Silva (www.arb-silva.de) updated for the taxonomic assignment of ruminal bacteria [30]. Details in bioinformatics analysis are described in Appendix A.

2.6. Statistical Analysis

The statistical analysis was conducted using GenStat 19th edition (VSN International, Hemel Hempstead, UK, 2017) and significance was declared at $p < 0.05$. The data of liveweight, ADG, carcass weight, dressing percentage, feed intake, digestibility and blood biochemical parameters were analyzed with a one-way ANOVA. Rumen fermentation parameters were analyzed using a mixed model (REML) with repeated measurements. Treatment, sampling time and the interaction of treatment and sampling time were defined to have fixed effects in the model, sampling date, and animal as random effects. Parameters obtained from each lamb at different sampling times were treated as repeated measures. A meta-analysis was conducted for ADG with experimental site using block and dietary treatment, sex, and their interaction as fixed effects. As the interaction was not significant, it was omitted from the model for analysis. For the data of ruminal bacterial community, only when average relative abundance of a species in any treatment group was over 0.5%, the species and its corresponding genus and phylum were statistically analyzed with a one-way ANOVA.

3. Results

3.1. Growth Performance

In Exp 1, the initial liveweight was similar for the two treatments ($p = 0.826$), but after 68 days the final liveweight tended to be higher for the LPL treatment than for the control ($p = 0.084$; Table 2). The LPL treatment was 12% higher in ADG than the control, however, the difference did not reach the level of significance ($p = 0.165$). In Exp 2, the supplementation of LPL tended to increase ADG ($p = 0.099$). In other experiments, a significant change in ADG with LPL supplementation was not observed ($p \geq 0.601$). The meta-analysis with all ADG data obtained in this study showed that the supplementation of LPL ($p = 0.471$) or the interaction of LPL and animal sex ($p = 0.400$) did not significantly change ADG, although male lambs (274 g/day) had higher ($p < 0.001$) ADG than female lambs (200 g/day).

Table 2. Growth performance of fattening lambs fed experimental diets supplemented without (CON) or with (LPL) lysophospholipids.

Experiment No.	Site	Sex	Item	Diet		*p* Value
				CON	LPL	
Exp 1	Jilin	Male	n	7	6	
			Days	68	68	
			Initial liveweight (kg)	24.6 ± 0.56	24.8 ± 0.58	0.826
			Final liveweight (kg)	44.6 ± 1.12	46.9 ± 0.86	0.084
			Average daily gain (ADG) (g/day)	294 ± 17.3	330 ± 8.1	0.165
Exp 2	Jilin	Male	n	8	8	
			Days	28	28	
			Initial liveweight (kg)	43.6 ± 1.34	44.7 ± 1.38	0.568
			Final liveweight (kg)	48.9 ± 1.65	52.4 ± 1.74	0.158
			Average daily gain (ADG) (g/day)	198 ± 38.9	282 ± 24.3	0.099
Exp 3	Tongyu	Female	n	18	17	
			Days	29	29	
			Initial liveweight (kg)	24.9 ± 0.75	25.0 ± 0.77	0.972
			Final liveweight (kg)	30.9 ± 0.93	30.7 ± 0.96	0.851
			Average daily gain (ADG) (g/day)	207 ± 13.2	197 ± 13.5	0.601
Exp 4	Chefeng	Male	n	23	26	
			Days	62	62	
			Initial liveweight (kg)	30.4 ± 1.17	31.6 ± 0.76	0.378
			Final liveweight (kg)	47.5 ± 1.56	48.5 ± 0.92	0.569
			Average daily gain (ADG) (g/day)	276 ± 12.3	275 ± 9.6	0.982
			Carcass (kg)	20.3 ± 0.64	20.2 ± 0.47	0.815
			Dressing percentage (%)	42.9 ± 0.38	41.5 ± 0.38	0.012
Exp 5	Chefeng	Female	Number of animals	18	27	
			Days	62	62	
			Initial liveweight (kg)	28.7 ± 0.76	29.8 ± 0.65	0.286
			Final liveweight (kg)	41.0 ± 1.14	42.4 ± 0.90	0.339
			Average daily gain (ADG) (g/day)	198 ± 13.7	203 ± 8.5	0.764
			Carcass (kg)	17.4 ± 0.48	18.1 ± 0.43	0.276
			Dressing percentage (%)	42.5 ± 0.38	42.8 ± 0.34	0.621

At the end of Exp 3 and Exp 4, all lambs were slaughtered. The supplementation of LPL did not significantly ($p \geq 0.276$) affect hot carcass weights for either male or female lambs. Dressing percentage was similar for the two treatments in female lambs. However, male lambs had a lower dressing percentage ($p = 0.012$) by 1.4 unit when the diet was supplemented with LPL than with the control.

3.2. Total Tract Nutrient Apparent Digestibility

During the period of digestibility trial, dry matter intake and other nutrient intakes were similar for the treatment and control animals ($p \geq 0.363$; Table 3). The supplementation of LPL slightly (by 2%), but statistically significantly ($p < 0.05$) increased DM and OM digestibilities. The increase in CP digestibility resulted from the supplementation of LPL was 4.3% ($p = 0.024$). However, NDF and ADF digestibilities had both statistically ($p = 0.012$ for NDF and $p = 0.001$ for ADF, respectively) and biologically (by −8% for NDF and −35% for ADF, respectively) significant decreases. The digestibility of EE did not significantly ($p = 0.263$) change with the supplementation of LPL.

Table 3. Intake and total tract apparent nutrient digestibility of fattening lambs offered a diet supplemented without (CON) or with (LPL) lysophospholipids.

Index	Diet		*p* Value
	CON (n = 7)	LPL (n = 6)	
Intake (g/day)			
Dry matter intake (DMI)	1328 ± 53.2	1382 ± 97.3	0.620
Organic matter intake (OMI)	1199 ± 48.1	1256 ± 88.3	0.566
Crude protein intake (CP intake)	207 ± 8.4	224 ± 15.7	0.363
Neutral detergent fiber intake (NDF intake)	562 ± 23.1	560 ± 46.9	0.969
Acid detergent fiber intake (ADF intake)	233 ± 9.3	240 ± 18.7	0.730
Ether extract intake (EE intake)	21.2 ± 1.83	22.2 ± 2.00	0.710
Digestibility (g/kg of DM)			
Dry matter (DM)	617 ± 3.3	629 ± 4.4	0.047
Organic matter (OM)	655 ± 3.1	669 ± 4.9	0.039
Crude protein (CP)	694 ± 7.0	723 ± 9.3	0.024
Neutral detergent fiber (NDF)	503 ± 8.3	463 ± 10.3	0.012
Acid detergent fiber (ADF)	285 ± 13.6	186 ± 17.3	0.001
Ether extract (EE)	941 ± 10.8	928 ± 1.3	0.263

3.3. Rumen Fermentation Parameters

Although ruminal pH value was higher before morning feeding than 3 h after morning feeding ($p < 0.001$), the supplementation of LPL did not significantly affect ruminal pH value ($p = 0.235$; Table 4). Both LPL supplementation ($p = 0.011$) and sampling time ($p < 0.001$) significantly affected ammonia-N concentration in the rumen and there was no significant interaction between them. The supplementation of LPL resulted in 37% higher ammonia-N concentration compared with the control. Supplementation, sampling time and their interaction significantly affected or tended to affect total SCFA concentration ($p \leq 0.083$). Although the total SCFA concentration at 3 h after morning feeding was similar for both LPL supplemented and unsupplemented lambs, the supplementation of LPL increased the total SCFA concentration before morning feeding by 61% ($p < 0.05$). In term of the molar proportion of individual SCFAs and the ratio of acetate to propionate, LPL supplementation and its interaction with sampling time did not have significant effects ($p \geq 0.172$) although sampling time did significantly affect the molar proportions of acetate and propionate and their ratio ($p \leq 0.026$).

Table 4. Ruminal fermentation parameters of fattening lambs (n = 7) fed experimental diets supplemented with (LPL) or without (CON) lysophospholipids.

Index	Diet				p Value		
	LPL (n = 7)		CON (n = 7)				
	Before [1]	After [1]	Before	After	Treatment (T)	Sampling Time (S)	T × S
Ruminal pH	6.75 ± 0.127 [a2]	5.94 ± 0.159 [b]	7.01 ± 0.048 [a]	5.98 ± 0.137 [b]	0.235	<0.001	0.372
Ammonia N (mmol/L)	8.92 ± 1.450 [b]	13.13 ± 1.328 [a]	6.92 ± 0.721 [b]	9.19 ± 0.911 [b]	0.011	0.006	0.384
Total SCFA [3] (mmol/L)	44.4 ± 6.93 [b]	65.9 ± 4.97 [a]	27.5 ± 2.28 [c]	66.0 ± 3.86 [a]	0.052	<0.001	0.083
Individual SCFA molar proportion (mol/100 mol)							
Acetate	60.5 ± 0.93 [a]	58.0 ± 1.47 [ab]	61.6 ± 1.86 [a]	54.9 ± 1.35 [b]	0.627	0.003	0.172
Propionate	22.3 ± 2.35 [b]	25.6 ± 2.35 [ab]	26.0 ± 2.63 [a]	31.8 ± 2.40 [a]	0.194	0.026	0.674
Butyrate	17.2 ± 1.88	16.4 ± 2.09	12.4 ± 1.37	13.3 ± 1.86	0.151	0.930	0.340
Acetate: Propionate	2.96 ± 0.322 [a]	2.50 ± 0.262 [ab]	2.77 ± 0.359 [a]	1.93 ± 0.216 [b]	0.443	0.011	0.447

[1] Before, rumen samples taken before morning feeding; After, rumen samples taken 3 h after morning feeding; [2] a, b, c: Different letters on the shoulder of values in a row mean a significant difference ($p < 0.05$); [3] SCFA, short chain fatty acid.

3.4. Blood Metabolites

Among serum biochemical parameters measured in this study (Table 5), only lipase activity was significantly ($p = 0.034$) altered by the supplementation of LPL and the activity of the enzyme decreased by 71% compared with the control.

Table 5. Serum biochemical parameters of fattening lambs fed experimental diets supplemented without (CON) or with (LPL) lysophospholipids.

Item	Diet		Normal Range	p Value
	CON (n = 7)	LPL (n = 7)		
α-Amylase (AMYL) (U/L)	21.7 ± 4.54	26.8 ± 4.50	1–30	0.439
Albumin (ALB) (g/L)	27.5 ± 1.31	27.2 ± 1.31	24–37	0.854
Alkaline phosphatase (ALKP) (U/L)	214 ± 18.0	186 ± 14.0	50–228	0.249
Alkaline transaminase (ALT) (U/L)	6.93 ± 0.913	8.84 ± 1.503	5–17	0.299
Aspartate transaminase (AST) (U/L)	54.4 ± 5.71	62.9 ± 8.01	40–96	0.403
Low density lipoprotein cholesterol (LDL) (mmol/L)	0.573 ± 0.0464	0.497 ± 0.0785	–	0.423
High density lipoprotein cholesterol (HDL) (mmol/L)	0.471 ± 0.0522	0.400 ± 0.0724	–	0.439
Total cholesterol (TC) (mmol/L)	1.29 ± 0.156	1.22 ± 0.095	1.14–2.12	0.725
Urea N (UREA) (mg/dL)	9.13 ± 0.861	8.66 ± 1.180	–	0.753
Creatinine (CREA) (μmol/L)	88.0 ± 18.87	53.5 ± 14.44	53–133	0.186
Triglyceride (TG) (mmol/L)	0.223 ± 0.0231	0.210 ± 0.0225	0.10–0.34	0.700
Total protein (TP) (g/L)	75.8 ± 1.60	74.5 ± 1.58	56–78	0.565
Glucose (GLU) (mmol/L)	4.10 ± 0.210	4.19 ± 0.238	2.78–4.45	0.778
Lipase (LIP) (U/L)	61.4 ± 16.29	17.8 ± 3.24	1–71	0.034
Globulin (GLOB) (g/L)	24.2 ± 1.01	25.6 ± 2.38	32–41	0.585

3.5. Rumen Bacterial Community

The supplementation of LPL altered the composition of bacterial community in the rumen (Figure 1). At the phylum level, the dominant phyla were *Bacteroidetes*, *Firmicutes*, and *Proteobacteria* in the rumen of lambs fed a diet containing either LPL or no LPL, accounting for 95.4% of total bacterial relative abundance (Figure S1). The relative abundance of these three phyla were 51.5%, 41.0%, and 2.8% in LPL treatment, 58.3%, 20.0%, and 17.1%, respectively. *Firmicutes* (CON 20.0% vs. LPL 41.0%;

$p = 0.055$) and *Fibrobacteres* (CON 1.87% vs. LPL 0.36%; $p = 0.011$) were significantly different between the two treatments (Figure S1; Table S1). On the average the relative abundance of *Proteobacteria* dropped over 83.5% when LPL was supplemented, although the difference was insignificant due to a large variation within treatment groups. At the genus level (Figure 2), *Christensenellaceae* ($p = 0.012$), *Lachnospiraceae* ($p = 0.090$), and *Saccharofermentans* ($p = 0.058$) were increased and *Fibrobacter* ($p = 0.011$), *Prevotella* 7 ($p = 0.037$), *Prevotella* 9 ($p = 0.031$), unclassified *Prevotellaceae* ($p = 0.011$), and *Syntrophococcus* ($p = 0.050$) were decreased with the supplementation of LPL (Supplementary Table S1). At the species level, the *Christensenellaceae* R-7 group increased from 0.6% to 2.4% ($p = 0.012$), *Erysipelotrichaceae* from 0.4% to 2.2% ($p = 0.078$), *Lachnospiraceae* from 2.7% to 11.9% ($p = 0.089$), and uncultured *Oribacterium* from 0% to 1.5% ($p = 0.088$) when LPL was supplemented, while *Fibrobacter* was decreased from 1.8% to 0.3% ($p = 0.013$) and uncultured *Syntrophococcus* from 0.5% to 0.1% ($p = 0.036$). The supplementation of LPL reduced the relative abundance of Gram-positive bacteria *Roseburia* spp. 0.11% to 0.02%.

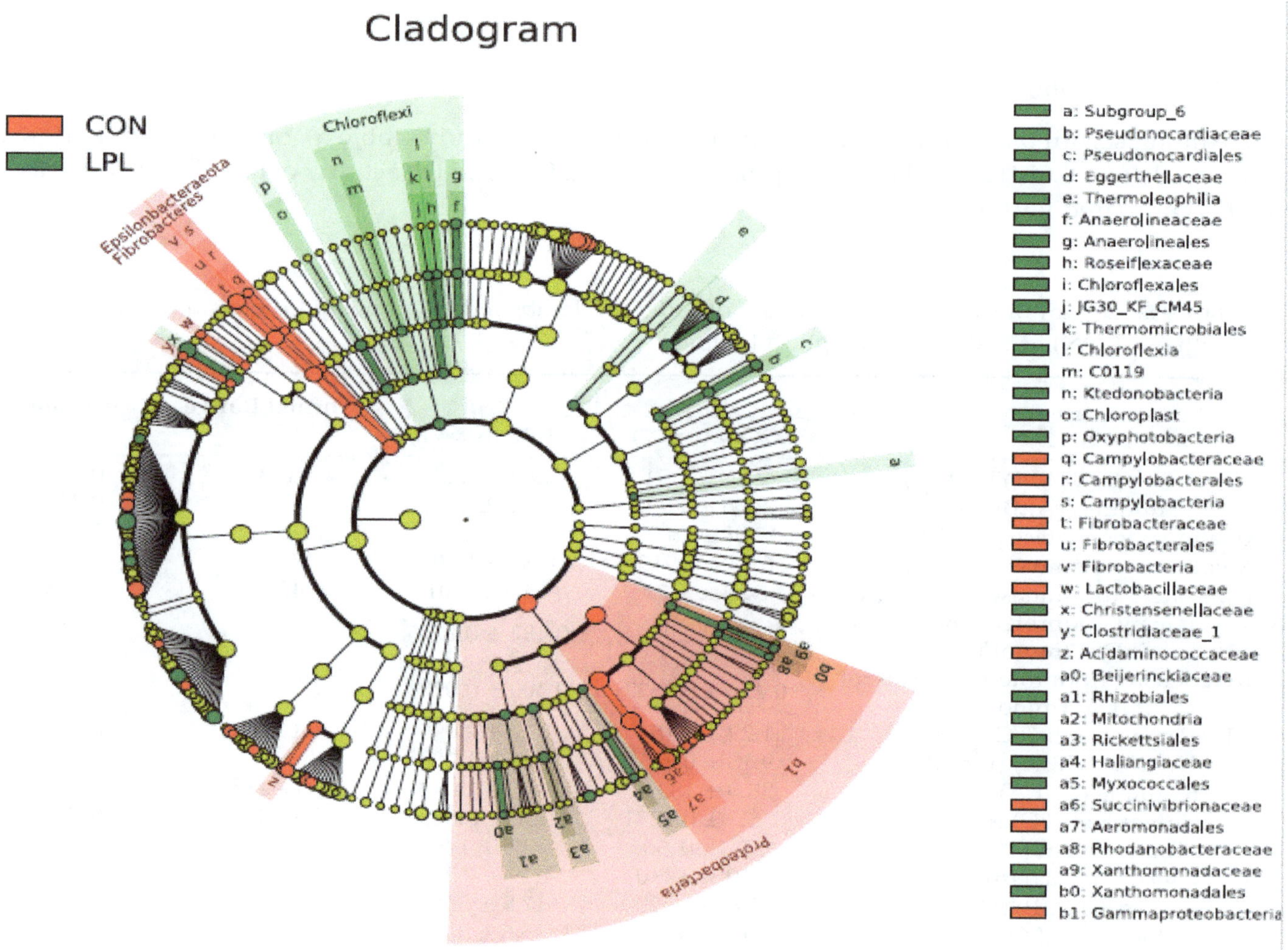

Figure 1. LEfSe cladogram of the composition of bacterial community in the rumen fattening lambs fed a diet supplemented without (CON) or with (LPL) 0.05% lysophospholipid. Differences are represented in the color of the group, where taxa are most abundant. Red: taxa abundant in CON; Green: taxa abundant in LPL.

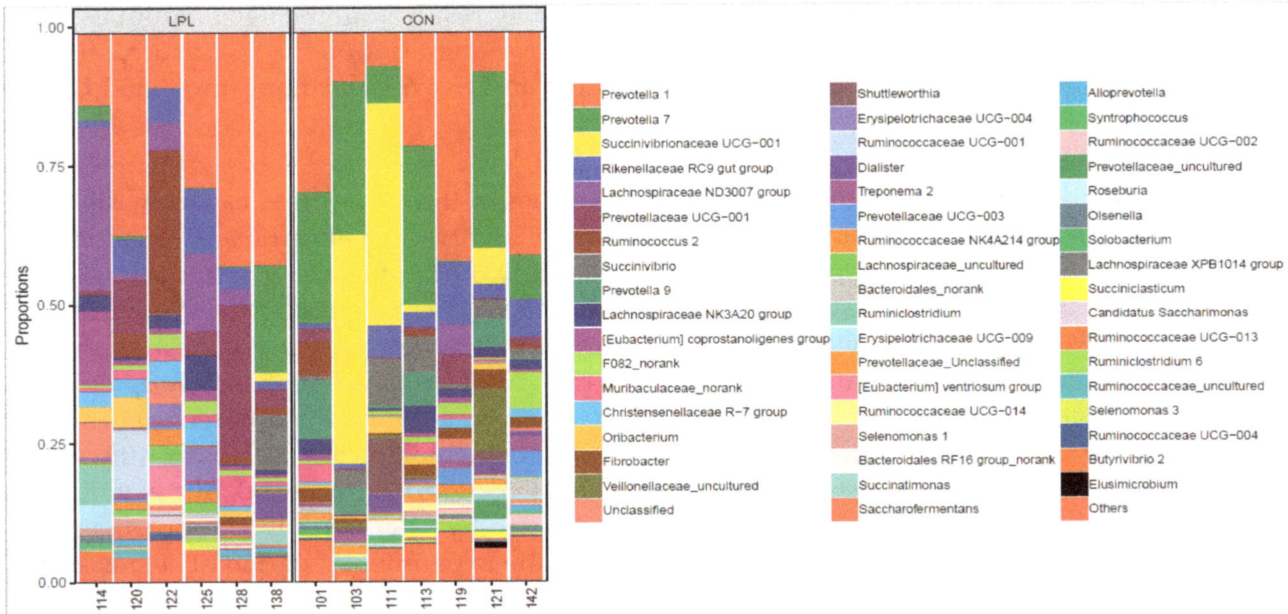

Figure 2. Distribution of the most dominant genera in the rumen of fattening lambs fed a diet supplemented without (CON) or with (LPL) 0.05% lysophospholipid (those with a relative abundance of less than 1% were combined as other).

4. Discussion

4.1. Animal Performance

In this study, the response of lambs to the supplementation of LPL at 0.05% of dietary DM was inconsistent. The ADG was increased in two experiments, but the difference did not reach the significance level (Exp 1) or just reached the tendency level (Exp 2). In the other three experiments, ADG did not change with the supplementation of LPL. There are no reported studies with LPL supplemented to sheep diet in the literature for us to directly compare with. In other ruminant species, the response of animal production to LPL supplementation was inconsistent as well. Song et al. [21] found that the supplementation of LPL at 0.3–0.5% of dietary DM to the diet of beef cattle did not improve growth performance. However, our preliminary trial showed that eleven 15.5-month-old Hanwoo beef cattle supplemented with LPL at 0.1% of the diet for 3 months had an average weight gain of 123.4 kg, being higher than the average weight gain of 117.5 kg from the control animals. Rico et al. [18] supplemented LPL at 10 g/day (approximately 0.035% of dietary DM) to lactating dairy cows for 10 days and did not observe an increase of milk yield. In contrast, a short-term supplementation of LPL at 0.05–0.075% of dietary DM did increase milk yield by 5.5%–6.5% [19].

The inconsistency in animal production in ruminants is contradictory to the studies on non-ruminants, in which LPL consistently increased growth and lactation performance and feed efficiency [14–16]. The inconsistent results among different ruminant studies might result from the variation in source of LPL (Sontakke et al. [20] vs. Lee et al. [19]), the dose of LPL (this study vs. Song et al. [21]), the duration of supplementation (e.g., this study vs. Lee et al. [19]), active compounds in LPL, the degradation of LPL in the rumen, etc. The degree of the removal of fatty acid chain from phospholipid during enzymatic hydrolysis affects the bioactivity of LPL [31]. The studies conducted previously might have used LPL with lower activity than the recent studies, which is expected to have different responses. It is well known that feed is subjected to degradation in the rumen before further digestion in a way similar to nonruminants. Phospholipids, from which LPL is produced, can be degraded in the rumen, but a considerable amount of LPL was able to bypass the rumen [32]. The degree of degradation may vary with the source of LPL and rumen environment which is moderated by diets and other factors [33]. We are not sure how much LPL can escape from the rumen. The discrepancy in growth response among experiments within this study may result from the age of animals, seasons and environment which all might affect rumen condition for degradation. In the future, studies are needed to determine the degradation of LPL in the rumen and to explore rumen-protected LPL.

4.2. Digestibility

To the best of our knowledge, there are only two studies in the literature measuring digestibility when LPL is supplemented to a diet for ruminants (Song et al. [21] in beef cattle and Lee et al. [19] in dairy cows). We found that LPL supplementation increased DM digestibility, which is consistent with the findings of Song et al. [21] who supplemented LPL to beef cattle diet at 0.3%–0.5% of dietary DM, but is contrasted with the results obtained from lactating dairy cows where DM digestibility slightly decreased when LPL was supplemented at 0.05−0.075% in the diet [19]. Although LPL supplementation resulted in an increase (this study) or a decrease [19] in DM digestibility, the difference between LPL and CON was only 12−14 g/kg. This small difference may not be biologically significant. However, in the study by Song et al. [21], the difference was small, only 8−17 g/kg on day 30, but increased with time. After 90 days, the difference in DM digestibility increased by 64 g/kg for LPL supplementation at 0.5% of the diet. Our digestibility trial started on day 15 of the experimental period. The measurement should be taken after LPL is supplemented for a longer time as well for better understanding in the effect of LPL on digestibility.

Song et al. [21] measured DM digestibility only. Lee et al. [19] is only a reference for comparison in nutrient digestibility. In our study, OM digestibility was increased, but decreased in the study conducted by Lee et al. [19]. Being similar to DM digestibility, the difference between LPL and CON was small as well and might be not relevant biologically at least for a short-term after LPL supplementation.

Our result showed that CP digestibility was increased, while it did not change in the study by Lee et al. [19]. The increase in CP digestibility was consistent with increased ruminal ammonia-N concentration 3 h after morning feeding with LPL supplementation. How LPL affect CP digestion is not known and warrants further study.

NDF digestibility decreased by around 40 g/kg DM in both Lee et al. [19] and our study, although the decrease was not statistically significant in the study by Lee et al. [19]. Lee et al. [19] did not measure ADF digestibility. Our result showed that LPL supplementation resulted in a large decrease in ADF digestibility by 99 g/kg DM. The great decrease in fiber digestibility is conflicted with our hypothesis proposed from the finding that fiber degradation is improved by emulsifying agents due to increased cellulolytic enzyme activity [34]. Further study is warranted for the mechanisms underlying the inhibition of fiber degradation by LPL.

4.3. Rumen Fermentation Parameters

Our study was contrasted with Lee et al. [19]'s findings on rumen fermentation parameters although ruminal pH value was not affected by LPL in both studies. We found ammonia and total SCFA concentrations were increased with LPL, but Lee et al. [19] did not. There was a lack of difference in the molar proportion of major individual SCFAs and the ratio of acetate to propionate in both studies. The reason for different effects of LPL on rumen fermentation is not known, but the chemical composition of the diet, especially EE content, was greatly different between the two studies, which might alter LPL reaction in the rumen. Our results did support the finding that LPL enhances feed digestion, but did not support the hypothesis that LPL results in rumen fermentation towards more propionate and less acetate formation that we proposed before.

4.4. Blood Biochemical Parameters

The major change observed in blood biochemical parameters was a decrease in lipase activity associated with the supplementation of LPL. Rico et al. [18] found that milk fat concentration was increased 5 days after dairy cows were supplemented to a higher fiber and lower unsaturated fatty acid diet. These authors attributed the increase in milk fat concentration to increased acetate supply. We suspect their findings might be associated with the decrease in lipase activity. How LPL affects lipase activity warrants further mechanistic investigation. We suspect LPL that escaped from the rumen is absorbed in the small intestines and stimulates the body to depress the expression of genes encoding lipase. From another point of view, the decrease in lipase activity might enhance fat deposition in muscle, which could be useful for the production of quality meat.

Most blood biochemical parameters tested did not alter with the supplementation of LPL, suggesting the status of energy and nitrogen in the body has not been improved, which supports the outcome of no or a slight difference in ADG between the two treatment groups.

4.5. Rumen Bacterial Community Profile

A significant change in bacterial community profile was observed when LPL was supplemented to the diet for lambs. Of the most abundant genera found in the study, *Prevotella* was most significantly decreased with LPL supplementation based on LDA effect size factors. *Prevotella* is well-known hemicellulose degrading bacteria [35]. The decreased abundance of *Prevotella* was consistent with our finding that fiber digestibility was decreased with LPL. It is worthy to mention that among the *Prevotella* bacteria, *Prevotella* 1 was increased, while *Prevotella* 7 and *Prevotella* 9 decreased with LPL, suggesting a wide diversity of *Prevotella* bacteria in their biochemical characteristics.

Lachnospiraceae ND3007 group was increased, having the largest LDA effect size factor, with LPL supplementation in the diet. The bacteria in this genus was found to be decreased with an increase in temperature and humidity index [36]. The genus has not been well described in the literature, but is a butyric acid producing strain and considered as a probiotics [36]. The increase of the genus might promote a healthy rumen environment.

To date, the six culturable lipolytic bacteria are *Anaerovibrio lipolytica* and some species in the genera *Butyrivibrio, Clostridium,* and *Propionibacterium* [37]. The populations of these bacteria were quite small in the rumen of lambs fed the two diets in this study. This reflected the low content of ether extract (1.6 g/kg) in the diet. The content is at the low end of the lipid content range of 20 to 100 g/kg in the diet for ruminants [38]. The abundance of these bacteria was not changed with LPL supplementation, suggesting lipid metabolism in the rumen is not affected. The decreased activity of lipase in serum in the lambs fed a diet supplemented with LPL might just result from the absorption of LPL into the blood and was not related to lipid metabolism in the rumen.

Roseburia spp., Gram-positive bacteria, can be selectively inhibited by LPL in vitro [11]. This is confirmed true in vivo as the relative abundance of the genus in the rumen dropped from 0.11% to 0.02% with the supplementation of LPL in this study although the genus might be not biologically important for fattening lambs due to the low abundance. However, the phylum *Firmicutes* belongs to Gram-positive bacteria, increasing from 20% to 41% in the relative abundance. It might be because LPL may not inhibit all Gram-positive bacteria or the LPL concentration is not high enough to have an inhibitory effect.

5. Conclusions

The supplementation of lysophospholipid did not consistently result in positive animal performance in fattening lambs in this study. Further studies on a large scale are warranted to confirm the effect of lysophospholipid in animal performance. Feed digestion and rumen fermentation were altered with lysophospholipid supplementation, which was in the opposite direction to the results reported in other ruminant species and contradictory with our hypothesis that lysophospholipid enhances fiber degradation. Mechanism studies are needed to explore the effects of lysophospholipid on rumen fermentation, ruminal degradation and bypass of lysophospholipid, and body metabolisms, especially fat metabolism.

Supplementary Materials:
Figure S1: Distribution of the most dominant phyla in the rumen of fattening lambs fed a diet supplemented without (CON) or with (LPL) 0.05% lysophospholipid (those with a relative abundance of less than 2% were combined as other), Figure S2: Linear discriminant analysis effect size of rumen bacterial populations of fattening lambs fed a diet supplemented without (CON) or with (LPL) 0.05% lysophospholipid. This plot represents the most differentially abundant taxa according to the logarithmic linear discriminant analysis (LDA) score cutoff of at 4.0. Table S1: Ruminal bacterial composition (% of total sequence reads) in fattening lambs fed experimental diets supplemented with (LPL) or without (CON) lysophospholipids.

Author Contributions: Conceptualization, P.Y. and S.S.; methodology, X.S.; formal analysis, X.S.; investigation, Q.H., B.L., T.W., Y.H., W.T., R.L., C.L., J.L., B.S., C.W. and X.S.; resources, P.Y. and Y.L.; data curation, X.S.; writing—original draft preparation, X.S.; writing—review and editing, X.S. and L.C.; supervision, X.S.; project administration, X.S.; funding acquisition, X.S. and S.S.

Acknowledgments: We thank Portal Agri-Industries Co., Ltd. for manufacturing pelleted feeds and Yue Yu for convenience during the experiment conducted at the Animal Experimental Station of Jilin Agricultural Science and Technology University.

Appendix A

Graphical representation of the relative abundance of bacterial diversity from phylum to species was visualized using Krona chart. Cluster analysis was preceded by principal component analysis (PCA) to reduce the dimension of the original variables using the QIIME software package [39]. QIIME calculates both weighted and unweighted unifrac distance, which are phylogenetic measures of beta diversity. We used unweighted unifrac distance for Principal Coordinate Analysis (PCoA) and Unweighted Pair Group Method with Arithmetic mean (UPGMA) Clustering.

To confirm differences in the abundances of individual taxonomy between the two groups, Metastats software was utilized. LEfSe was used for the quantitative analysis of biomarkers within different groups. This method was designed to analyze data in which the number of species is much higher than the number of samples and to provide biological class explanations to establish statistical significance, biological consistency, and effect-size estimation of predicted biomarkers. To identify differences of microbial communities between the two groups, ANOSIM [40] and MRPP (multi-response permutation procedure) [41] were performed based on the Bray-Curtis dissimilarity distance matrices.

References

1. Sun, X.; Song, B.; He, Y.; You, P. A review on pelleted complete feed for sheep and goats. *Mod. J. Anim. Husb. Vet. Med.* **2017**, *46*, 162–165. (In Chinese)
2. Toteda, F.; Facciolongo, A.M.; Ragni, M.; Vicenti, A. Effect of suckling type and PUFA use on productive performances, quanti-qualitative characteristics of meat and fatty acid profile in lamb. *Prog. Nutr.* **2011**, *13*, 125–134.
3. Salem, A.F.Z.M. *Nutritional Strategies of Animal Feed Additives*; Nova Science Publishers, Inc.: Hauppauge, NY, USA, 2013; pp. 1–221.
4. Guo, X.; Cheng, L.; Liu, J.; Zhang, S.; Sun, X.; Al-Marashdeh, O. Effects of licorice extract supplementation on feed intake, digestion, rumen function, blood indices and live weight gain of Karakul sheep. *Animals* **2019**, *9*, 279. [CrossRef] [PubMed]
5. Zhong, R.; Xiang, H.; Cheng, L.; Zhao, C.; Wang, F.; Zhao, X.; Fang, Y. Effects of feeding garlic powder on growth performance, rumen fermentation, and the health status of lambs infected by gastrointestinal nematodes. *Animals* **2019**, *9*, 102. [CrossRef]
6. Facciolongo, A.M.; De Marzo, D.; Ragni, M.; Lestingi, A.; Toteda, F. Use of alternative protein sources for finishing lambs. 2. Effects on chemical and physical characteristics and fatty acid composition of meat. *Prog. Nutr.* **2015**, *17*, 165–173.

7. Nagpal, R.; Shrivastava, B.; Kumar, N.; Dhewa, T.; Sahay, H. Microbial feed additives. In *Rumen Microbiology: From Evolution to Revolution*; Puniya, A.K., Singh, R., Kamra, D.N., Eds.; Springer: Berlin, Germany, 2015; pp. 161–175.
8. Bernal-Barragán, H.; Cerrillo-Soto, M.A.; García-Mazcorro, J.F.; Juárez-Reyes, A.S.; Salem, A.Z.M. Antibiotics in animal nutrition. In *Nutritional Strategies of Animal Feed Additives*; Salem, A.F.Z.M., Ed.; Nova Science Publishers, Inc.: Hauppauge, NY, USA, 2013; pp. 25–46.
9. Maron, D.F.; Smith, T.J.S.; Nachman, K.E. Restrictions on antimicrobial use in food animal production: An international regulatory and economic survey. *Glob. Health* **2013**, *9*. [CrossRef]
10. Mnasri, T.; Hérault, J.; Gauvry, L.; Loiseau, C.; Poisson, L.; Ergan, F.; Pencréac'H, G. Lipase-catalyzed production of lysophospholipids. *OCL Oilseeds Fats Crop. Lipids* **2017**, *71*. [CrossRef]
11. Van Rensburg, C.E.J.; Joone, G.K.; O'Sullivan, J.F.; Anderson, R. Antimicrobial activities of clofazimine and B669 are mediated by lysophospholipids. *Antimicrob. Agents Chemother.* **1992**, *36*, 2729–2735. [CrossRef]
12. Zhao, P.Y.; Li, H.L.; Hossain, M.M.; Kim, I.H. Effect of emulsifier (lysophospholipids) on growth performance, nutrient digestibility and blood profile in weanling pigs. *Anim. Feed Sci. Technol.* **2015**, *207*, 190–195. [CrossRef]
13. Brautigan, D.L.; Li, R.; Kubicka, E.; Turner, S.D.; Garcia, J.S.; Weintraut, M.L.; Wong, E.A. Lysolecithin as feed additive enhances collagen expression and villus length in the jejunum of broiler chickens. *Poult. Sci.* **2017**, *96*, 2889–2898. [CrossRef]
14. Zampiga, M.; Meluzzi, A.; Sirri, F. Effect of dietary supplementation of lysophospholipids on productive performance, nutrient digestibility and carcass quality traits of broiler chickens. *Ital. J. Anim. Sci.* **2016**, *15*, 521–528. [CrossRef]
15. Zhao, P.Y.; Kim, I.H. Effect of diets with different energy and lysophospholipids levels on performance, nutrient metabolism, and body composition in broilers. *Poult. Sci.* **2017**, *96*, 1341–1347. [CrossRef] [PubMed]
16. Zhao, P.Y.; Zhang, Z.F.; Lan, R.X.; Liu, W.C.; Kim, I.H. Effect of lysophospholipids in diets differing in fat contents on growth performance, nutrient digestibility, milk composition and litter performance of lactating sows. *Animal* **2017**, *11*, 984–990. [CrossRef] [PubMed]
17. Wang, Q.Q.; Long, S.F.; Hu, J.X.; Li, M.; Pan, L.; Piao, X.S. Effects of dietary lysophospholipid complex supplementation on lactation performance, and nutrient digestibility in lactating sows. *Anim. Feed Sci. Technol.* **2019**, *251*, 56–63. [CrossRef]
18. Rico, D.E.; Ying, Y.; Harvatine, K.J. Short communication: Effects of lysolecithin on milk fat synthesis and milk fatty acid profile of cows fed diets differing in fiber and unsaturated fatty acid concentration. *J. Dairy Sci.* **2017**, *100*, 9042–9047. [CrossRef]
19. Lee, C.; Morris, D.L.; Copelin, J.E.; Hettick, J.M.; Kwon, I.H. Effects of lysophospholipids on short-term production, nitrogen utilization, and rumen fermentation and bacterial population in lactating dairy cows. *J. Dairy Sci.* **2019**, *102*, 3110–3120. [CrossRef]
20. Sontakke, U.B.; Kaur, H.; Tyagi, A.K.; Kumar, M.; Hossain, S.A. Effect of feeding rice bran lyso-phospholipids and rumen protected fat on feed intake, nutrient utilization and milk yield in crossbred cows. *Indian J. Anim. Sci.* **2014**, *84*, 998–1003.
21. Song, W.-S.; Yang, J.; Hwang, I.H.; Cho, S.; Choi, N.-J. Effect of dietary lysophospholipid (LIPIDOL^{TM}) supplementation on the improvement of forage usage and growth performance in Hanwoo heifer. *J. Korean Soc. Grassl. Forage Sci.* **2015**, *35*, 232–237. [CrossRef]
22. Kamande, G.M.; Baah, J.; Cheng, K.J.; McAllister, T.A.; Shelford, J.A. Effects of tween 60 and tween 80 on protease activity, thiol group reactivity, protein adsorption, and cellulose degradation by rumen microbial enzymes. *J. Dairy Sci.* **2000**, *83*, 536–542. [CrossRef]
23. Ma, N. Introduction of breeding stock and protection and utilization of genetic resources of sheep. *J. Jilin Agric. Univ.* **2004**, *26*, 77–82. (In Chinese)
24. Alton, G.G.; Jones, L.M.; Pietz, D.E. *Laboratory Techniques in Brucellosis*; Monograph Series; World Health Organization: Geneva, Switzerland, 1975; pp. 1–163.
25. National Research Council. *Nutrient Requirements of Small Ruminants: Sheep, Goats, Cervids and New World Camelids*, 6th ed.; National Academy Press: Washington, DC, USA, 2007; p. 384.
26. Liang, J.; Zhu, L.; Xu, Z. Study on the determination of NH_4^+-N content in microbial fermentation liquor by indophenol blue spectrophotometric method. *Food Ferment. Ind.* **2006**, *32*, 134–137. (In Chinese)

27. Feng, Z.; Gao, M. A modified spectrophotometric method for the determination of ammonia concentration in ruminal liquor. *Anim. Husb. Feed Sci.* **2010**, *31*, 37. (In Chinese)
28. Magoč, T.; Salzberg, S.L. FLASH: Fast length adjustment of short reads to improve genome assemblies. *Bioinformatics* **2011**, *27*, 2957–2963. [CrossRef] [PubMed]
29. Edgar, R.C. UPARSE: Highly accurate OTU sequences from microbial amplicon reads. *Nat. Methods* **2013**, *10*, 996–998. [CrossRef] [PubMed]
30. Henderson, G.; Yilmaz, P.; Kumar, S.; Forster, R.J.; Kelly, W.J.; Leahy, S.C.; Guan, L.L.; Janssen, P.H. Improved taxonomic assignment of rumen bacterial 16S rRNA sequences using a revised SILVA taxonomic framework. *PeerJ* **2019**, *2019*. [CrossRef]
31. Fan, K.; Yi, Y.; Liu, Y.; Li, M.; Wang, J.; Wang, Z.; Zhang, H.; Gu, K. Preparation of soybean lysophospholipids and its biosafety analysis. *China Oils Fats* **2019**, *44*, 124–127, 132. (In Chinese)
32. Jenkins, T.C.; Gimenez, T.; Cross, D.L. Influence of phospholipids on ruminal fermentation in vitro and on nutrient digestion and serum lipids in sheep. *J. Anim. Sci.* **1989**, *67*, 529–537. [CrossRef]
33. Russell, J.B.; Rychlik, J.L. Factors that alter rumen microbial ecology. *Science* **2001**, *292*, 1119–1122. [CrossRef]
34. Hwang, I.H.; Lee, C.H.; Kim, S.W.; Sung, H.G.; Lee, S.Y.; Lee, S.S.; Hong, H.; Kwak, Y.C.; Ha, J.K. Effects of mixtures of Tween80 and cellulolytic enzymes on nutrient digestion and cellulolytic bacterial adhesion. *Asian Australas. J. Anim. Sci.* **2008**, *21*, 1604–1609. [CrossRef]
35. Rubino, F.; Carberry, C.; Waters, S.M.; Kenny, D.; Mccabe, M.S.; Creevey, C.J. Divergent functional isoforms drive niche specialisation for nutrient acquisition and use in rumen microbiome. *ISME J.* **2017**, *11*, 932–944. [CrossRef]
36. Zhong, S.; Ding, Y.; Wang, Y.; Zhou, G.; Guo, H.; Chen, Y.; Yang, Y. Temperature and humidity index (THI)-induced rumen bacterial community changes in goats. *Appl. Microbiol. Biotechnol.* **2019**, *103*, 3193–3203. [CrossRef] [PubMed]
37. Privé, F.; Newbold, C.J.; Kaderbhai, N.N.; Girdwood, S.G.; Golyshina, O.V.; Golyshin, P.N.; Scollan, N.D.; Huws, S.A. Isolation and characterization of novel lipases/esterases from a bovine rumen metagenome. *Appl. Microbiol. Biotechnol.* **2015**, *99*, 5475–5485. [CrossRef] [PubMed]
38. Harfoot, C.G.; Hazlewood, G.P. Lipid metabolism in the rumen. In *The Rumen Microbial Ecosystem*; Hobson, P.N., Stewart, C.S., Eds.; Blackie Academic and Professional Publishers: London, UK, 1997; pp. 382–426.
39. Caporaso, J.G.; Kuczynski, J.; Stombaugh, J.; Bittinger, K.; Bushman, F.D.; Costello, E.K.; Fierer, N.; Peña, A.G.; Goodrich, J.K.; Gordon, J.I. QIIME allows analysis of high-throughput community sequencing data. *Nat. Methods* **2010**, *7*, 335–336. [CrossRef] [PubMed]
40. Philippot, L.; Spor, A.; Hénault, C.; Bru, D.; Bizouard, F.; Jones, C.M.; Sarr, A.; Maron, P.A. Loss in microbial diversity affects nitrogen cycling in soil. *ISME J.* **2013**, *7*, 1609–1619. [CrossRef] [PubMed]
41. Price, M.N.; Dehal, P.S.; Arkin, A.P. FastTree 2—Approximately maximum-likelihood trees for large alignments. *PLoS ONE* **2010**, *5*, e9490. [CrossRef] [PubMed]

PERMISSIONS

All chapters in this book were first published in MDPI; hereby published with permission under the Creative Commons Attribution License or equivalent. Every chapter published in this book has been scrutinized by our experts. Their significance has been extensively debated. The topics covered herein carry significant findings which will fuel the growth of the discipline. They may even be implemented as practical applications or may be referred to as a beginning point for another development.

The contributors of this book come from diverse backgrounds, making this book a truly international effort. This book will bring forth new frontiers with its revolutionizing research information and detailed analysis of the nascent developments around the world.

We would like to thank all the contributing authors for lending their expertise to make the book truly unique. They have played a crucial role in the development of this book. Without their invaluable contributions this book wouldn't have been possible. They have made vital efforts to compile up to date information on the varied aspects of this subject to make this book a valuable addition to the collection of many professionals and students.

This book was conceptualized with the vision of imparting up-to-date information and advanced data in this field. To ensure the same, a matchless editorial board was set up. Every individual on the board went through rigorous rounds of assessment to prove their worth. After which they invested a large part of their time researching and compiling the most relevant data for our readers.

The editorial board has been involved in producing this book since its inception. They have spent rigorous hours researching and exploring the diverse topics which have resulted in the successful publishing of this book. They have passed on their knowledge of decades through this book. To expedite this challenging task, the publisher supported the team at every step. A small team of assistant editors was also appointed to further simplify the editing procedure and attain best results for the readers.

Apart from the editorial board, the designing team has also invested a significant amount of their time in understanding the subject and creating the most relevant covers. They scrutinized every image to scout for the most suitable representation of the subject and create an appropriate cover for the book.

The publishing team has been an ardent support to the editorial, designing and production team. Their endless efforts to recruit the best for this project, has resulted in the accomplishment of this book. They are a veteran in the field of academics and their pool of knowledge is as vast as their experience in printing. Their expertise and guidance has proved useful at every step. Their uncompromising quality standards have made this book an exceptional effort. Their encouragement from time to time has been an inspiration for everyone.

The publisher and the editorial board hope that this book will prove to be a valuable piece of knowledge for researchers, students, practitioners and scholars across the globe.

LIST OF CONTRIBUTORS

Raffaella Rossi, Sabrina Ratti and Carlo Corino
Dipartimento di Medicina Veterinaria, Università Degli Studi di Milano, Via dell'Università 6, 26900 Lodi, Italy

Francesco Vizzarri
Department of Agricultural and Environmental Science, Università di Bari Aldo Moro, Via G. Amendola 165/A, 70126 Bari, Italy

Marisa Palazzo and Donato Casamassima
Department of Agricultural, Environmental and Food Sciences, Università Degli Studi del Molise, Via F. De Sanctis 1, 86100 Campobasso, Italy

Janghan Choi and Woo Kyun Kim
Department of Poultry Science, University of Georgia, Athens, GA 30602, USA

Ahmed A. AL-Sagan
King Abdulaziz City for Science & Technology, Riyadh 11442, Saudi Arabia

Shady Khalil
AlzChem Trostberg GmbH, 83308 Trostberg, Germany

Elsayed O. S. Hussein
Department of Animal Production, College of Food and Agriculture Sciences, King Saud University, Riyadh 11451, Saudi Arabia

Youssef A. Attia
Arid Land Agriculture Department, Faculty of Meteorology, Environment, and Arid Land Agriculture, King Abdulaziz University, Jeddah 21589, Saudi Arabia

Francesco Bordignon and Angela Trocino
Department of Comparative Biomedicine and Food Science (BCA), University of Padova, Viale dell'Università 16, I-35020 Legnaro, Italy

Ana Tomás-Vidal, Maria C. Milián Sorribes, Miguel Jover-Cerdá and Silvia Martínez-Llorens
Institute of Animal Science and Technology, Group of Aquaculture and Biodiversity, Universitat Politècnica de València, Camino de Vera, 14, 46071 València, Spain

Lingyan Li, Xiaoyue Xin, Shuxin Yin and Yongli Qu
Key Laboratory of Efficient Utilization of Feed Resources and Nutrition Regulation in Cold Region of Heilongjiang Province, College of Animal Science and Veterinary Medicine, Heilongjiang Bayi Agricultural University, Daqing 163319, China

Jiachen Qu
College of Animal Science and Technology, Inner Mongolia Agricultural University, Hohhot 010018, China

Fuguang Xue, Dengke Hua, Yue Wang, Xuemei Nan, Yiguang Zhao and Benhai Xiong
State Key Laboratory of Animal Nutrition, Institute of Animal Science, Chinese Academy of Agricultural Sciences, Beijing 100193, China

Fuyu Sun
State Key Laboratory of Animal Nutrition, Institute of Animal Science, Chinese Academy of Agricultural Sciences, Beijing 100193, China
College of Animal Science and Technology, China Agricultural University, Beijing 100193, China

Linshu Jiang
Beijing Key Laboratory for Dairy Cow Nutrition, Beijing University of Agriculture, Beijing 102206, China

Juan Carlos Moyano Tapia
Facultad de Ciencias de la Vida/Centro de Investigación, Posgrado y Conservación Amazónica, Universidad Estatal Amazónica, Puyo 160150, Ecuador
Centro Latinoamericano de Estudios de Problemáticas Lecheras (CLEPL), Casilda 2170, Argentina

Simon Alexander Leib
Facultad de Ciencias de la Vida/Centro de Investigación, Posgrado y Conservación Amazónica, Universidad Estatal Amazónica, Puyo 160150, Ecuador

Pablo Roberto Marini
Centro Latinoamericano de Estudios de Problemáticas Lecheras (CLEPL), Casilda 2170, Argentina
Facultad de Ciencias Veterinarias, Universidad Nacional de Rosario, Casilda 2170, Argentina
Consejo de Investigaciones (CIC-UNR), Rosario 2000, Argentina

Maria Laura Fischman
Centro Latinoamericano de Estudios de Problemáticas Lecheras (CLEPL), Casilda 2170, Argentina
Facultad de Ciencias Veterinarias, Universidad de Buenos Aires, CABA 1427, Argentina
Instituto de Investigación y Tecnología en Reproducción Animal (INITRA), Universidad de Buenos Aires, CABA 1427, Argentina

Wen Yang Chuang and Yun Chen Hsieh
Department of Animal Science, National Chung Hsing University, Taichung 402, Taiwan

Tzu-Tai Lee
Department of Animal Science, National Chung Hsing University, Taichung 402, Taiwan
The iEGG and Animal Biotechnology Center, National Chung Hsing University, Taichung 402, Taiwan

Valentina Panettieri, Andrea Ariano, Fulvia Bovera, Lorella Severino and Giovanni Piccolo
Department of Veterinary Medicine and Animal Production, University of Napoli Federico II, Via F. Delpino 1, 80137 Napoli, Italy

Stavros Chatzifotis
Institute of Marine Biology, Biotechnology and Aquaculture, Hellenic Centre for Marine Research, Gournes Pediados Heraklion, Crete, Greece

Concetta Maria Messina and Simona Manuguerra
DiSTeM, Marine Biochemistry and Ecotoxicology Laboratory, University of Palermo, Via G. Barlotta 4, 91100 Trapani, Italy

Ike Olivotto and Basilio Randazzo
Department of sea science, University Polytechnic of Marche, via Brecce Bianche, 60100 Ancona, Italy

Andrea Santulli
DiSTeM, Marine Biochemistry and Ecotoxicology Laboratory, University of Palermo, Via G. Barlotta 4, 91100 Trapani, Italy
Consorzio Universitario della Provincia di Trapani, Institute of Marine Biology, Via G. Barlotta 4, 91100 Trapani, Italy

Christian Alvarado-Gilis and Juan Pablo Keim
Instituto de Producción Animal, Universidad Austral de Chile, Valdivia-Chile, Valdivia 5090000, Chile

Rodrigo A. Arias
Instituto de Producción Animal, Universidad Austral de Chile, Valdivia-Chile, Valdivia 5090000, Chile
Centro de Investigación de Suelos Volcánicos, Universidad Austral de Chile, Valdivia 5090000, Chile

Gonzalo Guajardo and Stefan Kunick
Escuela de Graduados, Facultad de Ciencias Agrarias y Alimentarias, Universidad Austral de Chile, Valdivia 5090000, Chile

Marianna Pauletto, Mery Giantin and Mauro Dacasto
Department of Comparative Biomedicine and Food Science, University of Padua, Viale dell'Università 16, 35020 Legnaro (PD), Italy

Ramy Elgendy
Department of Comparative Biomedicine and Food Science, University of Padua, Viale dell'Università 16, 35020 Legnaro (PD), Italy
Department of Pharmacology, Faculty of Veterinary Medicine, Suez Canal University, Ismailia 41522, Egypt
Department of Immunology, Genetics and Pathology, Uppsala University, Uppsala 75185, Sweden

Elettra Marone, Lisa Grotta, Solange Ramazzotti, Francesca Bennato and Giuseppe Martino
Faculty of Bioscience and Technology for Food, Agriculture and Environment, University of Teramo, 64100 Teramo (TE), Italy

Chen Zheng, Junjun Ma and Ting Liu
College of Animal Science and Technology, Gansu Agricultural University, No. 1 Yingmen Village, Anning District, Lanzhou 730070, China

Bingdong Wei and Huaming Yang
Branch of Animal Husbandry, Jilin Academy of Agricultural Sciences, No.186 East Xinhua Road, Gongzhuling 136100, China

Wen Yang. Chuang, Wei Chih. Lin, Yun Chen. Hsieh and Chung Ming. Huang
Department of Animal Science, National Chung Hsing University, Taichung 402, Taiwan

Shen Chang. Chang
Kaohsiung Animal Propagation Station, Livestock Research Institute, Council of Agriculture, Pingtung 912, Taiwan

Marco Iannaccone
Faculty of Bioscience and Technology for Food, Agriculture, and Environment, University of Teramo, Via R. Balzarini 1, 64100 Teramo, Italy

Andrea Ianni
Department of Medical, Oral and Biotechnological Sciences, "G. d'Annunzio" University Chieti-Pescara, Via dei Vestini 31, 66100 Chieti, Italy
Faculty of Bioscience and Technology for Food, Agriculture and Environment, University of Teramo, 64100 Teramo (TE), Italy

Camillo Martino
Istituto Zooprofilattico Sperimentale dell'Abruzzo e del Molise "G. Caporale" Via Campo Boario, 64100 Teramo, Italy

Lorenzo Cerretani
Pizzoli SPA, Via Zenzalino Nord, 40054 1 Budrio (BO), Italy

Mohammed A. Al-Harthi
Arid Land Agriculture Department, Faculty of Meteorology, Environment and Arid Land Agriculture, King Abdulaziz University, Jeddah 21589, Saudi Arabia

Francesco Iannaccone
Department of Veterinary Medicine and Animal Production, University of Napoli Federico II, via F. Delpino 1, 80137 Napoli, Italy

Abdulaziz A. Alaqil
Department of Animal and Fish Production, King Faisal University, Al-Hufof, Al-Hassa 31982, Saudi Arabia

Hassan S. Zeweil
Department of Animal and Fish Production, Faculty of Agriculture, Saba Basha, Alexandria University, Alexandria 21527, Egypt

Ali E. Mansour
Ministry of agriculture, Animal Production section, Behiri Governorate 22951, Egypt

Qin Huo, Bo Li, Tingting Wu, Yuhua He, Wannian Tian, Rongquan Li, Changsheng Li, Jianping Li, Baijun Song, Chunqing Wang and Xuezhao Sun
The Innovation Centre of Ruminant Precision Nutrition and Smart and Ecological Farming, Jilin Agricultural Science and Technology University, Jilin 132109, China

Long Cheng
Faculty of Veterinary and Agricultural Sciences, Dookie Campus, The University of Melbourne, Victoria 3647, Australia

Peihua You and Yiyong Li
Portal Agri-Industries Co., Ltd., Nanjing 211803, China

Shuanghua Shen
Floor 15, Building B, 650 Xinzhuang Avenue, Xinqiao Town, Songjiang District, Shanghai 201612, China

Index

www.ingramcontent.com/pod-product-compliance
Lightning Source LLC
LaVergne TN
LVHW060534210926
848485LV00005B/1199
* 9 7 8 1 6 4 7 4 0 4 1 1 6 *